DIGITAL THEORY AND PRACTICE
USING INTEGRATED CIRCUITS

PRENTICE-HALL SERIES IN ELECTRONIC TECHNOLOGY

Charles M. Thomson and Joseph J. Gershon,
Consulting Editors

DIGITAL THEORY AND PRACTICE USING INTEGRATED CIRCUITS

Morris E. Levine

The College of Staten Island

PRENTICE-HALL, INC. *Englewood Cliffs, New Jersey* 07632

Library of Congress Cataloging in Publication Data

LEVINE, MORRIS E.
　　Digital theory and practice using integrated circuits.

　　Bibliography: p.
　　Includes index.
　　1. Digital integrated circuits.　I. Title.
TK7868.D5L49　　　621.381'73　　　77-23177
ISBN　0-13-212613-3

10　9　8　7　6　5　4　3　2

Printed in the United States of America

PRENTICE-HALL INTERNATIONAL, INC., *London*
PRENTICE-HALL OF AUSTRALIA PTY. LIMITED, *Sidney*
PRENTICE-HALL OF CANADA, LTD. *Toronto*
PRENTICE-HALL OF INDIA PRIVATE LIMITED, *New Delhi*
PRENTICE-HALL OF JAPAN, INC., *Tokyo*
PRENTICE-HALL OF SOUTHEAST ASIA PTE. LTD., *Singapore*
WHITEHALL BOOKS LIMITED, *Wellington, New Zealand*

To Michelle

Contents

Chapter Two

SEMICONDUCTOR PRINCIPLES   44

Chapter Three

THE INTEGRATED-CIRCUIT LOGIC FAMILIES   70

Chapter Four

BINARY NUMBERS AND BINARY ARITHMETIC   87

Chapter Five

EXCLUSIVE-OR AND ARITHMETIC OPERATIONS   112

Chapter Six

THE BISTABLE   134

Chapter Ten

BINARY CODES, ENCODING, DECODING, MULTIPLEXING   245

Chapter Eleven

MEMORY   273

Preface

Electronic equipment and electronic systems employing digital techniques are being developed and are appearing at an ever increasing rate. The major force behind these developments is the digital integrated circuit. Initially, digital ICs were quite simple and were based upon circuits that had been successful with discrete components. To obtain improvements in such areas as speed, noise immunity, dissipation, loading and complexity, improved circuits and logic families were developed and introduced, and manufacturing techniques were improved. This led to higher and higher degrees of integration and to more complex ICs. Simultaneously the language of digital electronics was growing and such conventions as logic symbols, logic terminology, rules for IC usage, and methods of data presentation were being developed and improved. Minimization techniques such as mapping were also being discovered. Surprisingly, these developments both simplified digital system design and made it more complex. Systems were made simpler because discrete circuit design was no longer necessary and, by following prescribed rules, these trouble free ICs could be coupled and combined. Simultaneously they became more involved because such combinations permitted the design and operation of systems that were much more complex.

   The object of this book is to develop and present the theory and principle of logic design, digital electronics and digital ICs. It is a companion text to the author's laboratory manual *Digital Theory and Experimentation Using Integrated*

Circuits and the chapter sequence and text material is coordinated with the laboratory manual. The emphasis is on ICs and IC applications and is directed towards the interpretation, understanding, and usage of IC manufacturers data sheets. It is intended for use as a basic text in digital electronics in technology, technical and engineering programs. A minimum knowledge of algebra is required as a prerequisite. All other mathematics required are developed in the text.

The text is written around the author's personal experience working in industry with digital systems and working in the semiconductor industry. The language and style are based on that used by engineers and technicians in working with digital equipment and that used by the applications and sales engineers of the IC manufacturers. Many of the illustrations and problems used in the text are derived from situations encountered in actual equipments. A primary goal of this text and its accompanying laboratory manual is to prepare the student for employment in industry in the digital electronics field.

A major objective of this book is to show how ICs have developed and are used in digital systems. This is done by means of illustrative examples and problems using ICs. As one reads IC data sheets and the instruction manuals of digital equipment, one realizes that the timing of waveforms is of major importance. For that reason and because of their importance in the design and maintenance of such types of equipment, numerous examples and problems showing waveforms and timings for many building blocks and systems are developed and interpreted.

In all areas of electronics, mastery of the theory can be improved with as much laboratory experience as possible. This has been of major consideration in the chapter and subject organization. To facilitate this, the text begins with laboratory oriented theory. Chapter One, "Logic and Boolean Algebra", discusses the basic logic functions, gates, minimization both by Boolean Algebra and the Karnaugh Map, and introduces IC gates and their operation.

Chapters Two on semiconductors and Three on integrated circuits and the IC logic families discuss the circuit components inside the IC gates discussed in Chapter One and explore the properties and limits of these IC gates. If time does not permit discussion of these chapters at this point in the course, as may occur in attempting to keep pace with the laboratory, these two chapters have been written to allow them to be deferred to some later time in the semester.

Chapter Four discusses the binary system and binary arithmetic and concludes with a discussion of the octal and hexadecimal numerical system. Chapter Five, on the Exclusive-OR and arithmetic operations, discusses how the arithmetic of Chapter Four is accomplished with IC gates.

Chapter Six discusses the bistable. Starting with the simple two-inverter bistable, the NAND gate *R-S* flip-flop is developed. Then, using this NAND gate FF as a block and using its properties, other IC FFs are developed. In each case, two principles are followed. The truth table is generated and the timing waveform discussed. The object is to provide a format so that an IC data sheet can be understood and to provide basic building blocks for succeeding chapters.

Chapter Seven on Series Counters and Chapter Eight on Shift Registers follow since much of their operation and timing is based upon the application of the truth tables and timing of the FFs of Chapter Six. In these chapters, the count state concept is developed and its relationship to the timing waveforms discussed.

Chapter Nine discusses the two-transistor astable multivibrator at some length. This is done to give the student an understanding of how transistors can be made to charge and discharge capacitors. Also discussed is the UJT, which is frequently used as a simple clock in digital systems. The principles of the IC Schmitt Trigger are discussed, showing how the trip levels, logic levels, and hysteresis occur.

Chapter Ten covers codes, decoding, encoding, and multiplexing. Codes are discussed at this point in the text rather than earlier because many of the codes were previously developed as count states of counters.

Chapter Eleven discusses memory, both semiconductor and magnetic.

Chapter Twelve discusses the operational amplifier, a subject of enormous importance in today's digital electronics, and developes its properties and basic operation. A major application of the operational amplifier is its use in D/A and A/D converters, the subject of Chapter Thirteen.

Chapter Fourteen consists of illustrations of the digital systems used in a wide variety of industrial and entertainment applications.

Chapter Fifteen is an introduction to the simple-complex digital system of the microprocessor. It discusses the basic operation blocks of the microprocessor and gives some simple programming examples.

Chapter Sixteen discusses how to read and interpret an IC data sheet.

I wish to express my thanks to those who aided in the preparation of this book. To Dr. Irving Kosow for his assistance in the organization of the material, for his extremely helpful suggestions, comments, and critical review of sections of the manuscript; to Mr. Joseph Rickard of the Media Production Center of The College of Staten Island for his photographic illustrations; to the semiconductor manufacturers for their technical data; and, in particular, to Mrs. Jean Johnson for her invaluable assistance in the preparation, organization, and typing of the manuscript.

MORRIS E. LEVINE

ONE

Logic and Boolean Algebra

1-1 INTRODUCTION

This text is concerned with the transmission and processing of digital information. In this chapter we shall develop some of the *rules of digital operations*. As we shall see, these operations follow a set of basic rules, the *rules of logic*. They have their own language, are expressed in their own dual level mathematics known as *Boolean algebra*, and have a set of *graphic symbols* that are used in *logic diagrams* to provide a symbolic description of logic flow.

1-2 FUNDAMENTAL OPERATORS: AND, OR, AND NOT†

The fundamental *logic* operations are AND, OR, and NOT. Although not readily apparent, we have made common use of these operators, if somewhat subconsciously. For example, consider a punch-press operator. For safety reasons, two push buttons are required to operate the press, one for each hand. Both the right

†M. E. Levine, *Digital Theory and Experimentation Using Integrated Circuits.* Englewood Cliffs, N.J.: Prentice-Hall, Inc., 1974, Expt. 1, Basic Logic Functions.

push button AND the left push button are required to operate the press. We can write

$$L \text{ AND } R = P$$

to express the operation of the punch press. This equation says that both the left (L) *and* right (R) hand switches must be closed to make the press (P) operate.

Another typical simple system is shown in Fig. 1-1a. To light the lamp, switches A *and* B must both be closed. Table 1-1a is a tabulation of all the possible

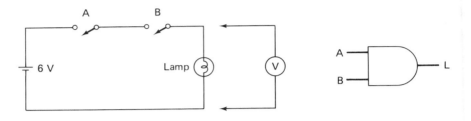

(a) 2 switch AND   (b) Logic symbol 2-input AND

Figure 1-1   (a) Two-switch AND circuit; (b) logic symbol

Table 1-1

(a) Two-switch AND

Switch A	Switch B	Lamp
Open	Open	OFF
Open	Closed	OFF
Closed	Open	OFF
Closed	Closed	ON

(b) Two-switch AND voltage levels

Switch A	Switch B	Lamp Voltage
Open	Open	0 V
Open	Closed	0 V
Closed	Open	0 V
Closed	Closed	6 V

(c) Two-switch AND symbolic levels

A	B	L
0	0	0
0	1	0
1	0	0
1	1	1

conditions of the two switches and the lamp condition. Table 1-1b is a similar tabulation except that, rather than tabulate the lamp condition, we have now tabulated the voltage read across the lamp. As far as the AND operation is concerned, whether we tabulate the lamp condition or measure the lamp voltage, the results are basically the same. To indicate this in the most general manner, the Boolean dual level logic value symbols 1 and 0 are universally used. This is shown in Table 1-1c. Open is represented by the symbol 0, closed by symbol 1, lamp OFF or 0 volts (V) by a 0 and lamp ON or 6 V by a 1.

The tabulations of Tables 1-1a, 1-1b, and 1-1c, which list all the possible combinations of the switches, are called *truth tables*.

The symbols 1 and 0 used in Table 1-1c are the logic levels of Boolean algebra. Boolean algebra is a two-valued algebra using two levels, 1 and 0, to represent the two possible states that can occur.

Since the equivalent of the circuit of Fig. 1-1a occurs in many different ways, such as in switches, relays, and transistors, the basic AND function has been given the graphic logic symbol shown in Fig. 1-1b. This symbol is the distinctively shaped symbol of the military standard MIL STD 806-C. Although many different graphic symbols have been used, the military standard symbols are the most popular; they are used by all integrated circuit (IC) manufacturers in their data sheets to describe the logic function being generated and throughout this text.†

We may write the AND operations of Fig. 1-1 in the form of an algebraic equation:

$$A \text{ AND } B = L \tag{1-1a}$$

or
$$A \times B = L \tag{1-1b}$$

or
$$(A)(B) = L \tag{1-1c}$$

or
$$A \cdot B = L \tag{1-1d}$$

or
$$AB = L \tag{1-1e}$$

In Boolean algebra, the multiplication symbol of ordinary algebra represents the AND function. Note the economy of expression used in Eq. (1-1e).

Now let us look at Table 1-1c as though it were a multiplication table for $A \times B$. It looks exactly like a conventional multiplication table. Boolean AND is just like multiplication.

Another AND circuit similar to that of Fig. 1-1 can be obtained using relays. This is shown in Fig. 1-2. The truth table can now be expressed in terms of the voltages applied to the relays and the lamp voltage (Table 1-2a), and in logic levels (Table 1-2b). In Tables 1-2a and 1-2b, logic level 0 is the equivalent of 0 V, and logic level 1 is the equivalent of 6 V. As we shall see in Chapter 3, in using integrated circuits the logic levels will turn out to be the voltages at transistor inputs and outputs.

†This standard has been superseded by the American National Standard ANSI Y32.14-1973, IEEE Std. 91-1973. See the Appendix.

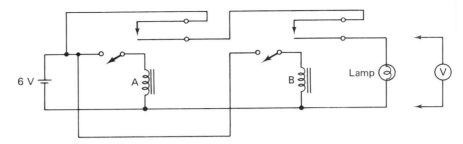

Figure 1-2   Two-relay AND

Table 1-2

(a) Two-relay AND voltages

Relay A	Relay B	Lamp
0 V	0 V	0 V
0 V	6 V	0 V
6 V	0 V	0 V
6 V	6 V	6 V

(b) Two-relay AND logic levels

A	B	L
0	0	0
0	1	0
1	0	0
1	1	1

Consider the simple system shown in Fig. 1-3a. To light the lamp requires closing switch *A or* switch *B*. Table 1-3a is the truth table for this two-switch OR circuit, and Table 1-3b is the truth table using Boolean algebra logic levels. Figure 1-3b is the MIL STD 806-C logic symbol for the OR logic function.

We may write the OR operation of Fig. 1-3a also in the form of an algebraic equation:

$$A \text{ OR } B = L \tag{1-2a}$$

or

$$A + B = L \tag{1-2b}$$

In Boolean algebra the $+$ sign of ordinary algebra represents the OR function.

Now let us look at Table 1-3b as though it were an addition table for $A + B$. Rows 1, 2, and 3 follow addition, but row 4 differs; $1 + 1 = 1$, and this is a case in which Boolean algebra differs from conventional algebra or arithmetic. We shall see that there will be other exceptions.

Truth Tables 1-1c, 1-2b, and 1-3b have important applications to the transmission of digital data. Consider Tables 1-1c and 1-2b for the AND circuits. In both tables, if $A = 0$, $L = 0$ and the level of B does not affect L. However, if $A = 1$, L and B are alike. Suppose that B was the input of a stream of digital data, 0s and 1s. We could control the transmission of B information to L by means of the level of A. If $A = 0$, no data are transmitted, whereas if $A = 1$, data are transmitted. A

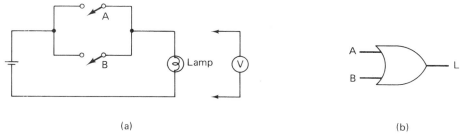

(a)   (b)

Figure 1-3   (a) Two-switch OR circuit; (b) logic symbol

Table 1-3

(a) Two-switch OR   (b) Two-switch OR symbolic levels

A	B	Lamp
Open	Open	OFF
Open	Closed	ON
Closed	Open	ON
Closed	Closed	ON

A	B	L
0	0	0
0	1	1
1	0	1
1	1	1

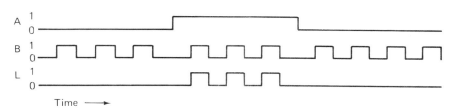

Time ⟶

Figure 1-4   AND gating of data

acts to gate the B information. We can show this on a time scale as in Fig. 1-4. Note that when $A = 0$ (no data transmission) the output level is at 0.

Truth Table 1-3b is for the OR function. Similarly, if $A = 0$, L and B are alike. If $A = 1$, $L = 1$ and the level of B does not affect the output L. A therefore controls the data stream by means of its level. With an OR circuit, $A = 0$ when it is desired to transmit, and $A = 1$ to not transmit. In the OR circuit, the output level $= 1$ when data are not being transmitted. This is shown in Fig. 1-5.

Integrated circuits (ICs) called gates are used to perform the logic functions of Figs. 1-1b, 1-4, 1-3b, and 1-5. Figure 1-1b is a two-input AND gate, and Fig. 1-3b is a two-input OR gate.

Suppose that the switches in Fig. 1-1a were toggle switches mounted on a vertical panel, so that to close the switch the toggle arm had to be pushed upward.

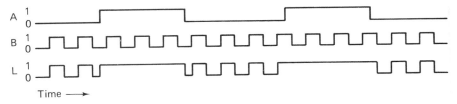

Figure 1-5   OR gating of data

This is indicated in Fig. 1-1a by the arrow location. The truth table for switches mounted in this way is given in Table 1-4.

Table 1-4   Switches of Figure 1-1a

Switch Position	Switch Action
Down	Open
Up	Closed

Suppose that the assembler of the panel inadvertently mounts switch *B* upside down, so that its truth table becomes that of Table 1-5 and the circuit of Fig. 1-1a becomes that of Fig. 1-7a.

Table 1-5   Switch *B* of Figure 1-7a

(a) Switch positions and action

Switch Position	Switch Action
Down	Closed
Up	Open

(b) Symbolic levels

B	$\bar{B}$
0	1
1	0

B ———▷○— $\bar{B}$ **Figure 1-6**   NOT logic symbol

Switch $\bar{B}$ of Fig. 1-7a acts like a switch with a NOT action, which is the inverse of switch *B* of Fig. 1-1a. To indicate this, the graphic symbol of Fig. 1-6, a triangle followed by a small circle, is used. The *small circle* indicates NOT. The symbol of Fig. 1-6 is used when an operator *B* is available and then has NOT action applied to it. In the same way, the *bar* above the *B* in $\bar{B}$ indicates that *B* has been subjected to the NOT function. The NOT function indicates that a 0 becomes a 1 and a 1 becomes a 0. Other terms used for NOT are negate, complement, and invert.

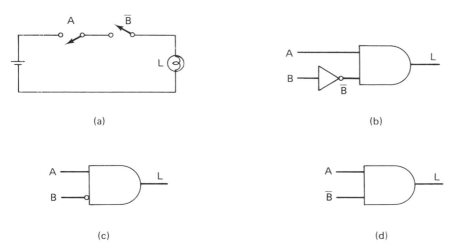

(a)   (b)

(c)   (d)

Figure 1-7 Two-switch AND with switch *B* NOT: (a) circuit diagram; (b) logic diagram; (c) logic symbol; (d) with $\bar{B}$ input

      Figure 1-7a is the now modified form of Fig. 1-1a. Figure 1-7b is a logic diagram for the modified circuit. Note that the NOT symbol converts *B* to $\bar{B}$ where it enters the AND. Rather than having to draw the triangle, Fig. 1-7c† is a logic symbol for Fig. 1-7a. The small circle at the input from *B* indicates a NOT on *B*. Figure 1-7d is used when $\bar{B}$ is available.

Table 1-6 Circuit of Figure 1-7

A	B	$\bar{B}$	L
0	0	1	0
0	1	0	0
1	0	1	1
1	1	0	0

      Table 1-6 is the truth table for the circuit of Fig. 1-7. Note that we first tabulated the *A* and *B* logic levels in the same manner as in Table 1-1c. *B* is then inverted to $\bar{B}$ by 0 and 1 substitutions. The AND, as indicated by the logic symbols of Figs. 1-7b and 1-7c, is activated by inputs *A* and $\bar{B}$ and gives a 1 *only* when *A* and $\bar{B}$ are *both* 1. If we now look at Fig. 1-7a, we can see that the lamp will be lit *only* if *A* is up and *B* is down, and this is what the *A*, *B*, and *L* columns give for row 3.

      †The use of and difference between Figs. 1-7b and 1-7c will become clearer when the implementation of these functions with integrated circuits is discussed in Chapter 3.

1-3 DIODE AND AND OR GATES

Figures 1-1a and 1-3a show how two-input AND and OR gates are made using switches. Simple but very useful electronic AND and OR gates can be made by using the switching properties of diodes. A diode, Fig. 1-8, is an electronic device that conducts current in the forward direction when its anode is made positive with respect to its cathode. Where conducting, the voltage across it is low (0.3 V for germanium, 0.7 V for silicon). When the voltage applied to its anode is made negative with respect to its cathode, it does not conduct. The diode therefore acts like a switch. The switch is closed when the anode is made positive, and the switch is open when the anode is made negative.

Forward direction
of current

C, cathode   A, anode

Figure 1-8 Diode

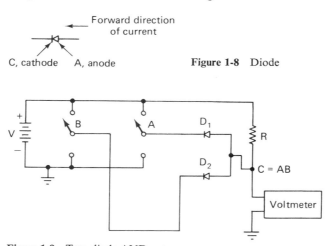

Figure 1-9 Two-diode AND gate

Consider the circuit with two diodes shown in Fig. 1-9. If switches A and B are both connected to ground, the cathodes of diodes D_1 and D_2 are connected to ground. Current will flow through resistor R and through the forward-biased diodes to ground. The voltmeter will measure the forward voltage drop across the diodes, a very low voltage. Now let switch A be connected to $+V$. This connects the cathode of diode D_1 to a positive voltage. Diode D_1 is reverse biased (cathode positive and anode negative), but diode D_2 is still forward biased and the voltmeter measures the low forward voltage across D_2. If we also connect switch B to $+V$, both diodes are connected across the resistor R. The diodes do not have their anodes connected to a positive voltage, and no current flows in resistor R. Hence the voltmeter will measure the battery voltage $+V$. Figure 1-9 is a diode AND gate. Both switches A *and* B have to be connected to $+V$ to make the voltage at point C go to the battery voltage. The Boolean equation for the circuit of Fig. 1-9 is $C = AB$.

Consider a second circuit with two diodes (Fig. 1-10). If switches A and B are both connected to ground, neither diode has a forward voltage applied to it and no current will flow in resistor R. The voltmeter will read 0 V. If switch A is connected

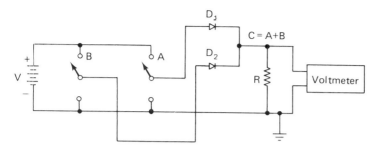

Figure 1-10 Two-diode OR gate

to $+V$, this will connect the anode of D_1 to a positive voltage. Diode D_1 conducts and the voltmeter reads the voltage V minus the voltage drop in the diode. By inspecting the circuit of Fig. 1-10, we can see that the voltage at point C will be equal to the voltage $+V$ (minus the diode voltage) if switch A *or* switch B is connected to $+V$. Figure 1-10 is a diode OR gate with logic equation $C = A + B$.

Table 1-7

(a) Diode AND gate of Figure 1-9

A	B	C
0	0	0
0	$+V$	0
$+V$	0	0
$+V$	$+V$	$+V$

(b) Diode OR gate of Figure 1-10

A	B	C
0	0	0
0	$+V$	$+V$
$+V$	0	$+V$
$+V$	$+V$	$+V$

Tables 1-7a and 1-7b give the voltage† truth tables for the circuits of Figs. 1-9 and 1-10.

1-4 MULTIPLE-INPUT AND AND OR

Figure 1-11 is the logic symbol for a three-input AND function. To describe its operation we have to write a truth table that includes all the possible combinations of inputs U, V, and W. For a logic circuit with N inputs, the number of combinations is given by Eq. (1-3):

$$\text{Number of combinations} = 2^N \qquad (1\text{-}3)$$

†Neglecting the voltage drops across the diodes.

We have already seen that for two inputs the number of combinations is $2^2 = 4$. For Fig. 1-11 the number of inputs is three, $N = 3$, so that there are $2^3 = 8$ possible combinations. Table 1-8 is the truth table. To tabulate all the combinations, a systematic procedure should be followed. In Table 1-8 the three inputs each head columns. Starting with the input whose column is farthest to the right alternately mark 0 and 1 until all the rows are completed. In the next column use two 0s and two 1s, double the groupings of the first row. In the third column use double the groupings of the middle row or four 0s and four 1s. We shall see in Chapter 4 that this organization of truth tables is compatible with binary counting and, in Chapter 7, with electronic counters.

In Table 1-8, note that the X is a 1 only when U, V, and W are all equal to 1.

Figure 1-11   Three-input AND

Table 1-8   Three-input AND

U	V	W	X
0	0	0	0
0	0	1	0
0	1	0	0
0	1	1	0
1	0	0	0
1	0	1	0
1	1	0	0
1	1	1	1

8 rows

Begin with this column.
Alternate 0s and 1s.

Let us consider another example, that of Fig. 1-12, which has three inputs, two of which are NOT inputs. Table 1-9 is the truth table for this logic system. Note that columns D, E, and F are listed in accordance with the previous rules. However, the three inputs to the OR are $\bar{D}$, E, and $\bar{F}$, and therefore we must list $\bar{D}$ and $\bar{F}$, the opposite of D and F. Output G is determined by the states of the inputs, E, $\bar{D}$, and $\bar{F}$, and is a 1 when E, $\bar{D}$, or $\bar{F}$ is at logic 1.

Figure 1-12 shows two possible ways of drawing the logic diagram. In Fig. 1-12a the inputs are NOT inputs; in Fig. 1-12b the inputs are direct but are NOTed by the small input circles. Output G, of course, is the same.

$$\bar{D} \quad \bar{E} \quad \bar{F} \quad G = \bar{D} + E + \bar{F}$$

(a) With NOT inputs

$$D \quad E \quad F \quad G = \bar{D} + E + \bar{F}$$

(b) With inverting circles

Figure 1-12 Three-input OR with two NOT inputs: (a) with NOT inputs; (b) with inverting circles

Table 1-9 Three-input OR of Figure 1-12

D	E	F	$\bar{D}$	$\bar{F}$	G
0	0	0	1	1	1
0	0	1	1	0	1
0	1	0	1	1	1
0	1	1	1	0	1
1	0	0	0	1	1
1	0	1	0	0	0
1	1	0	0	1	1
1	1	1	0	0	1

Output determined by E, $\bar{D}$, and $\bar{F}$ is $= 1$ when E, $\bar{D}$, or $\bar{F}$ is equal to 1

1-5 NAND AND NOR

Although AND, OR, and NOT operations are important and are used a great deal, the AND followed by a NOT resulting in the NAND and the OR followed by a NOT resulting in the NOR have considerably greater versatility. Moreover, they are readily implemented with transistor circuits. Figure 1-13 is the logic symbol for

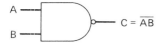

$$C = \overline{AB}$$

Figure 1-13 Two-input NAND

Table 1-10 Two-input NAND

A	B	AB	$C = \overline{AB}$
0	0	0	1
0	1	0	1
1	0	0	1
1	1	1	0

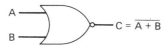

Figure 1-14 Two-input NOR

Table 1-11 Two-input NOR

A	B	$A + B$	$C = \overline{A + B}$
0	0	0	1
0	1	1	0
1	0	1	0
1	1	1	0

a two-input NAND, and Table 1-10 is its truth table. This truth table should be compared to that of Table 1-1c.

Figure 1-14 is the logic symbol for a two-input NOR, and Table 1-11 is its truth table. This truth table should be compared to that of Table 1-3b.

The NAND of Fig. 1-13 and the NOR of Fig. 1-14 are readily converted to AND and OR by the addition of an inverter. This is shown in Figs. 1-15a and 1-15b. This can be seen if we invert the C columns of Tables 1-10 and 1-11 and compare the results with Tables 1-1c and 1-3b.

Figure 1-15 (a) NAND to AND; (b) NOR to OR

1-6 BOOLEAN EXPRESSIONS AND LOGIC DIAGRAMS

Consider the Boolean expression

$$D = A + BC$$

How should this be interpreted? Which comes first, the OR part of the expression between A and B or the AND between B and C? Fortunately, it turns out that the interpretation and priority are exactly the same as in conventional algebra. The rules for interpreting such an expression are as follows:

1. AND separate and before OR; multiplication before addition.
2. Handle parentheses as in ordinary algebra. Complete any operations within parentheses first and then expand, if desired.
3. Inverted terms are considered distinct from noninverted terms and are treated separately. That is, A and $\overline{A}$ are separate terms. (See, however, Sec. 1-7 for simplification formulas.)

Let us again consider $D = A + BC$, draw a logic diagram, and write a truth table. Note that the AND term BC was treated separately and then OR combined with A. Two possible diagrams are shown in Fig. 1-16, and they are equivalent.

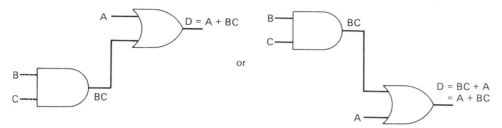

Figure 1-16   Logic diagrams for $D = A + BC$

The choice is essentially that of the draftsman. Justification for their equivalency is based upon one of the rules of Boolean algebra developed in Table 1-12. Note the procedure in organizing the table. The columns are arranged in alphabetical order. Column BC is developed separately and then ORed with column A.

Table 1-12 $D = A + BC$

A	B	C	BC	$D = A + BC$	
					←D is the OR of column A and column BC
0	0	0	0	0	
0	0	1	0	0	
0	1	0	0	0	
0	1	1	1	1	
1	0	0	0	1	
1	0	1	0	1	
1	1	0	0	1	
1	1	1	1	1	

Table 1-12 is an example of a careful systematic procedure for writing truth tables. If the columns are arranged alphabetically, if the 0s and 1s are set up in increasing order and in the doubling combinations illustrated, and if a separate column is set up for each step, it is simple to write correct truth tables. Attempts to short-cut these procedures will almost always give incorrect answers.

EXAMPLE 1-1 Draw a logic diagram for $D = (A + B)C$.

Solution

a. Complete operations within the parentheses first.

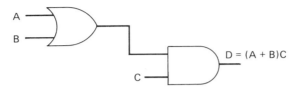

Figure 1-17 Logic diagram for $D = (A + B)C$

b. Expand as in algebra: $D = AC + BC$.

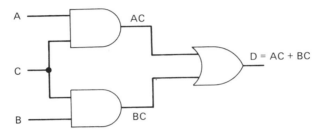

Figure 1-18 Logic diagram for $D = AC + BC$

Both sets of Boolean expressions and logic diagrams are correct and equivalent. Note that, on each point on the logic diagram, the Boolean expression has been given. This is good practice as it provides considerable assistance in understanding the operation of the circuit.

EXAMPLE 1-2 Draw a logic diagram for $E = \overline{A + BC + D\overline{C}}$.

Solution See Fig. 1-19. Note that $\overline{C}$ was treated as a separate term in drawing the logic diagram.

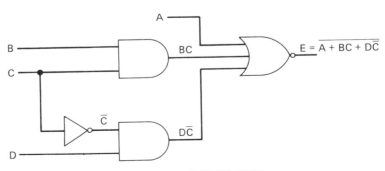

Figure 1-19 Logic diagram for $E = \overline{A + BC + D\overline{C}}$

EXAMPLE 1-3   Write a truth table for $D = \overline{AB + \overline{\overline{A}B} + C}$.

Solution

A	B	C	$\overline{A}$	AB	$\overline{A}B$	$\overline{\overline{A}B}$	$AB + \overline{\overline{A}B} + C$	$D = \overline{AB + \overline{\overline{A}B} + C}$
0	0	0	1	0	0	1	1	0
0	0	1	1	0	0	1	1	0
0	1	0	1	0	1	0	0	1
0	1	1	1	0	1	0	1	0
1	0	0	0	0	0	1	1	0
1	0	1	0	0	0	1	1	0
1	1	0	0	1	0	1	1	0
1	1	1	0	1	0	1	1	0

Note how each term is treated separately and combined for the final result. Only by following such a step-by-step procedure can errors be avoided.

1-7 BOOLEAN ALGEBRA FORMULAS†

This section develops some of the relationships and formulas of Boolean algebra and shows how these formulas are used to simplify and reduce logic circuitry. This process of reduction is called *minimization*. We shall find many similarities to ordinary algebra, but there will be some marked differences. Boolean algebra is a two-level algebra that, as we have already seen, can be represented by 0 and 1 or A and $\overline{A}$. It is when the ordinary algebraic operation of addition $(+)$ would result in something other than 0 or 1 or in multiplication $(\times)$ to an exponent that Boolean algebra differs from ordinary algebra. Magnitudes such as 2, 3, etc., exponents, negative, and fractions have no meaning in Boolean algebra. This is demonstrated in the following.

In Fig. 1-20a, two switches are mechanically coupled so that *both* act together. As far as circuit operation is concerned, one switch is not needed. We can therefore write

$$A \cdot A = A \tag{1-4a}$$

†M. E. Levine, *Digital Theory and Experimentation Using Integrated Circuits*. Englewood Cliffs, N.J.: Prentice-Hall, Inc., 1974, Expt. 2, Boolean Algebra and Simplification of Logic Equations.

Note that in conventional algebra the product of A and A leads to an exponential A^2.

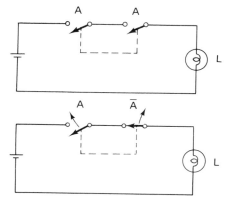

Figure 1-20a   Switches for $A \cdot A = A$

Figure 1-20b   Switches for $A \cdot \bar{A} = 0$

In Fig. 1-20b, two switches work together, but when A is closed $\bar{A}$ is open and when A is open $\bar{A}$ is closed. The circuit is *always* OPEN. We can write

$$A \cdot \bar{A} = 0 \qquad (1\text{-}4b)$$

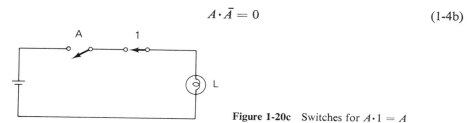

Figure 1-20c   Switches for $A \cdot 1 = A$

In Fig. 1-20c, there are two switches. One is *always* closed and represented by a 1. Whether the lamp is ON or OFF is determined by switch A, yielding

$$A \cdot 1 = A \qquad (1\text{-}4c)$$

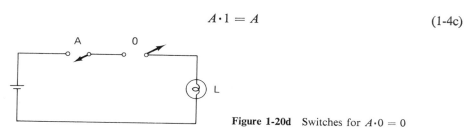

Figure 1-20d   Switches for $A \cdot 0 = 0$

In Fig. 1-20d, the circuit has a permanently *open* switch, represented by a 0, and the lamp can never light. This gives

$$A \cdot 0 = 0 \qquad (1\text{-}4d)$$

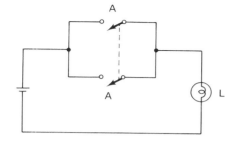

Figure 1-20e   Switches for $A + A = A$

In Fig. 1-20e, the two switches work together. One switch is not needed. We can therefore write

$$A + A = A \tag{1-5a}$$

By following a procedure similar to that of the previous sections we can show that

$$A + \bar{A} = 1 \tag{1-5b}$$

$$A + 1 = 1 \tag{1-5c}$$

$$A + 0 = A \tag{1-5d}$$

Equations (1-4) and (1-5) have considerably broader implications than are at first apparent. The As in these equations, although shown as single switches, really represent any circuitry or complex logic function. For example (and this particular case occurs frequently), Eq. (1-5c) says that 1 + *any logic function* is equal to 1. For example, $1 + ACD = 1$.

By using permanently open and closed switches as illustrations, we can demonstrate that

$$0 \cdot 0 = 0 \tag{1-6a}$$

$$0 \cdot 1 = 0 \tag{1-6b}$$

$$1 \cdot 1 = 1 \tag{1-6c}$$

and

$$0 + 0 = 0 \tag{1-7a}$$

$$0 + 1 = 1 \tag{1-7b}$$

$$1 + 1 = 1 \tag{1-7c}$$

Since the order of switch openings and closings is not a factor in logic circuit operation

$$A + B = B + A \tag{1-8a}$$

$$A \cdot B = B \cdot A \tag{1-8b}$$

The double NOT or double inversion returns a variable to its original state. This is shown in Fig. 1-20f:

A ——▷— $\overline{A}$ —▷— $\overline{\overline{A}} = A$ **Figure 1-20f** Double inversion

$$\overline{\overline{A}} = A \qquad (1\text{-}9)$$

The A in Boolean equations has a much more general interpretation. It can be interpreted as representing a combination of Boolean inputs. For example, Eq. (1-9) for the double conversion is applicable to any Boolean function:

$$\overline{\overline{AB}} = AB$$
$$\overline{\overline{A + B}} = A + B$$
$$\overline{\overline{A(B + C)}} = A(B + C)$$

What about a term $\overline{\overline{A} + \overline{B}}$? In this case, Eq. (1-9) does *not* apply and *cannot* be used. The bar above the complete expression is *wider* than the bar above the A. It is only when both bars have the *same width* that Eq. (1-9) can be used.

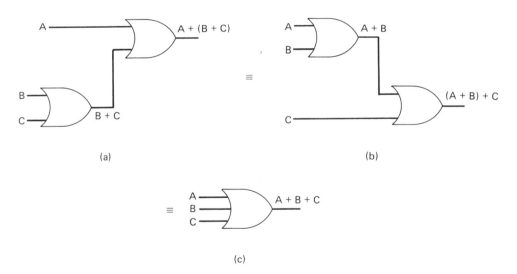

(a)

(b)

(c)

Figure 1-20g Grouping of variables

We can group our variables in any way we desire, as shown in Figs. 1-20g and 1-20h.

$$A + (B + C) = (A + B) + C \qquad (1\text{-}10a)$$
$$= A + B + C \qquad (1\text{-}10b)$$

and

$$A(BC) = AB(C) \qquad (1\text{-}11a)$$
$$= ABC \qquad (1\text{-}11b)$$

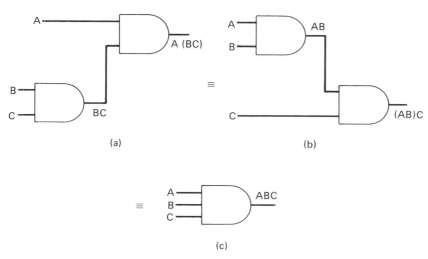

Figure 1-20h   Grouping of variables

Boolean complex expressions with many variables can be expanded in exactly the same manner as in ordinary algebra and with the same priority rules.

$$A(B + C) = AB + AC \tag{1-12}$$

We can prove that such a statement is valid by using a truth table (Table 1-13). If both sides of the equation are equal for all possible combinations of inputs, the equation must be true.

Table 1-13 $A(B + C) = AB + AC$

A	B	C	$B + C$	$A(B + C)$	AB	AC	$AB + AC$
0	0	0	0	0	0	0	0
0	0	1	1	0	0	0	0
0	1	0	1	0	0	0	0
0	1	1	1	0	0	0	0
1	0	0	0	0	0	0	0
1	0	1	1	1	0	1	1
1	1	0	1	1	1	0	1
1	1	1	1	1	1	1	1

EXAMPLE 1-4   Prove that $A + AB = A$.   (1-13)

Solution

$$A + AB = A(1 + B)$$ By using Eqs. (1-4c) and (1-12)

but $$1 + B = 1$$ By using Eq. (1-5c)

$$A(1) = A$$ By using Eq. (1-4c)

and $$A + AB = A$$

EXAMPLE 1-5   Prove that $A + \bar{A}B = A + B$.   (1-14)

Solution

$$A + B = A(1) + B(1)$$ By using Eq. (1-4c)

$$= A(B + \bar{B}) + B(A + \bar{A})$$ By using Eq. (1-5b)

$$= AB + A\bar{B} + AB + \bar{A}B$$

$$= AB + AB + A\bar{B} + \bar{A}B$$

but $$AB + AB = AB$$ By using Eq. (1-5a)

$$A + B = AB + A\bar{B} + \bar{A}B$$

$$= A(B + \bar{B}) + \bar{A}B$$

$$= A + \bar{A}B$$ By using Eq. (1-5b)

Of course, we could have proved this with a truth table.

1-8   DE MORGAN'S THEOREM†

Of major importance in the practical implementation of logic equations are the Boolean relationships known as De Morgan's theorem. They are used to convert OR logic to AND logic and AND logic to OR logic. As we shall see in Chapter 3, integrated-circuit (IC) families or systems are available in either OR–NOR or AND–NAND gates. De Morgan's relationships allow the realization of AND problems with OR circuits or OR problems with AND circuits. Frequently, the application of De Morgan's theorem results in circuit simplification and a reduction in package count (number of ICs needed).

De Morgan's theorem is expressed as follows:

$$A + B = \overline{\bar{A}\bar{B}}$$ (1-15a)

$$AB = \overline{\bar{A} + \bar{B}}$$ (1-15b)

†M. E. Levine, *Digital Theory and Experimentation Using Integrated Circuits*. Englewood Cliffs, N.J.: Prentice-Hall, Inc., 1974, Expt. 3, De Morgan's Theorem.

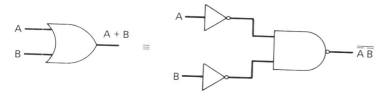

Figure 1-20i

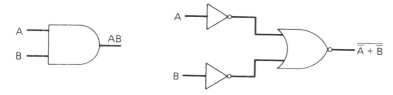

Figure 1-20j

Let us examine the circuits of Figs. 1-20i and j to see how they are obtained. The rules are as follows.

1   Where there is an OR, change to an AND.
2   Where there is an AND, change to an OR.
3   NOT each of the variables.
4   NOT the final result.
5   If necessary to simplify, apply Eq. (1-9) for the double NOT.

De Morgan's theorem is applicable to complex functions and they are handled in the same manner, applying these general rules. Examples follow.

EXAMPLE 1-6   Convert $A\bar{B}$ to an OR term using De Morgan's theorem.

Solution

$$A\bar{B} = \overline{\bar{A} + \bar{\bar{B}}} \quad \text{By using Eq. (1-15b)}$$
$$\bar{\bar{B}} = B \quad \text{By using Eq. (1-9)}$$
$$A\bar{B} = \overline{\bar{A} + B}$$

EXAMPLE 1-7   Convert $C + DE$ to AND/NAND logic by De Morgan's theorem.

Solution   The problem requests that the final solution have only AND or NAND terms. This means that only the OR term has to be converted. Since the DE term is already in AND form, it does not have to be converted. As we have discussed previously, Boolean equations are applicable to single terms or combinations of terms. In this case the term DE is treated as a single term:

$$C + DE = \overline{\bar{C}(\overline{DE})} \quad \text{By using Eq. (1-15a)}$$

This is shown in Fig. 1-21.

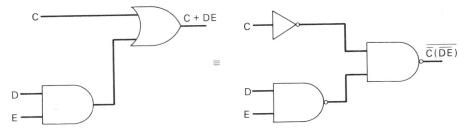

Figure 1-21 AND/NAND for $C + DE$

EXAMPLE 1-8 Convert $C + DE$ to OR/NOR logic.

Solution This is the same problem as in Example 1-7, but we now want the conversion to have either OR or NOR terms. Since there already is an OR term, it is to be left alone. Only the AND term has to be converted. This brings up another property of De Morgan's theorem. It can be applied to a complete logic equation or to any part of it. In this case, we need apply it only to the AND term:

$$C + DE = C + \overline{\overline{D} + \overline{E}} \qquad \text{By using Eq. (1-15b)}$$

In this case the wider bar has the same significance as parentheses, and the operation under the wider bar is to be performed separately. This is shown in Fig. 1-22.

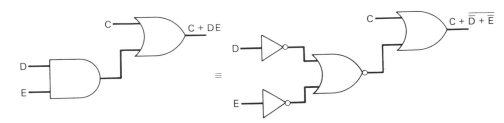

Figure 1-22 OR/NOR for $C + DE$

Examples 1-7 and 1-8 both required inverters. Inverters are considered valid implementations of the desired solution in terms of a common logic family. As we shall see in Chapter 3, IC inverters are available in all logic families. Furthermore, gates can be readily converted to inverters. For example, if we have two-input NAND or NOR gates, they can be converted in the following ways. The solutions of Examples 1-7 and 1-8 are equivalent, as shown in Figs. 1-23 and 1-24.

Figure 1-23 Invert using NAND

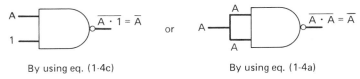

By using eq. (1-4c)   By using eq. (1-4a)

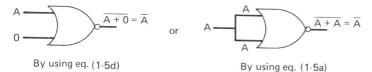

By using eq. (1-5d) | By using eq. (1-5a)

Figure 1-24 Invert using NOR

In Example 1-7, only ANDs and NANDs appear in the solution. In Example 1-8, only ORs and NORs appear.

EXAMPLE 1-9   Convert $A + B + C + \bar{D}$ to an AND/NAND term.

Solution

$$A + B + C + \bar{D} = \overline{\overline{A}\,\overline{B}\,\overline{C}\,D}$$

De Morgan's theorem can be applied to multiterm expressions. The rules are applied to each term in the equation.

The proof of De Morgan's theorem is readily obtained with truth tables (see Table 1-14). For example,

$$A + B = \overline{\overline{A}\,\overline{B}}$$

Table 1-14   Proof of $A + B = \overline{\overline{A}\overline{B}}$

A	B	$A + B$	$\bar{A}$	B	$\bar{A}\bar{B}$	$\overline{\bar{A}\bar{B}}$
0	0	0	1	1	1	0
0	1	1	1	0	0	1
1	0	1	0	1	0	1
1	1	1	0	0	0	1

$$=$$

1-9 MINIMIZATION BY BOOLEAN FORMULAS

Solutions to problems in logic are expressed in logic equation form. However, there may be a simpler solution that requires fewer components. Finding the simpler solution is called minimization. The Boolean equations of Sec. 1-7 and 1-8 are used in this minimization procedure. These equations are summarized as follows:

$$A \cdot A = A \tag{1-4a}$$

$$A \cdot \bar{A} = 0 \tag{1-4b}$$

$$A \cdot 1 = A \tag{1-4c}$$

$$A \cdot 0 = 0 \tag{1-4d}$$

$$A + A = A \tag{1-5a}$$

$$A + \bar{A} = 1 \tag{1-5b}$$

$$A + 1 = 1 \tag{1-5c}$$

$$A + 0 = A \tag{1-5d}$$

$$\bar{\bar{A}} = A \tag{1-9}$$

$$A + AB = A \tag{1-13}$$

$$A + \bar{A}B = A + B \tag{1-14}$$

$$\left. \begin{array}{l} A + B = \overline{\bar{A}\bar{B}} \\[6pt] AB = \overline{\bar{A} + \bar{B}} \end{array} \right\} \text{De Morgan's theorem} \qquad \begin{array}{l} \text{(1-15a)} \\[6pt] \text{(1-15b)} \end{array}$$

And from the basic Boolean concepts,

$$\left. \begin{array}{l} 0 + 0 = 0 \\ 0 + 1 = 1 \\ 1 + 1 = 1 \end{array} \right\} \text{OR} \qquad \begin{array}{l} \text{(1-7a)} \\ \text{(1-7b)} \\ \text{(1-7c)} \end{array}$$

$$\left. \begin{array}{l} 0 \cdot 0 = 0 \\ 0 \cdot 1 = 0 \\ 1 \cdot 1 = 1 \end{array} \right\} \text{AND} \qquad \begin{array}{l} \text{(1-6a)} \\ \text{(1-6b)} \\ \text{(1-6c)} \end{array}$$

Minimization is performed by the following systematic procedure.

1 If there are any parentheses in the equations, expand using conventional algebraic procedures and priorities. Treat variables and inverted variables as separate terms.

2 Look for simplification using the equations of Boolean algebra.

3 Look for common terms and factor as in conventional algebra. Treat variables and inverted variables as separate terms.

4 Look for simplification using the equations of Boolean algebra.

5 Repeat steps 1, 2, 3, and 4, continuing to look for minimization terms. Look for applications of De Morgan's theorem. Note that $(1 + \text{anything}) = 1$, and $(0 \times \text{anything}) = 0$.

EXAMPLE 1-10   Given $D = A + ABC + A\bar{B}\bar{C} + CB + C\bar{B}$.

a. Minimize.

b. Draw the logic diagram for the given equation.
c. Draw the reduced logic diagram.

Solution   See Fig. 1-25.

$$A + ABC + A\bar{B}\bar{C} + CB + C\bar{B}$$
$$= A(1 + BC + \bar{B}\bar{C}) + C(B + \bar{B}) \qquad \text{by factoring}$$
$$D = A + C \qquad \text{by using Eqs. (1-5c) and (1-5b)}$$

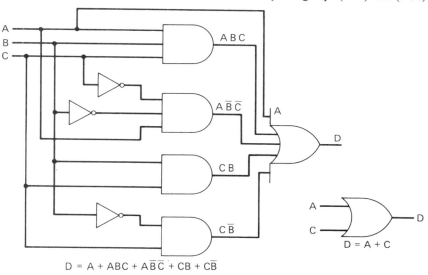

$$D = A + ABC + A\overline{B}\,\overline{C} + CB + C\overline{B}$$

Figure 1-25   Logic diagrams for Example 1-10

Note in Example 1-10 that minimization has both made the logic simple and reduced the system from three stages of logic to one. This stage reduction is also important when speed is a factor in a logic system.

EXAMPLE 1-11   Minimize $\bar{P}R\bar{Q}\bar{S} + \bar{P}RQ + \bar{P}RS + P\bar{R}Q + P\bar{R}\bar{Q} + S(1 + \bar{P}Q\bar{R})\bar{S}$.

Solution   Factoring, we obtain

$$= \bar{P}R(\bar{Q}\bar{S} + Q + S) + P\bar{R}(\underbrace{Q + \bar{Q}}_{= 1}) + \underbrace{S\bar{S}}_{= 0}(1 + \bar{P}Q\bar{R})$$

$$= \bar{P}R(\bar{Q}\bar{S} + \overline{\overline{Q}\overline{S}}) + P\bar{R}$$
$$\qquad\qquad \longrightarrow \text{by De Morgan's theorem}$$

$$= \bar{P}R(1) + P\bar{R}$$
$$\qquad \longrightarrow \text{equivalent to } A + \bar{A} = 1; \ \bar{Q}\bar{S} \equiv A$$

$$= \bar{P}R + P\bar{R}$$

Note that $\bar{P}$ and P and R and $\bar{R}$ in the final result cannot be combined but must be treated as separate inputs.

In Sec. 1-2, Fig. 1-1, we discussed a series circuit of two switches, a battery and a lamp. To light the lamp required that switch A *and* switch B be closed. Figure 1-1a was the circuit diagram. The truth tables were Tables 1-1a and 1-1c. This system could have been called a *positive logic* system.

Suppose instead that we really want the lamp to be OFF. We can call this a *negative logic* system. Then either switch A *or* B has to be open to *not* light the lamp. Using the NOT notation for open switches and a not-lit lamp, we can write Eq. (1-16) to describe the circuit operation:

$$\bar{A} + \bar{B} = \bar{L} \qquad (1\text{-}16)$$

Table 1-15a is the truth table for the switch positions, Table 1-15b is the truth table for the lamp voltages, and Table 1-15c is the table for the logic levels. In

Table 1-15

(a) Two-switch OR

Switch A	Switch B	Lamp
Closed	Closed	ON
Closed	Open	OFF
Open	Closed	OFF
Open	Open	OFF

(b) Voltage levels

Switch A	Switch B	Lamp
Closed	Closed	6 V
Closed	Open	0 V
Open	Closed	0 V
Open	Open	0 V

(c) Logic levels: negative logic

$\bar{A}$	$\bar{B}$	$\bar{L}$
0	0	0
0	1	1
1	0	1
1	1	1

Table 1-15c, a 0 corresponds to a closed switch and the ON lamp. A 1 is an open switch and an OFF lamp. These truth tables now have to be carefully compared with Tables 1-1a, 1-1b, and 1-1c, which are repeated now but arranged in reverse vertical order. If we now make the comparison, we can see that Table 1-15c is like Table 1-1c but with a 0 replacing a 1 and a 1 replacing a 0.

By changing the objective, we have changed from *positive* logic to *negative*

Table 1-1

(a) Two-switch AND (vertical reversed) : positive logic

Switch *A*	Switch *B*	Lamp
Closed	Closed	ON
Closed	Open	OFF
Open	Closed	OFF
Open	Open	OFF

(b) Two-switch AND voltage levels (vertical reversed) : positive logic

Switch *A*	Switch *B*	Lamp Voltages
Closed	Closed	6 V
Closed	Open	0 V
Open	Closed	0 V
Open	Open	0 V

(c) Two-switch AND logic levels (vertical reversed) : positive logic

A	*B*	*L*
1	1	1
1	0	0
0	1	0
0	0	0

logic. Comparing the voltage and logic levels of Tables 1-1b and 1-1c against Tables 1-15b and 1-15c, we can see the following:

Logic Level	Positive Logic	Negative Logic
0	0 V	6 V
1	6 V	0 V

We are therefore able to arrive at the following conclusions:

1 The same circuit can be used for positive and negative logic.

2 An AND circuit becomes an OR and an OR becomes an AND as we change.

3 In the truth table, all 0s are replaced by 1s and 1s by 0s as we change.

4 It is conventional to define positive logic as a logic system in which the 1 voltage level is a more positive voltage than the 0 level, and negative logic as a logic system in which the 1 logic level corresponds to the more negative voltage.

1-11 MAPPING TECHNIQUES: THE KARNAUGH MAP

Mapping techniques are powerful methods for simplifying Boolean logic expressions. They often quickly lead to simple and not obvious simplified expressions. Among the mapping methods are the Karnaugh Map, Venn and Vietch diagrams, and others. This discussion will concern itself with the Karnaugh map.

Consider a single variable A. It can exist as an $\bar{A}$ or an A. We can map it as in Table 1-16. If we have $\bar{A}$, we put a 1 in box a as in Table 1-17. If we have A, we put

Table 1-16 Karnaugh map for single variable

$\bar{A}$	A
a	b

Table 1-17 Single variable $\bar{A}$

$\bar{A}$	A
1 a	b

a 1 in box b as in Table 1-18. Suppose that we have $A + \bar{A}$. We would then map as shown in Table 1-19. But $A + \bar{A} = 1$. Also, the two boxes are filled. Hence, if the

Table 1-18 Single variable A

$\bar{A}$	A
a	1 b

Table 1-19 Single variable $A + \bar{A}$

$\bar{A}$	A
1 a	1 b

two boxes are filled (the complete area is filled), we have a 1. If both boxes were empty, we would have a 0.

In going to a more-than-one-variable function, certain rules must be followed. Enough boxes must be provided so that all possible combinations of variables can be mapped in single boxes: single variable, two boxes; two variables, four boxes; three variables, eight boxes; four variables, 16 boxes; etc. It is just like truth table combinations. How the boxes are organized will determine if simplifications are readily apparent.

In applying the Karnaugh map, it is necessary that the Boolean expression be expressed as an OR function, with each individual term of the OR function expressed as a simple AND term so that there are no combined or equivalent terms in parentheses. For example, $AB\bar{C}$ is allowed, but

$$A(\overline{B\bar{C}})$$

is not allowed.† Similarly, the expression $A(B + C)$ cannot be mapped, but its

---

†By De Morgan's theorem, it is equal to $\bar{A} + B\bar{C}$ or $A(\bar{B} + C) = A\bar{B} + AC$.

equivalent $AB + AC$ can be mapped. Each box on a multivariable map represents a *minterm*. A minterm of n variables is a simple product that includes all the variables once in either true or not true form. For example, in a four-variable logic system a typical minterm might be $A\bar{B}\bar{C}D$. The function must appear like a sum of minterms.

Suppose that we have a two-variable function; it can be mapped as in Table 1-20.

Table 1-20 Two-variable map

	$\bar{A}$	A
$\bar{B}$	a	b
B	c	d

Suppose that we wish to map AB. This is mapped by putting a 1 in box d. We would map $\bar{A}\bar{B}$ by putting a 1 in box a.

Suppose that we now have a function $A\bar{B} + AB$. We put 1 in two boxes as shown in Table 1-21. But $A\bar{B} + AB = A(\bar{B} + B) = A$. From this we see that, if

Table 1-21 $A\bar{B} + AB$

	$\bar{A}$	A	
$\bar{B}$		1	$\leftarrow A\bar{B}$
B		1	$\leftarrow AB$

the map is properly organized, two 1s in adjacent boxes can be simplified as shown in Table 1-22.

Table 1-22 $A\bar{B} + AB = A$

	$\bar{A}$	A	
$\bar{B}$		1	
B		1	$= A$

Let us now consider the following Boolean expression: $\bar{A}B + A\bar{B} + AB$. This can be simplified as $\bar{A}B + A(\bar{B} + B) = \bar{A}B + A$. This seems to be as far as one can go. Let us map it. By using the vertical 1s, we get A. If we use horizontal 1s, we get B. Hence we get $A + B$ if we use the box AB twice, and this is permitted (see Table 1-23).

Table 1-23 $\bar{A}B + A\bar{B} + AB$

	$\bar{A}$	A
$\bar{B}$		1
B	1	1

or

	$\bar{A}$	A	
$\bar{B}$	0	1	$= A + B$
B	1	1	

In examining the above expression we can see that one box, corresponding to $\bar{A}\bar{B}$, is missing. Let us represent a missing box such as this by

$$\overline{\overline{A}\overline{B}}$$

The wider overbar represents a missing or empty box. The box can either be left blank or have a 0 in it. Now, does $\overline{\overline{A}\overline{B}} = A + B$? It does by Eq. 1-15a. $\overline{\overline{A}\overline{B}}$ is a single NAND whereas the original expression had three ANDs and an OR. Mapping techniques therefore can often lead quickly to unexpected simplifications. Note that this really is De Morgan's theorem, which appears quickly as a property of the map.

Let us now consider the expression $A\bar{B} + \bar{A}B$. This plots as in Table 1-24.

Table 1-24 $A\bar{B} + \bar{A}B$ The EXCLUSIVE-OR function

	$\bar{A}$	A
$\bar{B}$		1
B	1	

or

	$\bar{A}$	A
$\bar{B}$	0	1
B	1	0

The 1s in the map are not adjacent. No further simplification is possible. We can change the form somewhat, as in the previous example, by using the missing boxes, and write by inspection

$$A\bar{B} + \bar{A}B = \overline{\overline{A}\overline{B} + AB} \qquad (1\text{-}17)$$

The wider overbar is placed over all the missing boxes. This can easily be proved to be correct by means of a truth table. The term $\overline{\overline{A}\overline{B} + AB}$ is the *complement* of $A\bar{B} + \bar{A}B$.

Three-variable functions can be plotted according to Table 1-25. Note that there are eight available boxes ($2^3 = 8$) and an area is available for all the possible combinations.

Table 1-25 Three-variable mapping

	$\bar{A}$ $\bar{B}$	$\bar{A}$ B	A B	A $\bar{B}$
$\bar{C}$				
C				

Suppose that we have the expression $A\bar{B}C + \bar{A}B\bar{C}$. It is plotted in Table 1-26.

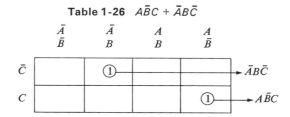

Table 1-26 $A\bar{B}C + \bar{A}B\bar{C}$

Since the boxes are not adjacent, no simplification is possible.

Suppose that we have the expression $X\bar{Y}Z + X\bar{Y}\bar{Z}$. It is plotted in Table 1-27.

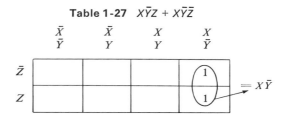

Table 1-27 $X\bar{Y}Z + X\bar{Y}\bar{Z}$

We have two adjacent boxes and a simplification is possible. $X\bar{Y}Z + X\bar{Y}\bar{Z} = X\bar{Y}(Z + \bar{Z}) = X\bar{Y}$. Thus two adjacent boxes on a three-variable map simplify to a two-variable expression. Note that the two 1s are in the $X\bar{Y}$ column.

Let us now consider the expression $\bar{X}\bar{Y}\bar{Z} + X\bar{Y}\bar{Z} = \bar{Y}\bar{Z}(\bar{X} + X) = \bar{Y}\bar{Z}$. This maps as shown in Table 1-28,

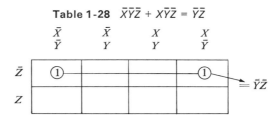

Table 1-28 $\bar{X}\bar{Y}\bar{Z} + X\bar{Y}\bar{Z} = \bar{Y}\bar{Z}$

We see that by organizing the map in this manner the opposite sides must be considered adjacent (as though the map were cylindrical). By looking carefully at the map one can see that the boxes used are common to $\bar{Y}$ and $\bar{Z}$.

Let us now consider mapping $\bar{L}\bar{M}N + \bar{L}MN + LMN + L\bar{M}N$. This is shown in Table 1-29.

Table 1-29 $\bar{L}\bar{M}N + \bar{L}MN + LMN + L\bar{M}N$

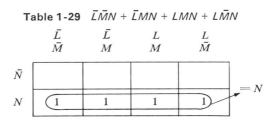

But

$$\bar{L}\bar{M}N + \bar{L}MN + LMN + L\bar{M}N = N(\bar{L}\bar{M} + \bar{L}M + LM + L\bar{M})$$
$$= N[(\bar{L})(\bar{M} + M) + L(M + \bar{M})]$$
$$= N[(\bar{L} + L)(\bar{M} + M)]$$
$$= N \quad \text{by using Eq. (1-5b)}$$

We see therefore that four adjacent boxes simplify to a single variable. We can also see that the four 1s all are in the N row.

Let us map $\bar{R}S\bar{T} + RS\bar{T} + \bar{R}ST + RST$. This is shown in Table 1-30. It is easy to show that this simplifies to S. We can see this also from the map because of the four adjacent boxes located in the common S area of the map.

Table 1-30 $\bar{R}S\bar{T} + RS\bar{T} + \bar{R}ST + RST = S$

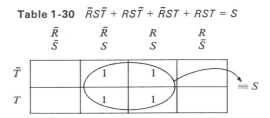

Let us map $\bar{A}B\bar{C} + \bar{A}B + AB\bar{C} + AC$. We can see that there are three variables and therefore we need an eight-box map. What about the terms $\bar{A}B$ and AC? They have two inputs. We have just seen that, in the process of simplification, two adjacent boxes reduce to a term with one less input. In the reverse procedure, terms like $\bar{A}B$ and AC plot as two 1s on adjacent boxes. The plot of this expression is given in Table 1-31. Had there been a single input such as C, it would have plotted as four 1s on the C row.

Table 1-31 $\bar{A}B\bar{C} + \bar{A}B + AB\bar{C} + AC$

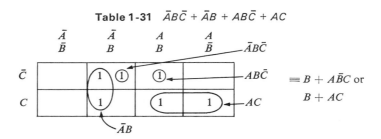

There are no restrictions on having a box duplicated as occurs in the box $\bar{A}B\bar{C}$. Inspection of the map shows that it can be simplified to $B + A\bar{B}C$. (As a guide to simplification, note that B is common to both $\bar{A}$ and A and $\bar{C}$ and C. Both $\bar{A}$ and A and $\bar{C}$ and C become equal to 1, and all four adjacent boxes simplify to B.) It can also be simplified to $B + AC$. This can be seen on the map; it also follows from Eq. (1-14).

We have just seen in Table 1-31 how a box can contain more than a single 1. This is allowed, since in Boolean algebra $1 + 1 = 1$. In the reverse procedure, a 1 in a box can be used as many times as needed. For example, plot $\bar{R}\bar{S}\bar{T} + \bar{R}S\bar{T} + RS\bar{T} + \bar{R}ST$ in Table 1-32.

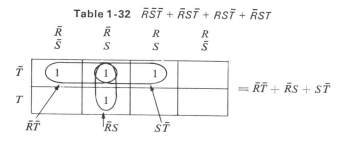

Table 1-32 $\bar{R}\bar{S}\bar{T} + \bar{R}S\bar{T} + RS\bar{T} + \bar{R}ST$

$$\equiv \bar{R}\bar{T} + \bar{R}S + S\bar{T}$$

This simplifies to $\bar{R}\bar{T} + \bar{R}S + S\bar{T}$. The box $\bar{R}S\bar{T}$ was used three times. It is quite apparent how the map has quickly enabled us to get from an expression of four three-input ANDs plus a four-input OR to a three two-input AND and one three-input OR.

Table 1-33   Four-variable map organization

		$\bar{A}$ $\bar{B}$	$\bar{A}$ B	A B	A $\bar{B}$
$\bar{D}$	$\bar{C}$				
D	$\bar{C}$				
D	C				
$\bar{D}$	C				

Four-variable maps use the organization shown in Table 1-33. This has 16 boxes ($2^4 = 16$), one box for each combination of the variables (corresponding to each line of a truth table). In this map

1   A four-input variable plots as a single box.
2   Two adjacent boxes reduce to a three-variable term. A three-variable term plots on two adjacent boxes.
3   Four adjacent boxes reduce to a two-variable term. A two-variable term plots as four adjacent boxes.
4   Eight adjacent boxes reduce to a one-variable term. A one-variable term plots as eight adjacent boxes.

EXAMPLE 1-12   Plot $\bar{A}\bar{B}\bar{C}\bar{D} + \bar{A}\bar{B}\bar{C}D + ABC$ and simplify.

Solution

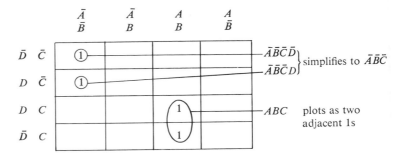

EXAMPLE 1-13

    a. Plot BC on a four-variable map whose inputs are A, B, C, and D.

    b. Plot $A\bar{D}$ on a four-variable map whose inputs are A, B, C, and D.

Solution

    a.   b.

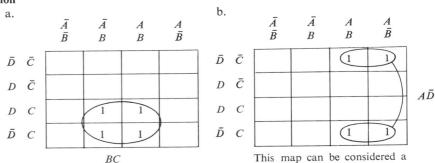

This map can be considered a cylinder also, both vertical and horizontal.

EXAMPLE 1-14

    a. Plot F on a four-variable map whose inputs are E, F, G, and H.

    b. Plot $\bar{F}$ on a four-variable map whose inputs are E, F, G, and H.

Solution

    a.   b.

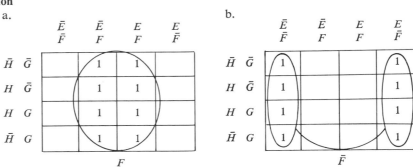

EXAMPLE 1-15 Plot and simplify $\bar{A}B\bar{C}\bar{D} + B\bar{C}\bar{D} + DCB + \bar{D}CB$.

Solution

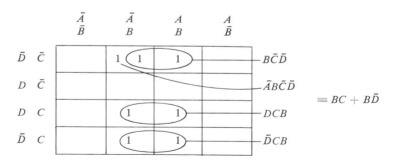

This simplifies to BC (combination of DCB and $\bar{D}CB$) + $B\bar{D}$ (combination of DCB and BCD, the bottom and top pairs of 1s).

1-12 THE RACE PROBLEM

Consider a two-input AND gate with waveforms A and B applied to its inputs (Fig. 1-26). What is waveform C? It appears that when A goes up B comes down, simultaneously. It would appear as though C would always be 0 since at no time are both A and B equal to 1. However, it is impossible practically to obtain waveforms that go up or come down in zero time. There is always some slope to the waveform.

In addition, another problem is present. We have seen that logic levels are represented by voltage levels. We can say that a 1 is the equivalent of $+5$ V and 0 is represented by 0 V. Then is $+4$ V or $+3$ V a 1? In IC logic circuits, there is a very sharp transition voltage around which logic 1 or logic 0 occurs. Its level is determined by circuit design. We shall call this the *threshold voltage*. Above this voltage, we have logic 1 and below it we have logic 0.

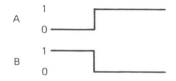

Figure 1-26 Race waveforms applied to a gate

Now let us apply gate input voltages with some slope, to a gate, as shown in Fig. 1-27. In Fig. 1-27a, the threshold voltage is above the crossover of A and B. At no time do A and B exceed logic 1 at the same time. C is always zero. In Fig. 1-27b

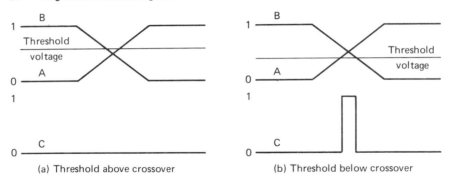

(a) Threshold above crossover

(b) Threshold below crossover

Figure 1-27 Race conditions and glitch generation: (a) threshold above crossover; (b) threshold below crossover

the threshold voltage is lower, and both *A* and *B* are above the threshold voltage for a short time. As a result, there is an undesired 1 level output at *C* for a short time. This is called a "glitch."

1-13 INTEGRATED-CIRCUIT GATES

In Chapter 3 we shall discuss in some detail various families of integrated circuits (ICs) that perform the logic functions described in this chapter. Each family has a basic function style, which is either NOR or NAND. The ICs are assembled in a sealed package, and gate inputs and outputs are brought to terminal pins for connection to external control signals. The degree of complexity within the package is to a great extent determined by the number of terminal pins in the package. With single gates, enough pins are available so that quite often more than one gate can be in an IC, but with complex circuits frequently the complexity is determined only by the number of available terminal pins.

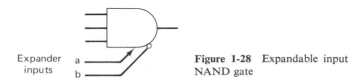

Expander inputs a
b

Figure 1-28 Expandable input NAND gate

Gates are available in some families with up to 10 inputs. However, some circuits are so complex that the number of available inputs may not be enough to provide additional input capacity. Some gates have additional input terminals for use as an expander input to provide additional input capability. The expander input logic symbol is shown in Fig. 1-28 as part of a NAND gate. Two input terminals, *a* and *b*, couple the expander inputs to this gate. Some families require only a single input, input *a*.

To provide an idea of the availability of gates within a given IC, Fig. 1-29 shows the pin connections for the IC type SN54/7400N or SN54/7400J. This IC is assembled in a plastic (N) or ceramic (J) container with 14 pins arranged in a dual-in-line arrangement of 7 pins on each long side. With 2 pins required for power, 12 pins are available for gating purposes. Since a two-input NAND gate requires three connections, a quadruple two-input positive logic NAND gate is available. For positive logic, the Boolean equation is $Y = \overline{AB}$; this can be performed in four different circuits.

J OR N DUAL-IN-LINE PACKAGE (TOP VIEW)

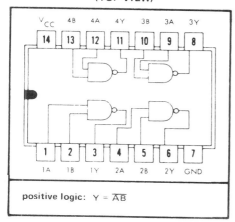

positive logic: $Y = \overline{AB}$

Figure 1-29 Typical IC gates: circuit types SN5400, SN7400 quadruple two-input positive NAND gates

In any one IC family many different types of gates are made. Figure 1-30 shows the various styles of gates and inverters that are available in one of the IC families, the TTL (transistor–transistor logic) 54/7400 family. The basic logic operation in this family is the NAND. These are inverters, NAND gates with inputs from two to eight. There are NOR gates and expanders. In addition, there are several styles of combinations of two-stage AND–OR invert gates for use in more complex circuitry.

To illustrate how these gating ICs can be used to perform circuit operations, Fig. 1-31 combines half the 7451, a two-wide, two-input AND–OR–INVERT gate, with two inverters (two sixths) of the 7404 to build an IC single-pole, double-throw switch. In Fig. 1-31, if $A = 1$, gate I is enabled and data X are transmitted to $\bar{Z}$ and Z. If $A = 1$, $\bar{A} = 0$, and gate II is disabled; data Y are not transmitted. If $A = 0$, gate I is disabled and gate II is enabled, transmitting Y data to $\bar{Z}$ and Z.

A practical way of using ICs is to mount them on a logic card. Figure 1-32 shows a logic card on which three 14-pin dual-in-line ICs are mounted. Each IC contains two four-input NAND gates with expander capability. This card therefore has six four-input expandable NAND gates. The card material is a glass epoxy and is very good electrically and strong mechanically. This card has 44 gold-plated connecting fingers along one edge for insertion into a socket.

NAND/NOR GATES

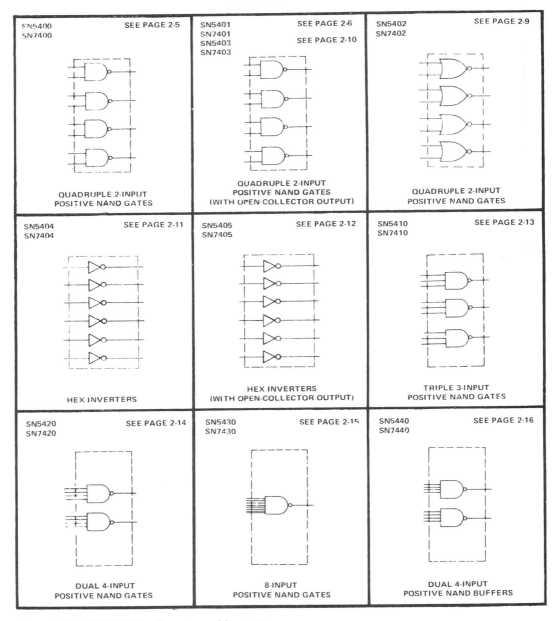

Figure 1-30 SN54/7400 family: gates and inverters

AND–OR–INVERT GATES

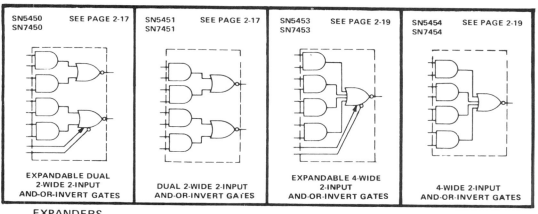

Figure 1-30 (continued)   SN54/7400 family: gates and inverters

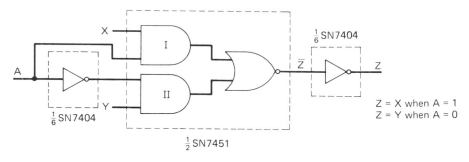

Figure 1-31   Integrated-circuit single-pole double-throw switch

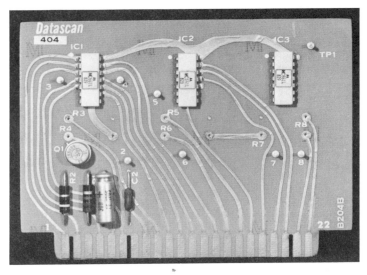

Figure 1-32 Integrated-circuit logic card with six four-input expandable NAND gates (Datascan Division of Amadex Instruments)

Problems

1-1. Write a truth table and draw the logic symbol for $X + Y = Z$.

1-2. Write a truth table and draw the logic symbol for $RS = T$.

1-3. The battery, relays, and lamp of Fig. 1-2 are replaced with a 120-V battery, relays, and lamp. Write a truth table for the circuit for the relay voltages and the lamp voltage. What voltage would correspond to logic level 0 and what voltage to logic level 1?

1-4. Write a truth table and draw the logic symbol for $K + \bar{L} = M$.

1-5. Repeat Prob. 1-4 for $\bar{K}L = M$.

1-6. Write a truth table and draw the logic symbol for $XY\bar{Z} = P$.

1-7. Write a truth table and draw the logic symbol for $X\bar{Y}\bar{Z} = R$.

1-8. Write truth tables and draw the logic symbols for (a) $\overline{A\bar{B}} = C$; (b) $\overline{A} + B = C$.

1-9. Write a truth table and give the logic equation for

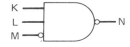

1-10. Write a truth table and give the logic equation for

1-11. Draw switch arrangements similar to Figs. 1-20a through 1-20e to demonstrate Eqs. (1-5b), (1-5c), and (1-5d).

1-12. Prove the equivalence of all three forms of Eqs. (1-10) and the logic diagrams of Fig. 1-20g by means of truth tables. For example, as in Fig. 1-20g, part (a), first do $B + C$ and then combine its result with A.

1-13. Repeat Prob. 1-12 for Eqs. (1-11) and Fig. 1-20h.

1-14. (a) Does $\bar{A} + \bar{B} = \bar{A}\bar{B}$? Prove or disprove.

(b) Use a truth table to show that $A\bar{B} + \bar{A}B = \overline{AB} + \overline{\bar{A}\bar{B}}$.

1-15. Use De Morgan's theorem to convert to AND/NAND logic only. Draw logic diagrams for the given and converted expressions: (a) $A + \bar{B} + C + \bar{D} + E$; (b) $XY + Y\bar{Z}\bar{X} + RS\bar{T}\bar{U}$. (*Hint:* do not change the AND terms.)

1-16. Repeat Prob. 1-15 for (a) $\bar{X} + Y + \bar{Z}$; (b) $A + \bar{B}E + \overline{KLM}$.

1-17. Use De Morgan's theorem to convert to OR/NOR logic only. Draw logic diagrams for the given and converted expressions: (a) $RS\bar{T}$; (b) $(K\bar{L})(\overline{A\bar{B}D})$; (c) $A\bar{B}D + (UV)(A + B)$.

1-18. Repeat Prob. 1-17: (a) $\bar{K}L\bar{M}P$; (b) $(\overline{\overline{XYZ}})(B)$; (c) $(X + \bar{Y})(\overline{L + N})(\overline{\bar{A}\bar{B}})$.

1-19. Draw the waveforms you would expect at X and Y.

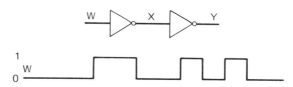

1-20. Draw the waveform at C.

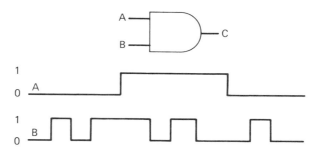

1-21. Draw the waveform at C.

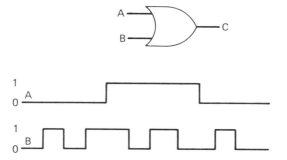

1-22. Draw the waveform at *D*.

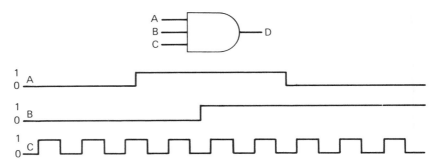

1-23. At each point on the logic diagram, write the Boolean expression in terms of *A*, *B*, and *C*.

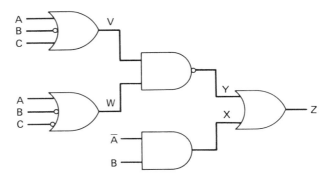

1-24. At each point on the logic diagram, write the Boolean expression in terms of *A*, *B*, *C*, and *D*.

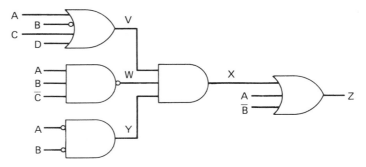

1-25. Draw a logic diagram for (a) $(A\bar{B}C + \bar{A}C\bar{D})DA\bar{C}$; (b) $(XY)(X + Y) + AB + AC$.

1-26. Draw a logic diagram for $(K + L + \bar{M})(KL + \bar{M}) + \bar{K}LM$.

1-27. Minimize by Boolean formulas and draw the logic diagrams: (a) $(A + B)(A + \bar{B})$; (b) $AB + AB\bar{C}D + AB\bar{D}$.

1-28. Simplify using Boolean formulas and draw the logic diagrams: (a) $X\bar{Y} + XY\bar{Z} + \bar{X}Z + \bar{X}X$; (b) $ABC\bar{D} + CDBA + AB$.

1-29. Minimize using Boolean formulas and draw the logic diagrams: (a) $(A + B + \bar{C})(\bar{A} + B + \bar{C})$; (b) $(A + \bar{B})(A + \bar{C})(B + \bar{C})$.

1-30. Show that $\bar{A}\bar{B} + \bar{A}B + A\bar{B} = \overline{AB} = \bar{A} + \bar{B}$. Use both Karnaugh mapping techniques and Boolean formulas.

1-31. Map $\bar{A}B + AB$ and simplify.

1-32. Map $\bar{A}\bar{B}C + \bar{A}B\bar{C} + ABC$. Can this be simplified?

1-33. Map $\bar{A}\bar{B}C + A\bar{B}C$ and simplify. Explain this simplification.

1-34. Map $\bar{A}B\bar{C} + AB\bar{C} + \bar{A}BC + ABC$. Can this be simplified? If so, give the simplified result.

1-35. (a) Map $S\bar{T} + \bar{R}S\bar{T} + R\bar{S}\bar{T} + RST$. Simplify.
(b) Express the result in complementary form.

1-36. Map $S + \bar{R}S\bar{T} + RST$ and simplify.

1-37. Map $\bar{A}\bar{B}\bar{C}\bar{D} + \bar{A}\bar{B}\bar{C}D + ABCD + A\bar{B}C\bar{D}$.

1-38. Map $\bar{A}\bar{B}\bar{C}D + AB\bar{C}D + ABC\bar{D} + A\bar{B}C\bar{D}$. Simplify.

1-39. Map $\bar{A}B\bar{C}\bar{D} + \bar{A}BC\bar{D}$. Can this be simplified? Explain.

1-40. Map $A\bar{B}CD + \bar{A}\bar{B}CD$. Can this be simplified? Explain.

1-41. Map B on a four-variable map in which the variables are A, B, C, and D.

1-42. Map $\bar{K}\bar{L} + K\bar{L}$ on a four-variable map in which the variables are K, L, M, and N. Simplify. Explain.

1-43. Map and show that $\bar{A}\bar{B} + \bar{A}BC\bar{D} + \bar{A}B\bar{C}D + \bar{A}BC + AB\bar{C} + ABC\bar{D} + A\bar{B}\bar{C} + A\bar{B}C = \overline{ABCD} = \bar{A} + \bar{B} + \bar{C} + \bar{D}$.

1-44. Al, Bob, and Chuck have decided to form a stock investment club. The decisions on buy or sell are to be based upon a majority vote. They wish to vote using an electronic logic system. Draw the simplest logic diagram with logic symbols, and use the Karnaugh map to show how you would do this.

1-45. Given that $X = \bar{A}B$. Determine which TTL gates you would need to implement this problem. Use the gates given in Fig. 1-30. Use ICs whose numbers begin with SN74. If inverters are needed, use the SN7404. If two-input NAND gates are needed, use the SN7400. Draw a logic diagram in the style of Fig. 1-31. *Note:* De Morgan's theorem can be used if needed.

1-46. Repeat Prob. 1-45 for $\overline{(AB)(CD)} = X$.

1-47. Repeat Prob. 1-45 for $AB\bar{C}\bar{D}E\bar{F} = X$.

1-48. Repeat Prob. 1-45 for $\overline{AB + CD + AC} = X$.

1-49. Repeat Prob. 1-45 for $\overline{(\bar{A} + \bar{B})(E + \bar{A})(B + D)} = X$.

1-50. Repeat Prob. 1-45 for $A + B + \bar{C} + \bar{D} = X$.

TWO

Semiconductor Principles

2-1 INTRODUCTION

In this chapter the principles of semiconductors are discussed. This establishes a basis for understanding the operation of diodes, transistors, and their combination into integrated circuits. The emphasis is on those characteristics which are applicable to digital integrated circuits. This chapter serves as an introduction to Chapter 3, in which the various types and families of integrated circuits are discussed.

2-2 SEMICONDUCTOR PHYSICS

The chemical elements silicon (Si) and germanium (Ge), if carefully allowed to solidify, form crystals. Both elements have four electrons in their outermost shell. In this crystal lattice, the four outermost or valence electrons are tightly held or bonded to the electrons of adjacent atoms in a type of bonding known as covalent (electron-pair) bonding. This is shown in Fig. 2-1.

In the structure of Fig. 2-1, which occurs at low temperatures, the electrons are held very tightly by the bonding and there are no charge carriers available to conduct current. This material is a very poor conductor. At high temperatures or with strong electric fields, it is possible to break the bonding and have current flow.

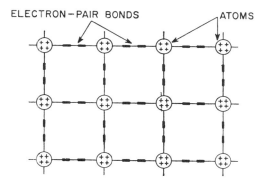

ELECTRON – PAIR BONDS   ATOMS

Figure 2-1 Crystal Lattice structure
(Radio Corporation of America)

If very small amounts of impurities (materials that have five electrons in their outer shell) are added to the material (doping), it enters into the lattice as shown in Fig. 2-2. The covalent bonding is still maintained, but there is now present in the crystal an "extra" electron available to carry current. It is held very loosely and can readily carry current. This type of material is called *N* material (for the negative charge on the electron) and is a much better conductor than the material of Fig. 2-1.

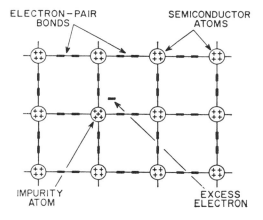

ELECTRON – PAIR
BONDS   SEMICONDUCTOR
                        ATOMS

IMPURITY   EXCESS
ATOM   ELECTRON

Figure 2-2 Lattice structure of *N*-type (−) material (Radio Corporation of America)

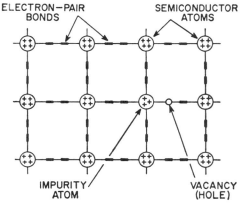

ELECTRON – PAIR
BONDS   SEMICONDUCTOR
                        ATOMS

IMPURITY   VACANCY
ATOM   (HOLE)

Figure 2-3 Lattice structure of *P*-type (+) material (Radio Corporation of America)

In a similar manner, if a material with three electrons in its outer shell is added to the crystal in very small amounts, it also enters into the crystal structure, as shown in Fig. 2-3. The bonding is still maintained, but an electron is now "missing" to complete the covalent bonding. This vacancy is called a *hole* and is considered to have a positive charge (absence of electron). The material is called *P* material. Current flows in the following manner. Under an electric field an electron from an adjacent atom moves to complete the covalent bonding at the impurity atom. But this leaves a hole in the adjacent atom. Electrons move in one direction and holes in the opposite direction. This material is also a much better conductor than the material of Fig. 2-1.

In Fig. 2-1, at higher temperatures or with high electric fields, electron bonding can be broken. This leaves an electron–hole pair available for the conduction of current. As temperature increases, the number of electron–hole pairs generated doubles for approximately every 10°C rise in temperature.

2-3 DIODE

If we add *N*-type impurities to one side of a crystal lattice and *P*-type impurities to the other side, as shown in Fig. 2-4a, we form a *diode* or two-element device. The region where the two materials meet is called a *junction*. Some of the free electrons from the *N*-type material diffuse across the junction and recombine with the holes in the *P* material. This leaves the *N* material with a slight positive charge. Similarly, some of the holes diffuse across the junction to combine with electrons in the *N* material. This leaves the *N* material with a slight negative charge. The net result is a region near the junction that has no charge carriers, as shown in Fig. 2-4b. This region is known as the *space-charge region* or *transition region* or *depletion layer*. The resultant positive charge in the *N* region and the negative charge in the *P* region act to both limit the number of charge carriers that diffuse across the junction and to form a voltage barrier of approximately 0.3 V for germanium and 0.7 V for silicon, which has to be overcome if current is to flow.

Let us now apply a voltage to the crystal, positive to the *P* region and negative to the *N* region. The polarity of the voltage applied attracts electrons and holes

Figure 2-4 Crystal with *P-N* junction: (a) before diffusion; (b) after diffusion; (c) diode symbol

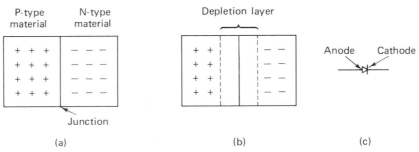

toward and across the junction. After overcoming the barrier voltage, current flows easily. The diode is forward biased. The diode symbol is given in Fig. 2-4c. The direction of the arrow corresponds to that for conventional current flow, and the bar side of the symbol corresponds to the cathode or *N*-material side. The voltage across the diode is approximately equal to the barrier voltage. The forward voltage across the diode is temperature sensitive and decreases by approximately 2.2 millivolts (mV) per °C increase in junction temperature. For example, if the voltage across the crystal is 0.7 V at 25°C, at a temperature of 125°C we would expect a voltage of

$$[0.7 - 0.0022(125°C - 25°C)] = 0.48 \text{ V}$$

The voltage across the diode in the forward direction depends to some extent upon the current through the diode. The incremental (small ac signal) resistance of the diode in the forward direction is expressed by Eq. (2-1), where I_d is in milliamperes of direct current.

$$r_{\text{inc}} = \frac{26}{I_d} \, \Omega \qquad (2-1)$$

For example, if the diode current is 2 mA, we can expect the small-signal incremental resistance to be approximately $26/2 = 13 \, \Omega$.

If we now reverse the polarity of the voltage applied to the diode by connecting the *P* material to the negative voltage terminal and the *N* material to the positive voltage terminal, the voltage polarity will be such that the charge carriers will be attracted away from the junction and current will not flow. The diode is reverse biased. Actually, a small amount of leakage current will flow owing to thermally generated electron–hole pairs in the space-charge region. These reverse currents flow as electrons in the *P* region and holes in the *N* region; the current carriers are called minority current carriers. This current doubles for approximately every 10°C increase in junction temperature.

If we increase the reverse voltage across the diode, there will be a sudden large increase in current as a critical reverse breakdown voltage is reached. If the amount of added impurities is high, the critical voltage is low. The electric field can reach within the crystal lattice and break the bond, releasing electrons. This is called the Zener effect. Zener-effect diodes have a negative temperature coefficient of voltage. If the doping is lower, the breakdown voltage is higher. The reverse leakage electrons in the depletion region acquire kinetic energy and strike another atom, releasing another electron–hole pair. This avalanches, and there is a sudden sharp increase in current. Avalanche breakdown has a positive temperature coefficient of voltage. The dividing line between the Zener and avalanche effect is at about 5 V. Nevertheless, all diodes designed to use this reverse breakdown characteristic are usually called Zener diodes. Avalanche diodes have a much sharper onset (knee), and the increase in current is much more rapid than for the Zener effect.

Figure 2-5a shows the voltage current characteristics of a silicon diode for both forward (*F*) and reverse (*R*) voltage biasing. Note the change in scale between the two regions. Figure 2-5b shows these characteristics idealized.

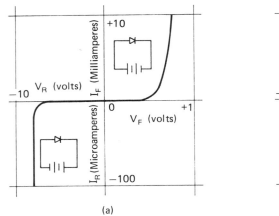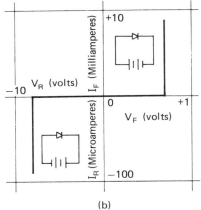

(a) (b)

Figure 2-5 Diode characteristics: (a) actual; (b) idealized

Figure 2-6 shows the symbol of a Zener diode. The positive temperature characteristic of the Zener avalanche diode can be compensated by connecting a forward-biased diode in series with it. Its negative temperature characteristic will balance that of the avalanche Zener diode. Such diodes are called *reference diodes* (Fig. 2-7). Reference diode voltages are typically 6–8 V.

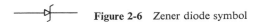 **Figure 2-6** Zener diode symbol

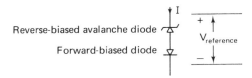

Figure 2-7 Reference diode

When a diode is reverse biased, the *P* region and the *N* region are separated by the depletion layer, which is free of current carriers and acts like a dielectric. Hence capacitance exists between the terminals of a reverse-biased diode. The depletion layer widens as the reverse bias applied to the diode is increased and the capacitance is reduced.

2-4 BIPOLAR TRANSISTOR

2-4.1 Basic Transistor Operation and Current Gain

If, in a single crystal lattice, *N* and *P* regions are made as shown in Fig. 2-8a, two back-to-back diodes are formed around two junctions. This forms an *NPN* transistor whose symbol is shown in Fig. 2-8b. A *PNP* transistor can be similarly formed

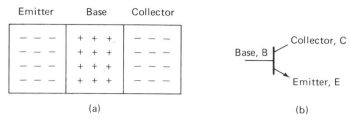

Figure 2-8   Bipolar transistor: (a) N and P regions in NPN transistor;
(b) symbol

by making the emitter and collector of P material and the base of N material. The symbol for the PNP transistor is the same as that for the NPN transistor, with the exception of the arrow on the emitter lead, which is reversed from that shown in Fig. 2-8b. In the discussion on transistors that follows, the characteristics of both types, NPN and PNP, are identical except for the polarity of the voltages applied to the transistor. In the fabrication of transistors, the emitter and collector are heavily doped and the base lightly doped, which makes the base resistance much higher than that of the collector and emitter.

If the emitter–base diode of the transistor of Fig. 2-8 is forward biased, base positive with respect to the emitter, current will flow in this diode. This injects electrons (minority carriers) into the base region. If we now make the collector–base diode reverse biased, collector positive with respect to the base, the positive potential on the collector will attract most of the emitter-originated electrons from the base region into the collector.

We can write the following equation for the dc conditions of the transistor:

$$I_E = I_C + I_B \tag{2-2}$$

Current gain collector to emitter (common base):

$$h_{FB} \quad \text{or} \quad \alpha_{\text{dc}} = \frac{I_C}{I_E} \approx 1 \tag{2-3}$$

Current gain collector to base (common emitter):

$$h_{FE} \quad \text{or} \quad \beta_{\text{dc}} \quad \text{or} \quad B = \frac{I_C}{I_B} \tag{2-4}$$

Current gain emitter to base (common collector):

$$h_{FC} = \frac{I_E}{I_B} = \frac{I_C + I_B}{I_B} = h_{FE} + 1 \tag{2-5a}$$

$$\approx h_{FE} \tag{2-5b}$$

For small-signal conditions, we can write, in a similar manner,

$$i_e = i_c + i_b \tag{2-6}$$

Current gain collector to emitter (common base):

$$h_{fb} \quad \text{or} \quad \alpha = \frac{i_c}{i_e} \approx 1 \tag{2-7}$$

Current gain collector to base (common emitter):

$$h_{fe} \quad \text{or} \quad \beta = \frac{i_c}{i_b} \tag{2-8}$$

Current gain emitter to base (common collector):

$$h_{fc} = \frac{i_e}{i_b} = \frac{i_c + i_b}{i_b} = h_{fe} + 1 = \beta + 1 \tag{2-9a}$$

$$\approx h_{fe} \approx \beta \tag{2-9b}$$

Typical values of h_{FB} and h_{fb} vary from 0.90 to 0.999. Typical values of h_{FE} and h_{fe} vary from 10 to 1000.

There is a relationship between h_{fe} and h_{fb}, which is expressed as

$$h_{fe} = \frac{h_{fb}}{1 - h_{fb}} \tag{2-10a}$$

or

$$\beta = \frac{\alpha}{1 - \alpha} \tag{2-10b}$$

The three possible configurations for operating the transistor are shown in Fig. 2-9. In the common-base configuration, a signal current (I_E or i_e) is applied into the emitter. The output is a current (I_C or i_c) in the collector. Equations (2-3) and (2-6) are applicable.

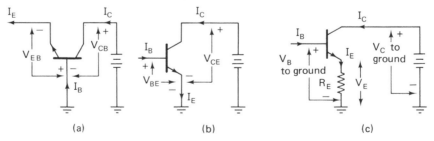

(a)   (b)   (c)

Figure 2-9   Three transistor operating configurations: (a) common base; (b) common emitter; (c) common collector-emitter follower

In the common-emitter configuration, a signal current (I_B or i_b) is applied into the base. The output is a current (I_C or i_c) in the collector. Equations (2-4) and (2-8) are applicable.

In the common-collector configuration, a signal current (I_B or i_b) is applied into the base. The output is a current (I_E or i_e) in the emitter. Equations (2-5) and (2-9) are applicable. In many applications of this configuration a resistor R_E, as shown in Fig. 2-9c, is connected between the emitter and ground. The emitter current develops a voltage V_E or v_e across R_E. This circuit is commonly called an emitter follower. The voltage gain (V_E/V_B or v_e/v_b) is approximately unity, and the output voltage V_E or v_e is in phase with the base signal voltage. We say that the emitter voltage *follows* the base voltage.

2-4.2 Input Resistance

The input resistance r_i of the transistor is of considerable importance.

1. In the common-base configuration $r_{i_b} = v_{eb}/i_e$ and is essentially equal to the resistance of a forward-biased diode, which we have given previously as $r = 26/I_E \ \Omega$ [Eq. (2-1)].

2. In the common-emitter configuration $r_{ie} = v_{be}/i_b$. But, from Eq. (2-8), we have

$$h_{fe} = \frac{i_c}{i_b}$$

In most transistors i_b is small compared to i_c and i_e, so that $i_c \approx i_e$. We can write, using Eq. (2-8),

$$r_{ie} = \frac{v_{be}}{i_c/h_{fe}} \approx \frac{v_{be}}{i_e/h_{fe}} \approx h_{fe}\frac{v_{be}}{i_e} \approx h_{fe}r_{ib} \tag{2-11}$$

For example, if $I_E \approx I_C = 1$ mA, and $h_{fe} = 100$, then

$$r_{ie} \approx 100\left(\frac{26}{I_E}\ \Omega\right) \approx 2600\ \Omega \tag{2-12}$$

3. In the common-collector emitter-follower configuration the input resistance is

$$r_{ic} = \frac{v_b}{i_b} = \frac{v_{be} + v_e}{i_b} = \frac{v_{be}}{i_b} + \frac{v_e}{i_b}$$

But
$$v_e = i_e R_e = (i_c + i_b)R_E = (h_{fe}i_b + i_b)R_E$$
$$= (h_{fe} + 1)i_b R_E$$

Since in most transistors $i_c \gg i_b$ and $h_{fe} \gg 1$,

$$v_e \approx h_{fe}i_b R_E$$

Substituting, we obtain

$$r_{ic} = \frac{v_{be}}{i_b} + h_{fe}R_E$$

$$= r_{ie} + h_{fe}R_E \tag{2-13a}$$

In most cases, $r_{ie} \ll h_{fe}R_E$, and we obtain

$$r_{ic} \approx h_{fe}R_E \qquad\qquad (2\text{-}13b)$$

2-4.3 Comparison of the Three Circuit Configurations

In common-base operation, the input resistance is low and the output resistance is high. Current gain is approximately unity, but there can be voltage gain. The input and output voltages are in phase. Transistors are infrequently used in this manner.

In common-emitter operation, the input resistance is moderate and the output resistance is moderately high. There is both current and voltage gain, with phase inversion between input and output. Transistors are used most frequently in this manner. In digital work, the transistor used as an inverter is of great importance.

In common-collector operation, more commonly known as emitter-follower operation, the input impedance is very high and the output impedance low. This type of operation is used as an impedance converter. Current gain is high and voltage gain is approximately unity. The output voltage at the emitter is in phase with the input voltage to the base. This is a fairly common type of operation.

2-4.4 Cutoff, Saturation, and Active Regions

Transistor characteristics are commonly displayed by means of families of curves in which the currents in the terminals are plotted versus applied voltage or currents. Although both base current and collector voltage do have some effect upon the input characteristics, the effect is small; in most cases it can be safely assumed that the input voltage of the transistor $V_{BE} \approx 0.3$ V (germanium) or 0.7 V (silicon).

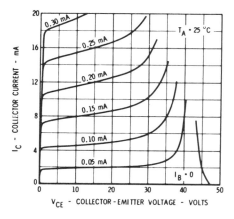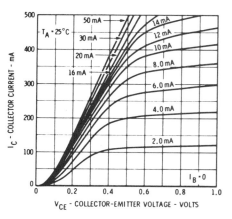

Figure 2-10 Common emitter-collector characteristics (Fairchild Semiconductor type 2N2845)

Figure 2-10 plots the collector characteristics of a transistor in the common-emitter connection. This is a plot of collector current I_C against the collector-to-emitter voltage V_{CE}, with the base current I_B as a parameter.

In the characteristics of Fig. 2-8, there are three regions of interest.

1. Cutoff, where $I_B = 0$. The collector current is ≈ 0.

2. Saturation: the region where $V_{CE} < V_{BE}$. If $V_{CE} < V_{BE}$, the collector–base junction becomes forward biased and the collector starts to inject minority carriers also into the base in a direction such as to oppose the minority carriers that are emitter derived. Hence the collector current falls rapidly. With both emitter and collector injecting minority carriers into the base region, minority carriers are stored in the base region (and collector region) and it requires time to remove them. This has important consequences for transient response and switching time.

3. Active: the remainder of the region in the collector family.

2-4.5   Circuit Conditions for Saturation

Consider the transistor in the circuit of Fig. 2-11, which appears frequently in digital work. We can write in this circuit for the base

$$V_{CC} = I_B R_B + V_{BE} \tag{2-14a}$$

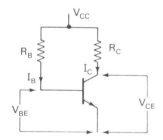

Figure 2-11   Transistor in common-emitter circuit

If V_{CC} is 5 V or larger, the effect of V_{BE} can be neglected, and we can write

$$V_{CC} = I_B R_B \tag{2-14b}$$

$$I_B = \frac{V_{CC}}{R_B} \tag{2-14c}$$

and for the collector

$$V_{CC} = I_C R_C + V_{CE} \tag{2-15}$$

The most current we can obtain will occur when $V_{CE} \approx 0$.

$$V_{CC} = I_{C\,max} R_C \tag{2-16a}$$

But when $V_{CE} \approx 0$, the transistor is in saturation, and we can call the collector current $I_{C\,sat}$.

$$V_{CC} = I_{C\,sat} R_C \tag{2-16b}$$

But $I_C = h_{FE}I_B$, and we can now write

$$I_{C \text{ sat}} = h_{FE}I_{B \text{ sat}} \tag{2-17}$$

where $I_{B \text{ sat}}$ is the base current just needed for saturation. Any base current in *excess* of this will drive the transistor further into saturation. Therefore, combining Eqs. (2-14b), (2-16b), and (2-17),

$$V_{CC} = h_{FE}I_{B \text{ sat}}R_C = I_{B \text{ sat}}R_B$$

we obtain

$$h_{FE}R_C = R_B \tag{2-18a}$$

$$h_{FE} = \frac{R_B}{R_C} \tag{2-18b}$$

These equations show that to just get a transistor into saturation make $R_B = h_{FE}R_C$, or, alternatively, if

$$\frac{R_B}{R_C} < h_{FE} \tag{2-18c}$$

the transistor is in saturation. The term (R_B/R_C) is frequently called the *circuit* h_{FE}. Equation (2-18c) says that, for the transistor to be in saturation, the *circuit* h_{FE} must be *lower* than the transistor h_{FE} in the circuit of Fig. 2-11.

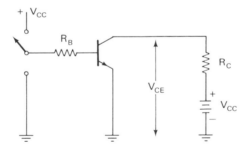

Figure 2-12   Transistor as a switch

In digital work the transistor is used in the configuration of Fig. 2-11 to act as an inverter and maintain two levels. One level is high and equal to V_{CC}; the other is low and equal to $V_{CE \text{ sat}} \approx 0$ V. This is illustrated in Fig. 2-12. The resistance R_B is switched between two voltages (usually electronically) between V_{CC} and ground. R_B and R_C are selected in accordance with Eq. (2-18c). If R_B is at ground,

$$I_B = 0, \qquad I_C = 0, \qquad V_{CE} = V_{CC}$$

If R_B is at V_{CC},

$$I_B \approx \frac{V_{CC}}{R_B}, \qquad I_C = I_{C \text{ sat}}, \qquad V_{CE} \approx 0$$

This, of course, is inverter action.

EXAMPLE 2-1   The data sheet for a silicon transistor specifies $h_{FE\,min} = 20$. (This means that no transistor using this type number will have an h_{FE} lower than 20.) The transistor is to be used in the illustrated inverter circuit. Determine R_B for saturated operation.

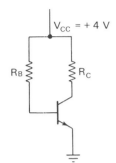

Example 2-1   Transistor inverter circuit

Solution

a. Using Eq. (2-18c), to just get the transistor into saturation

$$R_B = h_{FE}R_C$$
$$= 20(1000\ \Omega) = 20{,}000\ \Omega$$

If $R_B > 20{,}000\ \Omega$, the transistor will not be in saturation. This solution neglects the effect of V_{BE}.

b. To include the effect of V_{BE},

$$I_{C\,sat} = I_{C\,max} = \frac{V_{CC}}{R_C} = \frac{4\ V}{1000\ \Omega} = 4\ mA$$

This requires an

$$I_{B\,sat} = \frac{I_{C\,sat}}{h_{FE}} = \frac{4\ mA}{20} = 0.2\ mA$$

Using Eq. (2-14a),

$$4 = 0.2 \times 10^{-3}\ R_B + 0.7$$
$$R_B = 16{,}700\ \Omega$$

c. The h_{FE} of transistors varies with temperature and aging. At lower temperatures transistors have a lower h_{FE}. To ensure saturation, a safety factor is used. Using a 2 to 1 safety factor, the circuit would be designed for an $h_{FE} = 10$. Neglecting the effect of V_{BE},

$$R_B = 10(1000) = 10{,}000\ \Omega$$

2-4.6 Transistor Frequency Response

It takes time for the minority carriers to cross the base region. As a result, both h_{fb} and h_{fe} have frequency dependency, and their gain decreases as frequency increases.

The frequency at which h_{fb} has decreased to 70 percent (-3 dB) of its low-frequency value is called

$$f_{h_{fb}} \quad \text{or} \quad f_\alpha$$

Since h_{fb} and h_{fe} are related [Eq. (2-10a)], there is a corresponding frequency at which h_{fe} has decreased to 70 percent (-3 dB) of its low-frequency value. This is

$$f_{h_{fe}} = f_\beta = \frac{f_{h_{fb}}}{h_{fe}} = \frac{f_\alpha}{\beta} \qquad (2\text{-}19)$$

For example, if $f_\alpha = 200$ megaherz (mHz) and $\beta = 50, f_\beta = 4$ mHz.

The common-emitter gain continues to decrease as frequency increases and eventually reaches a gain of unity at a much higher frequency than f_β. The frequency at which this takes place is

$$f_T = \text{frequency for } h_{fe} = 1$$

and is an important figure of merit for transistors. It is approximately equal to 80 percent of f_α. Most modern small-signal silicon planar transistors have an $f_T \approx 250$ mHz.

2-4.7 Switching Time

There is a delay between the time of application of a signal to the control element in a transistor and the time at which the transistor output current reaches its final value. This is due to three causes:

1 Time to charge interelectrode capacitances.
2 Transit time of minority carriers to cross the base region.
3 Removal of stored charge in base (and collector) regions when the transistor is coming out of saturation.

When, for example, a transistor in the common-emitter configuration is turned *on* from *cutoff*, the following occurs:

1 Delay time, t_d, due to the need to charge the emitter–base diode capacitance to this diode conduction voltage (0.3 V for Ge, 0.7 V for Si).
2 Because of transit time for minority carriers to cross the base region, the collector current takes time to increase. The *rise time*, t_r, is the time for the collector current to virtually get to its final value. The higher f_T is, the shorter is t_r. We can write

$$t_{\text{on}} = t_d + t_r \qquad (2\text{-}20)$$

When a transistor is turned off after being in saturation, the following occurs:

1   Time, t_s, to remove the stored charges in the base (and collector) regions. The more deeply a transistor is in saturation, the longer t_s is. t_s is called the *storage time*.

2   Fall time, t_f, for the transistor current to drop to virtually cutoff. Like t_r, this is also dependent upon f_T. We can write

$$t_{off} = t_s + t_f. \tag{2-21}$$

The total switching time is equal to $t_{on} + t_{off}$.

Typical values in nanoseconds for the type 2N4401 are

$$\left.\begin{aligned} t_d &= 15 \text{ ns} \\ t_r &= 20 \text{ ns} \end{aligned}\right\} t_{on} = 35 \text{ ns}$$

$$\left.\begin{aligned} t_s &= 225 \text{ ns} \\ t_f &= 30 \text{ ns} \end{aligned}\right\} t_{off} = 255 \text{ ns}$$

It is apparent that for really high speed switching the transistor should be kept out of saturation. As we can see, it is the major factor in switching time.

2-4.8   Transistor Biasing

Biasing is the term given to the method used to set the operating conditions of a transistor in a circuit. Although in most cases biasing implies setting conditions in the active region, it can also refer to both cutoff and saturated conditions. Example 2-1 illustrated how to bias a transistor for saturation. There are many techniques used to bias bipolar junction transistors. The method used is determined by considerations of operating point stability with transistor-to-transistor parameter variation, temperature, and aging. Figure 2-13 shows several ways of biasing transistors in the active region.

Fixed bias: Fig. 2-13a

The base current is fixed and given by

$$I_B = \frac{V_{CC} - V_{BE}}{R_B} \tag{2-22}$$

and
$$I_C = h_{FE} I_B$$

This is the simplest biasing method. The operating point (I_C) is subject to considerable variation since it is dependent upon the value of h_{FE}, which may vary from transistor to transistor, with aging, and with temperature.

Emitter bias: Fig. 2-13b

We can write for the base current

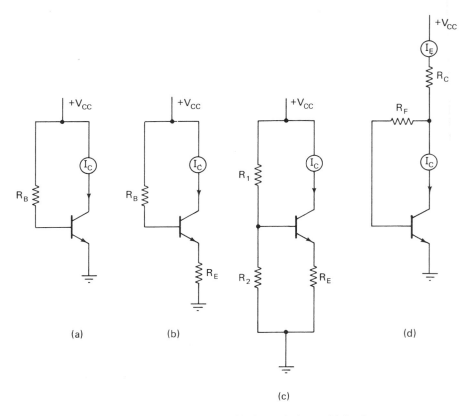

Figure 2-13 Junction transistor active region biasing techniques: (a) fixed bias; (b) emitter bias; (c) voltage divider bias; (d) collector feedback bias

$$V_{CC} = I_B R_B + V_{BE} + I_E R_E$$
$$= I_B R_B + V_{BE} + (I_B + h_{FE} I_B) R_E$$
$$I_B = \frac{V_{CC} - V_{BE}}{R_B + (h_{FE} + 1) R_E} \qquad (2\text{-}23)$$

and
$$I_C = h_{FE} I_B$$

There is compensation in I_B for changes in h_{FE}. Should h_{FE} increase, we see from Eq. (2-23) that I_B will decrease and I_C will decrease; I_C will not be as high as might be expected. In Eq. (2-23), if $(h_{FE} + 1)R_E \gg R_B$, R_B can be ignored in the denominator. I_B will now vary inversely with h_{FE}, and I_C will be constant. Practically, it is difficult to select resistors to meet these conditions.

Voltage divider bias: Fig. 2-13c

This common method of biasing transistors is the same biasing method of Fig. 2-13b if we apply Thevenin's theorem to the voltage divider $R_1 - R_2$. The base current is given by

$$I_B = \frac{V_{\text{Th}} - V_{BE}}{R_{\text{Th}} + (h_{FE} + 1)R_E} \qquad (2\text{-}24)$$

V_{Th} is the open-circuit Thevenin's voltage $\{V_{CC}[R_2/(R_1 + R_2)]\}$ at the junction of R_1 and R_2, and R_{Th} is the Thevenin's resistance $R_1 R_2/(R_1 + R_2)$ of the voltage divider. In this biasing method it is much easier to select resistors to satisfy the conditions $(h_{FE} + 1)R_E \gg R_{\text{Th}}$, and excellent compensation is obtained. When this is done, I_C can be kept very nearly constant.

Collector feedback bias: Fig. 2-13d

This circuit is a modification of Fig. 2-13b and is obtained by interchanging the battery and R_E as shown in Fig. 2-14.

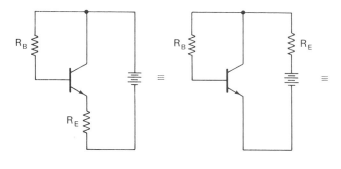

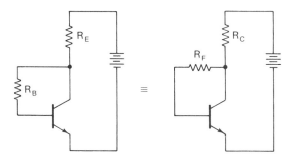

Figure 2-14 Equivalence of Figs. 2-14b and 2-14d obtained by interchanging R_E and the battery

If we replace R_E with R_C and R_F with R_B, we obtain Eq. (2-25) for the base current:

$$I_B = \frac{V_{CC} - V_{BE}}{R_F + (h_{FE} + 1)R_C} \tag{2-25}$$

and
$$I_C = h_{FE}I_B$$

In this circuit, the transistor must always stay in the active region. Cutoff is not possible, since cutoff means that $I_C R_C = 0$. If this were so, V_{CC} would force base current into R_F, with resultant I_C. Saturation is not possible since in saturation $V_{CE} < V_{BE}$. But $V_{CE} < V_{BE}$ implies that I_B will be zero, since there would be no voltage available to force base current into R_F. However, this contradicts the assumption of $I_{C\,sat}$.

2-4.9   Differential Amplifier

The differential amplifier shown in Fig. 2-15 is a frequently used dual-transistor amplifier circuit.

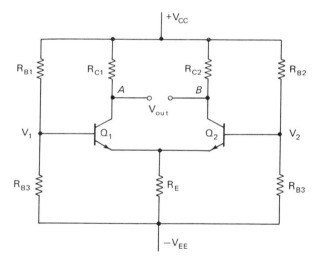

Figure 2-15   Differential amplifier

Transistors Q_1 and Q_2 are matched as closely as possible, usually by making them at the same time on a common substrate. If they do not match, component adjustments are made (not shown) to balance the circuit. If the circuit components are matched, the voltage at A equals the voltage at B; and $V_{out} = 0$. Suppose that voltage V_1 is made positive. This will increase the current into Q_1, causing the voltage at point A to go negative. The same is true for the voltage V_2, which will make point B go negative. If the voltages V_1 and V_2 are both equal in magnitude, points A and B go negative equally, and V_{out}, the *difference* in voltage between A and B, does not change. Hence the amplifier output V_{out} measures the difference in voltage between V_1 and V_2.

2-4.10 Parallel Transistor Operation

There are times when it is desired to operate two transistors in parallel, as shown in Fig. 2-16a. The base current is supplied through R_B; but since the bases are in parallel, it is the *voltage* from base to emitter, which is common to both transistors, that determines the parallel operation. Therefore, we have to plot the curve of collector current I_C (or base current I_B) versus V_{BE}. The collector current is of course equal to $h_{FE} \times I_B$, and the base–emitter section of the transistor operates as a diode. Unfortunately, the base–emitter characteristics of transistors are not quite the same, and small differences in V_{BE} can make large differences in base current or collector current. This is *current hogging* (shown in Fig. 2-16b). A common solution to this problem is to use separate base resistors, as shown in Fig. 2-16c.

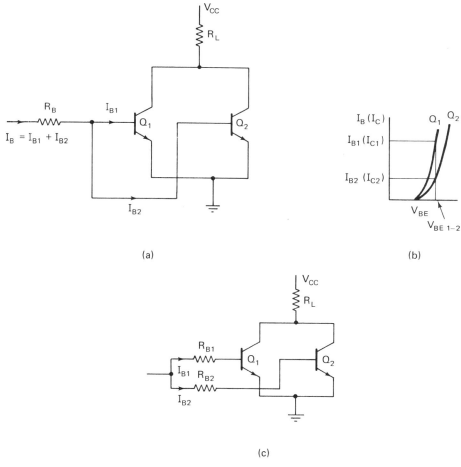

(a)

(b)

(c)

Figure 2-16 Parallel transistor operation: (a) transistors, common-base resistor; (b) $I_B(I_C) - V_{BE}$ characteristics; (c) two transistors, separate base resistors

2-5 JUNCTION FIELD EFFECT TRANSISTOR
(JFET)

There are two limitations to the bipolar transistor: (1) the input resistance is low since current must be driven into the input element to make the bipolar transistor operate, and (2) the collector offset voltage, although small, can in some switching applications cause problems.

These limitations are overcome by the field effect transistor. There are two types: the junction field effect transistor (JFET), which will be discussed in this section, and the metal oxide semiconductor field effect transistor (MOSFET), which will be discussed in the next section.

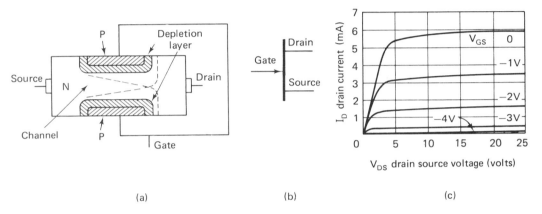

     (a)   (b)   (c)

Figure 2-17 *N*-channel junction field effect transistor: (a) construction; (b) symbol; (c) typical characteristics (Motorola Semiconductor, Type 2N4223)

Consider a bar of *N*-type silicon with *P* regions diffused into opposite sides of the bar as shown in Fig. 2-17a. The *P* regions form a diode with the *N* material of the bar and are surrounded by a depletion region. The *P* regions are connected together and become the *gate* terminal. One end of the bar is called the *source* and the other the *drain*. If the gate is connected to the source and the drain made positive with respect to the source, current flows in the bar. The section of the bar between the *P* regions is called the *channel*. If the gate is now made negative with respect to the source, the depletion region extends into the channel (as shown by the dashed lines in Fig. 2-17a) until finally no charge carriers exist in the channel. No current can now flow in the device. The gate voltage required to do this is called *pinch-off*.

Let the gate again be at zero voltage with respect to the source. Let the drain voltage begin at zero voltage with respect to the source and be gradually made more positive. Drain current begins at zero voltage and increases. There is a voltage drop in the channel between the depletion region and the source due to the bar resistance.

This makes the gate terminal (*P* regions) negative with respect to the depletion region, which now again extends into the channel. This continues until the transistor pinches itself off at some value of current.

Figure 2-17c shows the characteristics of an *N*-channel JFET. Pinch-off in the transistor occurs when the gate voltage is approximately −4 V with respect to the source. In the same way, it occurs when the drain is +4 V with respect to the source when the gate voltage V_{GS} is equal to 0.

Figure 2-17b shows the symbol for the *N*-channel JFET. The arrow points in the direction current would flow in the gate if it were made positive. This device operates only with the gate reverse biased with respect to the source; hence the input current is only the leakage current of a reverse-biased diode. Thus its input resistance is quite high. The JFET is controlled by gate *voltage* rather than by current as in bipolar transistors. The offset voltage V_{DS} of JFETs is zero. *P*-channel JFET transistors are made by interchanging the *P* and *N* regions of the device. The device symbol is the same but the arrow is reversed.

2-6 METAL OXIDE SEMICONDUCTOR FIELD EFFECT TRANSISTOR (MOSFET)

Figure 2-18a shows the construction features of an *N*-channel enhancement mode MOSFET. There are three terminals and the substrate or bulk. The gate is insulated from the channel by a layer of silicon dioxide, an almost perfect insulator. If the drain is made positive with respect to the source, and the gate is at zero potential with respect to the source, no charge carriers exist in the channel region between the source and gate regions. No current flows. If the gate is now made positive, it attracts electrons to the channel region below the gate, changing the channel from a *P* to an *N* region, and makes it conductive. We say that the channel is open until it is "enhanced" by application of the positive bias. Conduction does not begin in enhancement-type transistors until the gate is made several volts positive.

The primary advantages of this type of transistor are the almost infinite gate resistance and the enhancement-mode characteristics.

A *P*-channel MOSFET is made by interchanging the *N* and *P* regions in Fig. 2-18a.

By changing the impurity concentration in the channel and the location and size of the gate, a depletion *N*-channel-type MOSFET can be made, which operates with V_{GS} in the negative region.

Figure 2-18b shows the symbol for an *N*-channel enhancement-type MOSFET. The broken line indicates enhancement. It is solid for a depletion type. In a *P*-channel type the arrow is reversed.

Figure 2-18c shows a typical set of drain characteristics. As can be seen, the device operates by having the gate voltage control the drain current. In normal operation the source and substrate (bulk) are connected.

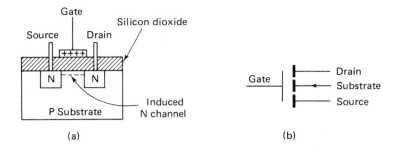

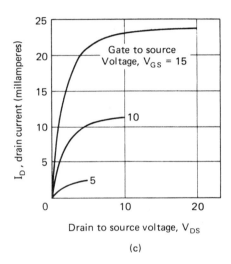

Figure 2-18 *N*-channel MOSFET transistor: (a) construction; (b) symbol; (c) characteristics

Problems

Unless otherwise noted, the transistors are silicon. Assume that $V_{BE} = 0.7$ V.

2-1. (a) $V_{BB} = +1$ V. Find V_{CE}. Include the effect of V_{BE}.

(b) Repeat (a), but assume that $V_{BE} = 0$.

(c) Repeat (a), but for $V_{BB} = +2$ V.

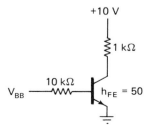

(d) What value of V_{BB} is required to just put the transistor into saturation? Include V_{BE}.

(e) Repeat (d), but assume that $V_{BE} = 0$.

2-2. Repeat Prob. 2-1 for $h_{FE} = 150$.

2-3. Repeat Prob. 2-1 for $R_B = 4\,\text{k}\Omega$.

2-4. In the transistor circuit of Prob. 2-1,

(a) For $V_{BB} = +10$ V, what is the minimum value of h_{FE} to make the transistor go into saturation? Include V_{BE}.

(b) Repeat (a), but assume that $V_{BE} = 0$.

(c) What is the circuit h_{FE} for Prob. 2-1?

(d) What is the ratio of transistor h_{FE} to circuit h_{FE} in Prob. 2-1?

2-5. Repeat Prob. 2-4 for $R_B = 4\,\text{k}\Omega$.

2-6. Assuming that $V_{BE} = 0$ in Prob. 2-1 and that $V_{BB} = +10$ V, how many additional 1-kΩ load resistors can be connected in parallel with the 1-kΩ collector load resistor before the transistor will begin to come out of saturation? (*Note:* Resistors from the collector to V_{CC} are called *pull-up* resistors.)

2-7. (a) The circuit shown gives the component values for the mw RTL (milliwatt resistor–transistor logic) inverter. What is the minimum value of h_{FE} required to saturate the transistor if $V_{BB} = +3.6$ V? Include V_{BE}.

(b) Repeat (a), but assume that $V_{BE} = 0$.

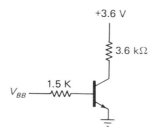

2-8. Repeat Prob. 2-7 for $R_B = 450\,\Omega$ and $R_C = 640\,\Omega$ (RTL 700/800 series).

2-9. Assume that the transistor of Prob. 2-7 has an $h_{FE} = 20$. What is the minimum value of V_{BB} required to just make the transistor go into saturation? Include the effect of V_{BE}.

2-10. Repeat Prob. 2-9 using the component values of Prob. 2-8.

2-11. Assume that the transistor of Prob. 2-7 has an $h_{FE} = 20$. Including the effect of V_{BE}, plot a smooth curve of V_{CE} versus V_{BB}. Plot V_{CE} on the vertical axis and V_{BB} on the horizontal axis. V_{BB} is to vary between 0 and $+3.6$ V. (*Note:* this is called a *transfer curve.*) The effect is quite sharp in the vicinity of $V_{BB} = 0.7$ V. Select about six values of V_{BB} that include the section of the curve where V_{CE} is changing.)

2-12. In the figure shown, the resistor R_L from a collector to ground is called a *pull-down* resistor.

(a) For $R_L = 1.5\,\text{k}\Omega$, what is V_{CE}?

(b) Repeat (a) for $R_L =$ two 1.5-kΩ resistors in parallel.

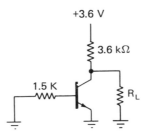

2-13. Repeat Prob. 2-12 for $R_B = 450\ \Omega$, $R_C = 640\ \Omega$, and $R_L = 450\ \Omega$.

2-14. Transistor Q_1 drives transistor Q_2 as shown. What is V_{CE} for transistor Q_1? Include V_{BE} of Q_2.

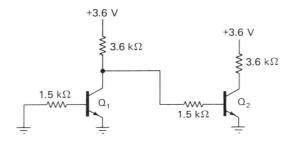

2-15. Repeat Prob. 2-14 for $R_B = 450\ \Omega$ and $R_C = 640\ \Omega$.

2-16. Transistor Q_1 drives two transistors, Q_2 and Q_3, in parallel. What is V_{CE} for Q_1? Include the effect of V_{BE} for Q_2 and Q_3.

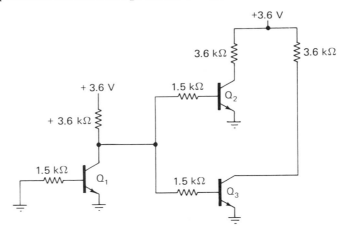

2-17. (a) Repeat Prob. 2-9, but with Q_1 driving three transistors in parallel.
(b) Repeat Prob. 2-9, but with Q_1 driving four transistors in parallel.

2-18. Using the results of Probs. 2-14, 2-16, and 2-17, plot a curve of V_{CE} versus the number of transistors being driven by Q_1. Plot V_{CE} vertically and transistor numbers horizontally. Start with 0 transistors.

2-19. Find I_C. Include V_{BE}.

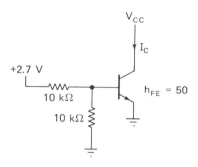

2-20. (a) Find I_B. Include the effect of V_{BE}.
   (b) Find I_B. Assume that $V_{BE} = 0$. (*Note:* I_B is the difference between the currents in R_1 and R_2.)

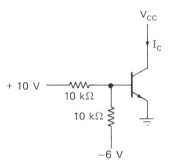

2-21. (a) If Q_1 is in saturation, what is V_{BE_2}?
   (b) What value of V_{CE_1} is needed to just make Q_2 conduct? Include the effect of V_{BE}.

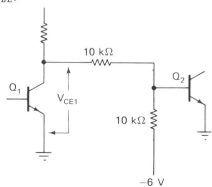

2-22. Find: (a) I_B; (b) I_C; (c) I_E; (d) V_{CE}. Include the effect of V_{BE}.

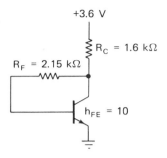

2-23. Repeat Prob. 2-22 for $R_C = 1\ \text{k}\Omega$ and $R_F = 2.75\ \text{k}\Omega$.

2-24. Find I_{C_2}. Include the effect of the diode voltage drop and the V_{BE} drop. (*Note:* this is the basic circuit of the DTL 930 series of digital logic. Use the results of Prob. 2-22.)

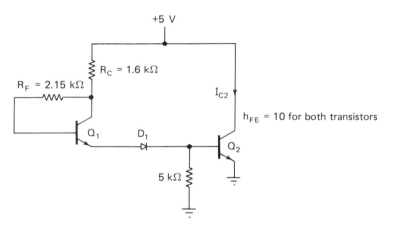

2-25. Transistor Q_2 in circuit (b) has to drive loads as shown in (a). How many such loads can it drive before it comes out of saturation? (*Note:* use the results of Prob. 2-24.)

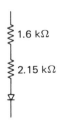

(a)

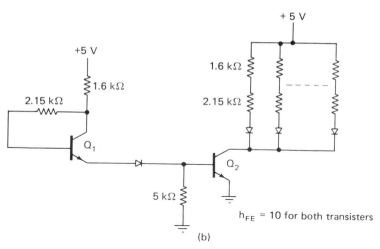

(b)

2-26. Repeat Prob. 2-24 with the following changes (DTL/HTL logic): $V_{CC} = +15$ V, $R_C = 3$ kΩ, $R_F = 12$ kΩ, and diode D_1 is a reverse-biased 7.5-V Zener diode.

THREE

The Integrated-Circuit Logic Families

3-1 INTRODUCTION

Many different forms of integrated-circuit (IC) digital logic have been developed. The first were based upon circuits that had been successful with discrete components. Initially, basic simple functions were provided, but as technology progressed, more complex functions became available. Concurrently, new families of ICs were being developed, which were better able to utilize the advancing state of the art. They could be manufactured more readily and were able to satisfy basic circuit requirements and applications in a more satisfactory manner. Moreover, these families had the capability of providing ICs of enormous complexity.

Complexity of integrated circuits is categorized as follows:

1 SSI: small-scale integration; up to 12 transistors.
2 MSI: medium-scale integration; 12–100 transistors.
3 LSI: large-scale integration; 100 or more transistors.

In circuit design with integrated circuits, the following requirements have to be considered:

1   Availability of logic functions.
2   Single or multiple power supply requirement.
3   Logic swing.
4   Noise immunity.
5   Noise generation.
6   Wired collector logic capability.
7   Complex function availability.
8   Interface with other logic families.
9   Dissipation.
10   Speed.
11   Cost.
12   Temperature range.

The IC families to be discussed are the following:

1   RTL: resistor–transistor logic.
2   DTL: diode–transistor logic.
3   TTL or T^2L: transistor–transistor logic.
4   ECL: emitter-coupled logic.
5   CMOS: complementary-symmetry logic.
6   PMOS, NMOS.

3-2   WIRED-COLLECTOR LOGIC

Consider what happens if the outputs of two gates, for example, two inverters, are wired together as shown in Fig. 3-1a. The output terminals of gates generally are at the collector of an output transistor, as shown in Fig. 3-1b. Hence the con-

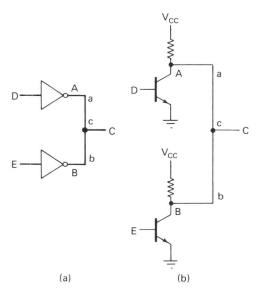

Figure 3-1   Two inverters in wired-collector logic: (a) logic diagram; (b) circuit diagram

(a)   (b)

nection a–b between the two output terminals is called *wired-collector logic*. The collector operates at either of two voltage and impedance levels:

1   V_{CC} and high impedance.
2   Ground and low impedance.

At point C, where the output collectors are wired together with the connector *acb*, if either A or B is at a low potential, the low impedance takes over. We can then write a truth table (Table 3-1). As can be seen, the logic relationship is that of

Table 3-1   Wired-collector logic at collectors

A	B	C
Low	Low	Low
Low	High	Low
High	Low	Low
High	High	High

an AND:

$$C = A \cdot B \tag{3-1}$$

Now let us add an inverter to the wired-collector logic of Fig. 3-1, as shown in Fig. 3-2, and again write a truth table (Table 3-2). As we can see, the relationship between D, E, and C is

$$C = \overline{D + E}, \quad \text{a NOR} \tag{3-2}$$

$$F = D + E, \quad \text{an OR} \tag{3-3}$$

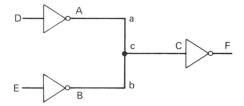

Figure 3-2 Wired-collector logic with added inverter

Because of these considerations, we obtain a form of logic known as *WIRED-OR, COLLECTOR-OR, PHANTOM-OR* or *IMPLIED-AND,* or *DOT-AND.* The most popular is WIRED-OR. Figure 3-3 shows the logic dashed symbol for this connection.

Table 3-2   Truth table for Figure 3-2

		If Connection *abc* Did Not Exist		With Connection *acb*	
D	*E*	*A*	*B*	*C*	*F*
Low	Low	High	High	High	Low
Low	High	High	Low	Low	High
High	Low	Low	High	Low	High
High	High	Low	Low	Low	High

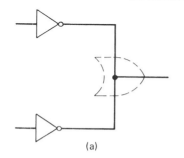
(a)

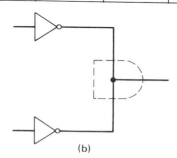
(b)

Figure 3-3   Wired-collector logic symbols: (a) wired-OR; (b) implied-AND

We must now discuss whether the connections of Figs. 3-1, 3-2, and 3-3 can safely be made. In the circuit of Fig. 3-1b we have made the connection *acb* without regard to the type of circuit in the collector of the transistors. Two types of loads exist, as shown in Fig. 3-4. In Fig. 3-4a, output transistors Q_1 and Q_2 have collector load resistors to V_{cc}. This is known as *passive pull-up* (pull-up resistor from collector to V_{cc}). If transistor Q_1 is on and Q_2 is off, transistor Q_1 has to carry (sink) the current of its own load resistor and that of Q_2. This is no problem.

In Fig. 3-4b, a totem-pole output, consider output *A*. When *A* is low, transistor Q_1 is on and in saturation, and transistor Q_3 is off. The impedance at point *A* is low. When *A* is high, transistor Q_1 is off, and Q_3 is on. The impedance at point *A* is low since it is an emitter follower. Hence, if we make the connection *acb*, we connect two low-impedance points and we can damage the transistors. This type of pull-up is known as an *active pull-up*, and wired-collector logic is not allowed. How logic families that have active pull-up outputs solve this problem will be discussed under each logic family.

Wired-collector logic is very important. It provides an extra level of logic free. It has a very important application in expansion of electronic memories.

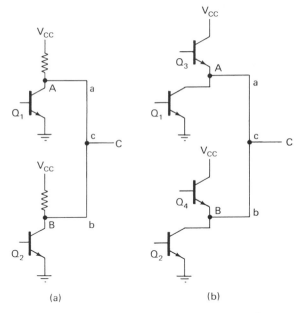

Figure 3-4 Type of collector loading: (a) passive pull-up;
(b) active pull-up

For example, by starting with a basic 4096 (4K) bit memory and connecting eight such memories in a wired-collector-logic manner, the memory is simply expanded to a 32K-bit memory.

3-3 NOISE IMMUNITY

Noise immunity or noise margin expressed in volts is an important consideration in the design and choice of ICs. Noise immunity is a measure of the ability of a logic circuit to prevent unwanted noise signals from changing a desired logic level to an undesired level.

Consider two inverters connected together as shown in Fig. 3-5. Noise interference due to (1) electromagnetic coupling to the wire *BC, or* (2) electrostatic coupling to the wire *BC*, or (3) power supply variation can couple an undesired voltage into the wire *BC*.

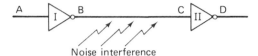

Figure 3-5 Two inverters subject to noise interference

In our discussion of transistor characteristics in Chapter 2, we noted that the emitter–base voltage, V_{BE}, was essentially equal to 0.7 V. Although not exactly constant at this value, V_{BE} changes very little with change in base current. As a result of this small change in V_{BE}, if we plot a transfer curve of input voltage versus output voltage for inverters such as those of Fig. 3-5, we obtain Fig. 3-6. As can be seen, the transition from high output level to low output level occurs over a very small change in input voltage. The *input voltage* at which this occurs is called the *threshold* voltage.

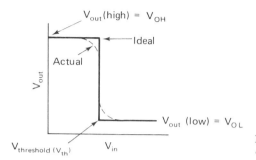

Figure 3-6 Transfer characteristics of inverters of Figure 3-5

Suppose that *A* is low and *B* is high. *C* should be high and *D* low. However, a downward noise voltage coupled into wire *BC* can make *C* go low, which will make *D* go high, an undesired logic level. How low a downward voltage can be tolerated before inverter II is adversely affected? This depends upon (1) the high logic voltage level of inverter I, and (2) the input voltage level at which the output level of inverter II changes from low to high.

We can see now that the ability of inverter II of Fig. 3-5 to discriminate against a downward-going noise voltage depends upon the *difference* in voltage between the high output level of inverter I, V_{OH}, and the threshold voltage, V_{th}, of inverter II. This is known as *high-level noise immunity*. Since the curve of Fig. 3-6 applies to both inverters, we can read the values directly from the curve and write

$$\text{High-level noise immunity} = V_{OH} - V_{th} \qquad (3\text{-}4)$$

Similar considerations apply if we start with input *A* high. This should make *B* and *C* low. Now an upward-going noise voltage can make point *C* go above the threshold voltage. We can again write

$$\text{Low-level noise immunity} = V_{th} - V_{OL} \qquad (3\text{-}5)$$

EXAMPLE 3-1 From a gate transfer characteristics curve, we obtain the following values:

$$V_{OH} = 14 \text{ V}$$
$$V_{th} = 8.2 \text{ V}$$
$$V_{OL} = 0.4 \text{ V}$$

Determine the noise immunity of the gate.

Solution High-level noise immunity $= 14 - 8.2 = 5.8$ V. A downward-going noise voltage in excess of this will cause an unwanted output level change.

Low-level noise immunity $= 8.2 - 0.4 = 7.8$ V. An upward-going noise voltage in excess of this will cause an unwanted output level change.

3-4 RESISTOR–TRANSISTOR LOGIC (RTL)

The basic circuit of the RTL gate is shown in Fig. 3-7. If *both* inputs A and B are at a low potential, the output X is high. If *either* A or B is high, either Q_1 or Q_2 is driven into saturation and X is low. The RTL gate of Fig. 3-7 is, therefore, a NOR gate for positive logic and a NAND gate for negative logic. The passive pull-up load resistor R_L permits wired-collector applications. High-level noise immunity is provided by the difference between the high output level and the threshold voltage of a silicon transistor. Ground-level noise immunity is provided by the difference between the low output level at point X and the threshold voltage of a silicon transistor, and is approximately equal to the offset voltage V_{BE} of a silicon transistor.

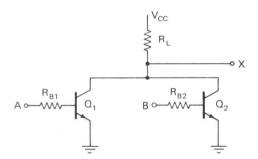

Figure 3-7 Resistor—transistor logic

In RTL, the output terminal X is required to act as a high level source for any RTL gate inputs driven from X. The high output voltage level drops, therefore, as the loading presented to point X increases (Prob. 2-18). Both the high-level noise immunity and the logic swing are therefore reduced with increased loading. This limits the output loading capability of the gate.

RTL logic operates with a single positive $+3.0$- or $+3.6$-V power supply. This makes it incompatible with other logic families; RTL is slow compared to other logic families. Complex functions are available in RTL.

3-5 DIODE–TRANSISTOR LOGIC (DTL)

The basic circuit of a DTL logic gate is shown in Fig. 3-8. Simple diode logic was discussed in Sec. 1-3. The circuit of Fig. 3-8 is a diode AND gate followed by an inverter. If input A or B is at a low potential, current can flow from V_{CC} through

R_1 and R_2 to ground. The voltage at node N is one diode drop (0.7 V) above ground. This voltage is not enough to overcome the offset voltage of the V_{BE} of Q_1, the diode drop of D_3, and the V_{BE} of Q_2, a total of 2.1 V. Now let input A be at a high potential equal to V_{CC}. Diode A is reverse biased. If the voltage at point B, starting from ground, is slowly raised, the voltage at N will be one diode drop higher than the voltage at B. When input B reaches 1.4 V, N reaches 2.1 V, and current flows into the base of Q_1, turning on Q_1, D_3, and Q_2 and driving Q_2 into saturation. Further increase of voltage at B reverse-biases diode D_2 since the voltage at point N is now clamped at 2.1 V. This occurs for any combination of input voltages at A and B or both. Only when *both* input diodes are reverse biased will Q_1, D_3, and Q_2 conduct.

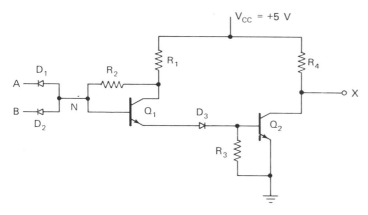

Figure 3-8 DTL gate

The gate of Fig. 3-8 is therefore a NAND gate for positive logic and a NOR gate for negative logic. The passive pull-up load resistor R_L allows wired-collector logic. At room temperature the threshold voltage occurs at 1.4 V. This makes the high-level noise immunity at room temperature equal to $V_{OH} - V_{th}$, which can be as high as several volts, and the ground-level noise immunity at room temperature approximately 1.4 V.

Load resistor R_L is typically 6 kΩ. Some gates have R_L lowered to 2 kΩ. This increases the dissipation and reduces the drive capability of the IC. This reduction in load resistance allows the IC to drive higher-capacity loads.

Consider an IC gate driving a capacitive load as shown in Fig. 3-9. Figure 3-9a shows the output circuit. Figure 3-9b shows the input gate waveform. Figure 3-9c shows the output waveform (inverted from b) with the capacitor $C = 0$. Figure 3-9d shows the output waveform with a capacitor. When Q_2 is driven into saturation, it is a low resistance and discharges C, so that the voltage across $C \approx 0$. When Q_2 is turned off, the voltage at X can increase only as fast as the capacitor C is charged through R_L. How fast it will charge is determined by the time constant $R_L C$.

For the gate to successfully drive another gate and cause it to switch, the

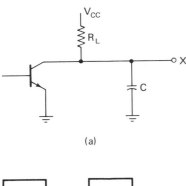

(a)

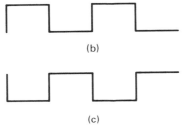

(b)

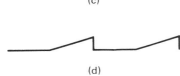

(c)

(d)

Figure 3-9 Integrated-circuit gate with capacitive load. (a) Gate output circuit; (b) input waveform; (c) output waveform $C = 0$; (d) output waveform with C

voltage at X in Fig. 3-9d must rise above the threshold voltage for a following gate. If C is too large, or the switching rate is too fast, the voltage at X as shown in Fig. 3-9d cannot reach the threshold voltage before the next cycle arrives, and the next gate will not have time to switch. Operation at a higher switching rate requires that the time constant $R_L C$ be reduced. C is fixed by the circuit configuration. Hence R_L must be reduced. Some versions of DTL have a reduced value for R_L within the IC, but this increases the internal dissipation and reduces the gate drive capability. Another solution that is used frequently is to parallel R_L with an external pull-up resistor, but this again subtracts from the useful drive capability of the IC gate. As we shall see in the next section, TTL solves this problem with an active pull-up; but active pull-ups do not permit wired-collector logic. In addition to this speed limitation, transistor switching time also limits the maximum frequency at which a gate can operate.

In a logic family, the outputs of logic elements drive inputs of following logic elements. A logic element can drive other units of the same logic family as well as compatible units. For example, DTL and TTL are compatible, but DTL and RTL are not.

Limits to the output drive capability of a logic family are called *fan-out*, and limits to the input capability are called *fan-in*. Let us consider these specifically for DTL. Similar considerations apply to other logic families.

In DTL, when a gate output is low and is coupled to other gates, the output transistor Q_2 of Fig. 3-8 has to sink the current of the gates that it is driving. As more inputs are connected to the output of Q_2, Q_2 must carry more current. As more current flows into Q_2, its $V_{CE\,sat}$ increases, and it may come out of saturation. This increase in $V_{CE\,sat}$ reduces ground-level noise immunity, increases dissipation, and may make the gate ineffective if the voltage at the collector of Q_2 exceeds the threshold voltage of the next stage. Hence there is an IC gate output drive limit called maximum fan-out. This gate output limit is generally expressed as a loading factor given in terms of the number of gates that can be driven (see, for example, Prob. 2-17).

The fan-in input capability of DTL is limited by switching speed requirements. In Fig. 3-8, let input A be high and input B low. The diode connected to input A is reverse biased and therefore acts like a capacitor. Suppose that input B is very rapidly switched to a high potential. Transistor Q_1 cannot conduct until its base input reaches the threshold voltage, and this cannot be reached until the charge stored in the capacitance of the diode at A has discharged enough to allow the voltage at the base of Q_1 to rise high enough so that Q_1 can conduct. More diode inputs, representing greater fan-in, add to the capacitance that has to be charged and therefore slow up the switching speed.

DTL operates with a single $+5$-V power supply. It has moderate speed and good noise immunity. Although some MSI is available in DTL, many complex functions are available in TTL with which it is compatible.

If diode D_3 in Fig. 3-8 is replaced with a Zener diode, the noise immunity of the gate can be improved. This type of logic is called HTL. V_{CC} is equal to 15 V, and the Zener diode is selected to make the low- and high-level noise immunities approximately 7.5 V. This is done at a sacrifice of speed capability.

3-6 TRANSISTOR–TRANSISTOR LOGIC (TTL)†

The basic circuit of a TTL gate is shown in Fig. 3-10. In many respects its operation is similar to that of DTL. Q_1 is a multiple emitter transistor. Each emitter acts like the input diode in DTL (Fig. 3-8). Figure 3-11 shows how a two-diode AND gate and inverter converts to the two-emitter input transistor of Fig. 3-10.

In Fig. 3-10, if either emitter X or Y or both are at low potential, current flows through R_1 and the base–emitter diode of Q_1. The base of Q_1 is 0.7 V above ground. This voltage prevents current from flowing through the base–collector diode of Q_1 into the base of Q_2. If both X and Y are at a high level, current can flow through resistor R_1, through the base–collector diode of Q_1 into the base of Q_2 and into the base of Q_4. This saturates Q_2 and Q_4, leaving the output at low potential ($V_{CE\,sat}$ of Q_4). When either X or Y is at a low potential so that no

---

†M. E. Levine, *Digital Theory and Experimentation Using Integrated Circuits.* Englewood Cliffs, N.J.: Prentice-Hall, Inc., 1974, Expt. 4, TTL NAND/NOR Gates—Definitions and Operation.

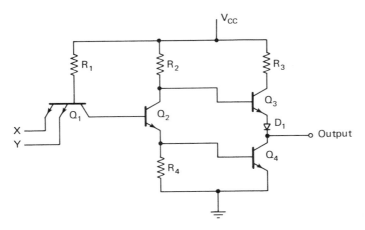

Figure 3-10   TTL gate

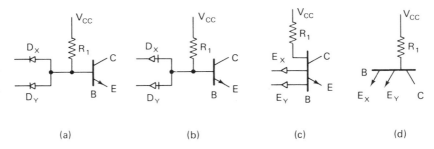

(a)   (b)   (c)   (d)

Figure 3-11   Deriving TTL from DTL: (a) two diodes and inverter; (b) interchange anode and cathode of diode symbol; (c) combine diode cathodes with transistor base; call diode anode emitters; (d) rotate and omit original emitter; change location of base and emitters

current flows into the base of Q_2, current flows through R_2 and causes Q_3 to conduct. The output level is now high, but is the low impedance of an emitter follower Q_3. Both the low-level and the high-level impedances are low, enabling the circuit to drive high-capacity loads. This type of output is called an active pull-up. Wired-collector logic is not possible. Diode D_1 is needed for biasing reasons.

For the condition in which transistor Q_4 is in saturation, resistors R_2 and R_4 have been selected so that the voltage at the base of Q_3 is not high enough to overcome the offset voltages of D_1 and the V_{BE} of Q_3. Q_3 is therefore cut off.

TTL logic is NAND for positive logic and NOR for negative logic. It is the fastest of all the saturated logic systems, because of the gain of transistor Q_1. In DTL, the input diode has no gain. In Fig. 3-10 let both inputs X and Y be high. Now quickly return an input to a low potential, ground for example. For Q_1, this makes the conditions $V_E =$ ground, $V_B = +0.7$ V, and $V_C = +1.4$ V. Q_1 is in its active state, and it quickly removes charges stored in the base of Q_2 (and Q_4), turning them off rapidly. This transistor action yields a very short gate turn-off time. Diode–transistor logic is slower because the input logic diodes of Fig. 3-8 can have no gain.

When the gate of Fig. 3-10 goes from the high to the low state, transistor Q_4 turns on faster than Q_3 turns off. As a result, for a short time both Q_3 and Q_4 are on, resulting in a momentary low impedance across the power supply. There is a short-duration current spike from the power supply, with the resultant short-duration voltage spike (glitch) in the power supply leads. Active pull-ups in TTL therefore generate switching noise problems. Power supply leads must be carefully bypassed with TTL to avoid spurious switched responses. Common practice is to use 0.1 microfarad (μF) per 20 gates.

TTL is the fastest form of saturated logic and has become very popular. It operates with a single $+5$-V power supply. It has good low-frequency noise immunity, but its speed makes it sensitive to short-duration noise pulses. The active pull-up generates power supply switching transients. Many complex functions are available. It is compatible with DTL.

Several other versions of TTL are available:

1   Schottky-clamped TTL: a very high speed modification in which a diode with a different type of doping, called a Schottky barrier diode, is placed across the collector–base junction of the transistors in TTL. This diode is free of minority carriers and has a lower voltage drop than a silicon *PN* junction. As a result, base current is diverted away from the base, which reduces most of the stored charge in the base region.
2   L or low dissipation.
3   H or high speed, a modification of the standard TTL.
4   Low-power Schottky. It has the same speed as standard TTL but at lower power. It is supplanting the standard TTL.

Because of the active pull-up, TTL cannot be used in wired-collector logic application. This is overcome in two ways.

a. Open collector: in the IC of Fig. 3-10, resistor R_3, transistor Q_3, and diode D_1 are omitted. This permits a passive pull-up resistor to be connected from the output to V_{CC}. This resistor can be common to all the gates that are being "wire collectored."

b. Tristate logic: a modification of the circuit of Fig. 3-10 in which provisions are made to turn off both Q_3 and Q_4. There are therefore three states: high, low, and off or high impedance. In the high-impedance state the outputs can be wire-ORed.

3-7 EMITTER-COUPLED LOGIC (ECL)

Previously discussed logic families were of the saturated logic type. Because of the stored charge that occurs in saturation, the switching speed of the RTL, DTL, and TTL is limited. Emitter-coupled logic (also called current-mode logic) is a nonsaturating type of logic. The basic circuit of an ECL gate is shown in Fig. 3-12.

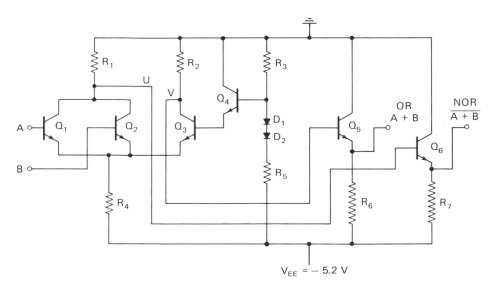

Figure 3-12 Emitter-coupled logic

The input circuit, which consists of transistors Q_1 and Q_2, and transistor Q_3 form a differential amplifier configuration in combination with R_4. Resistors R_3 and R_5 in combination with Q_4 and diodes D_1 and D_2 set a reference voltage at the base of Q_3. Transistors Q_1 and Q_2 form a NOR gate. As the logic levels are changed at inputs A and B, current is switched between Q_1, Q_2, and Q_3 to change the voltage levels at points U and V. U and V in turn drive low-impedance emitter followers Q_5 and Q_6. The logic levels involved are such that none of the transistors go into saturation; hence the switching time is low. The collector load resistors are grounded and the emitter resistors connected to $-V_{CC}$ to reduce the effects of noise in the power supply bus. It operates with a single -5.2-V power supply and is not compatible with other forms of IC digital logic.

3-8 COMPLEMENTARY-SYMMETRY LOGIC (CMOS)

Complementary-symmetry IC logic uses both P- and N-channel metal oxide (MOS) field effect (FET) transistors in complementary-symmetry arrangements. The transistors are enhancement types. CMOS is also known as COS/MOS.

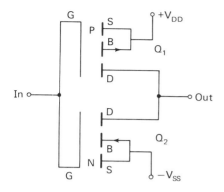

Figure 3-13 Complementary-symmetry inverter with MOSFET transistors

Figure 3-13 shows a CMOS inverter circuit using P- and N-channel MOSFET transistors. The upper transistor Q_1 is a P-channel transistor with source S connected to the bulk or substrate B. The lower transistor Q_2 similarly is an N-channel transistor. The two gates G are connected together to form the input, and the two drains D are connected together to provide an output.

If the input voltage is made $= -V_{SS}$, Q_2 is cut off, but Q_1 is on (P-channel with gate voltage equal to supply voltage). Q_1 has low impedance, and the output voltage is very nearly equal to $+V_{DD}$ (inversion) with low source impedance. If the output requires no current such as would occur if the output were driving MOS transistors, the current drawn from $+V_{DD}$ is extremely low (leakage current). When the input voltage is high and equal to $+V_{DD}$, the conditions are reversed. Q_1 is off and Q_2 is on, and again the power supply current requirement when driving MOS transistors is extremely low. The output voltage is now very nearly equal to V_{SS} (inversion).

Under either condition, one transistor is on and the other off. The output voltages are very close to V_{SS} or V_{DD}, and the unit acts as an inverter. The quiescent dissipation is extremely low.

Threshold voltage occurs at approximately 50 percent of the supply voltage; this results in both high- and low-level noise immunities equal to approximately 50 percent of the supply voltage.

When the output terminal of Fig. 3-13 drives a capacitive load, the dissipation increases as the switching speed increases for the following reasons. When a gate is switched from a low output level to a high output level, the load capacity must charge through the upper transistor. It is therefore on and conducting during the time the capacitor is charging. Current is drawn from V_{DD} and the gate dissipation increases with the switching rate of the signal.

Logic functions are built up by combining N-channel and P-channel transistors to form any desired logic function. A typical example of a two-input NAND gate is shown in Fig. 3-14. If A and B are both low, Q_1 and Q_2 are both *on*, Q_3 and Q_4 are both *off*, and the output is *high*. If A is *high* and B low, Q_2 is still *on* and Q_4 is still *off*, and the output is still *high*. Only if A and B are *both high* do *both* Q_3 and Q_4, now *both on* and in series, provide a low impedance path between the

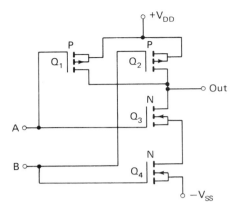

Figure 3-14   Two-input complementary-symmetry MOS NAND gate (positive logic)

output terminal and V_{SS}. At this time also, both Q_1 and Q_2 in parallel are *both off*, and the output impedance to V_{DD} is *high*. The gate of Fig. 3-14 is a NAND gate for positive logic.

Complementary-symmetry MOSFET gates have distinct advantages. The quiescent dissipation is extremely small. Fan-in and fan-out capabilities are quite high compared to other logic families. Noise immunity is very good. The first designs required high voltages and level converters to interface with other logic families. They also had switching speed shortcomings. These difficulties have been overcome. Logic elements are available that can operate at power supply voltages from $+1.3$ to $+15$ V. Standard ICs can be operated over a wide range (3 to 18 V) of supply voltage. In this respect they have a considerable advantage over the other forms of logic we have discussed, which require quite narrow power supply tolerances. This logic family is not directly compatible with other logic families and needs level or impedance converters to interface with other logic families.

3-9   PMOS, NMOS

Transistors made with these technologies can be made very small and require very few processing steps in their manufacture. This is particularly applicable in ICs where many identical functions are required, such as in memories and shift registers. PMOS ICs require power supplies between -8 and -27 V. They have limited current capability and are slow. This makes it awkward to interface with TTL. NMOS is a more recent development and is compatible with TTL. These are discussed in greater detail in the chapters on shift registers (Chapter 8) and memories (Chapter 11).

A comparison of the major IC digital logic families is given in Table 3-3.

Table 3-3 Comparison of the integrated logic circuit families

Family	RTL	DTL	TTL	ECL	CMOS	HTL
Circuit	Resistor–transistor	Diode–transistor	Transistor–transistor	Nonsaturating differential–transistor	Complementary MOS	Diode–transistor
Basis logic function positive logic	NOR	NAND	NAND	NOR	NOR or NAND	NAND
Wired-collector capability	Yes	Yes	With passive pull-up or with tri-state output	Yes	With tri-state output	Yes
Fan-out	4	8	10	25	50	10
Nominal supply voltage (V)	$3.0 \pm 10\%$ $3.6 \pm 10\%$	$5.0 \pm 10\%$	$5.0 \pm 10\%$	$-5.2 \pm 10\%$	1.3 to 18	15 ± 1
Gate dissipation (mW)	2.5 or 12	8 or 12	12 or 22	55	0.01 (dc) 1 (1 MHz)	55
Noise immunity	Fair	Good	Good	Good	Very good to excellent	Excellent
Noise generation	Good	Good	High	Good	Low	Good
Gate propagation delay (ns)	12	30	4–12	2	50	90
Flip-flop clock rate (MHz)	8	12–30	15–200	400	5	4
Number of functions	High	Moderate, new functions in TTL	Very high, increasing rapidly	High, growing	High, increasing rapidly	Moderate
Compatibility with other families	No	With TTL	With DTL	No	No	No

Problems

3-1. What is meant by wired-collector logic?

3-2. Where can wired-collector logic be used?

3-3. Define: (a) passive pull-up; (b) active pull-up.

3-4. Give some of the popular names for wired-collector logic.

3-5. What type of logic occurs at the point where the wired-collector logic is made?

3-6. When wired-collector logic is made with two inverters, what is the type of logic referred to the inverter inputs?

3-7. What is the Boolean equation at point X in the following wired-collector logic?

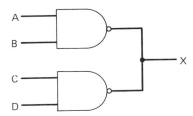

Figure P3-7   Wired-collector of two NANDs

3-8. What is the Boolean equation at point X?

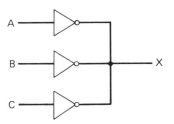

Figure P3-8   Wired-collector of three inverters

3-9. In a logic family, $V_{OH} = 9.2$ V, $V_{OL} = 0.8$ V, and $V_{th} = 4.0$ V. Find the noise immunity of this logic family.

3-10. What is meant by a pull-up resistor? Why is it needed?

3-11. What is meant by loading factor?

3-12. Define fan-out and fan-in.

3-13. Why is TTL faster than DTL?

3-14. Why is ECL faster than RTL, DTL, and TTL?

FOUR

Binary Numbers and Binary Arithmetic

4-1 INTRODUCTION

One major achievement of modern technology is the development of the electronic computers and electronic calculators able to perform arithmetic operations at incredible speeds. In this chapter we shall develop the methods employed by such computers to do arithmetic.

4-2 BASE 10 NUMBER SYSTEM

Before becoming involved with computer arithmetic, we should review the basic concepts of our base 10 number system and then extend it to the computer. For example, what do we mean by the three digit number 683? To begin, our number system is a base 10 system. In this decimal system we have 10 different symbols: 0, 1, 2, 3, 4, 5, 6, 7, 8, and 9. Each symbol brings a different concept. For example, the symbol 5 represents a defining quantity such as in the following examples: five fingers, five objects, five apples, five eggs, five cents, five dollars. We automatically associate the symbol 5 with the concept of "how many." Our decimal system has 10 symbols from 0 through 9. What happens if we have more than nine objects, say ten? We write this as 10. When we go to amounts in excess of 9, we make use of the position concept by beginning again in the next column to the left. A 1 in the second column represents ten objects, a 2 represents twenty. Now, what do we mean by twenty and how do we express it with a more scientific and yet a more basic representation? We can write

$$\text{Twenty} = 20 = 2 \times 10$$

The $\times$ in arithmetic is the multiplication symbol and indicates a repetitive process. 2×10 means that 10 is repeated 2 times for the final quantity. What happens if we have ninety-nine objects and add 1 (symbol $+$)† to the group? We can express this as $99 + 1 = 100$. One additional unit to the first column on the right exceeds the 9, and it now goes to a 0, with a 1 in the middle column. This in combination with its 9 exceeds its list of symbols, and we write 100. In scientific notation, we can write $100 = 1 \times 10^2$; the 2 is an exponent and represents the number of zeros in the number, and 10 is the base.

To get back to the decimal number 683, we can now write it as

$$683 = 600 + 80 + 3$$
$$= 6 \times 100 + 8 \times 10 + 3 \times 1$$
$$= 6 \times 10^2 + 8 \times 10^1 + 3 \times 10^0$$

remembering that any number with an exponent of 0 has the value 1. 10 is the *radix* or *base* of the decimal system.

What about a number like 67.832? We interpret this as

$$6 \times 10 + 7 \times 1 + 8 \times \tfrac{1}{10} + 3 \times \tfrac{1}{100} + 2 \times \tfrac{1}{1000}$$
$$= 6 \times 10^1 + 7 \times 10^0 + 8 \times 10^{-1} + 3 \times 10^{-2} + 2 \times 10^{-3}$$

The decimal point is a reference point. To the left all positions are multiplied by increasing multiples of 10, and to the right they are multiplied by decreasing multiples of 10. In scientific notation for the decimal number, note that negative exponents represent the process of multiplying by decreasing multiples of 10 or its equivalent of division.

In general, we can write that any decimal number may be represented as

$$X_{10} = D_n \times 10^n + D_{n-1} \times 10^{n-1} + \ldots D_0 \times 10^0 + D_{-1} \times 10^{-1}$$
$$+ D_{-2} \times 10^{-2} \ldots \tag{4-1}$$

---

†In this chapter the symbol $+$ is the arithmetic symbol for addition, as opposed to $+$ in Chapter 1, which indicated OR logic.

In X_{10}, the subscript 10 for X indicates the base or number system. The D's are the digits. We shall see later that this is required to eliminate confusion between number systems.

4-3 BINARY NUMBER SYSTEM

Why not use the decimal number system in computers? Unfortunately, it does not conveniently lend itself to available electronic components. In the computer we have to make use of the dual physical properties of devices such as transistors operated with a high- or low-level of voltage or current, relays that are open or closed, magnets polarized in one direction or the other, etc. In the computer we use logic decisions such as we previously described in Chapter 1.

Table 4-1   Decimal–binary equivalents

Decimal	Binary
0	0
1	1
2	10
3	11
4	100
5	101
6	110
7	111
8	1000
9	1001
etc.	etc.

In the computer we use a number system based upon the two symbols 1 and 0. This is a binary system.† It starts with 0 representing zero objects. One object is represented by 1. For two objects there is no symbol. As in the decimal system, we use the position concept. Two then is represented by beginning the column to the right over again with a 0 and putting a 1 in the next column to the left. This continues, and we can write an equivalent binary system and decimal system, such as in Table 4-1. Note that in going from 3 to 4, both binary columns have been completed, and we go from 11 to 100. In going from 7 to 8, three binary columns have been completed, and we go from 111 to 1000. In decimal we went from 99 to 100.

---

†Although we may not realize it, we make use of number systems other than decimal quite frequently. For example, the dozen and the gross comprise a system using 12 as the base.

4-4 BINARY-TO-DECIMAL CONVERSION

As in decimal, we can write that any binary number can be represented as

$$X_2 = B_n \times 2^n + B_{n-1} \times 2^{n-1} + \ldots B_0 \times 2^0 + B_{-1} \times 2^{-1} + B_{-2} \times 2^{-2} \ldots$$

(4-2)

In this case the subscript 2 for X indicates the binary base or binary number system. It can be seen by comparing numbers in both base systems that identical numbers can appear in both. For example, we could have 111_{10} and 111_2. Both represent considerably different numbers of objects. Without the subscripts it would be impossible to know what 111 really means. The B's are binary digits and are called *bits*. Groups of 8 bits occur frequently and are called *bytes*.

Equation (4-2) written in this basic manner will permit conversion from binary notation to decimal equivalent.

EXAMPLE 4-1   Convert the binary number 110011_2 to its decimal equivalent.

Solution

$$110011_2 = 1 \times 2^5 + 1 \times 2^4 + 0 \times 2^3 + 0 \times 2^2 + 1 \times 2^1 + 1 \times 2^0$$
$$= 1 \times 32 + 1 \times 16 + 0 \times 8 + 0 \times 4 + 1 \times 2 + 1 \times 1$$
$$= 51_{10}$$

EXAMPLE 4-2   Convert the binary number 0.0101_2 to its decimal equivalent.

Solution

$$0.0101_2 = 0 \times 2^{-1} + 1 \times 2^{-2} + 0 \times 2^{-3} + 1 \times 2^{-4}$$
$$\underset{\text{binary point}}{\uparrow} = 0 + \tfrac{1}{4} + 0 + \tfrac{1}{16}$$
$$= 0.25 + 0.0625$$
$$= 0.3125_{10}$$

EXAMPLE 4-3   Convert the combined binary number 1101.011_2 to its decimal equivalent.

Solution

$$1101.011_2 = 1 \times 2^3 + 1 \times 2^2 + 0 \times 2^1 + 1 \times 2^0 + 0 \times 2^{-1}$$
$$+ 1 \times 2^{-2} + 1 \times 2^{-3}$$
$$= 8 + 4 + 1 + \tfrac{1}{4} + \tfrac{1}{8}$$
$$= 13.375_{10}$$

The point which acts as the indicator between positive and negative exponents is called the *binary point*.

The values of 2^n and 2^{-n} occur frequently. Table 4-2 gives the values for n from 0 to 10.

Table 4-2 Values of 2^n and 2^{-n}

2^n	n	2^{-n}
1	0	1
2	1	0.5
4	2	0.25
8	3	0.125
16	4	0.0625
32	5	0.03125
64	6	0.015625
128	7	0.0078125
256	8	0.00390625
512	9	0.001953125
1024	10	0.0009765625

4-5 DECIMAL-TO-BINARY CONVERSION: THE 2S METHOD

The previous discussion has dealt with the problem of binary-to-decimal conversion. How do we go in the opposite direction, from decimal to binary? One way is by repeated application of the base 2 to the decimal number. Which way the 2 application is applied depends upon which side of the decimal point is being considered. For whole numbers (left of decimal point), the procedure is to continually divide by 2 and record the remainders next to the result of the division until a final division resulting in a zero is reached. The binary equivalent is the binary number reading the remainders upward.

EXAMPLE 4-4 Convert the decimal number 47_{10} to binary.

Solution

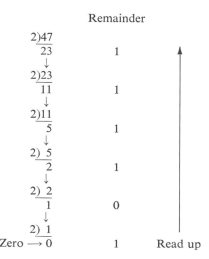

We can then write that $47_{10} = 101111_2$. The equivalency can easily be verified.

For fractional numbers, the procedure is reversed; instead of dividing by 2, we continually multiply by 2. When a carry occurs, we tabulate it, but we do not use it in the next multiplication.

EXAMPLE 4-5   Convert the number 0.4375_{10} to binary.

Solution   Reading the carry column downward,

	Carry	
$0.4375 \times 2 = 0.875$	0	Read down
$0.875 \times 2 = 1.75$	1	
$0.75 \times 2 = 1.5$	1	
$0.5 \times 2 = 1.0$	1	
$0 \times 2 = 0$	0	

We can now write $0.4375_{10} = 0.0111_2$. The equivalency can readily be verified.

EXAMPLE 4-6   Convert the decimal number 0.6_{10} to its binary equivalent.

Solution

	Carry	
$0.6 \times 2 = 1.2$	1	Read down
$0.2 \times 2 = 0.4$	0	
$0.4 \times 2 = 0.8$	0	
$0.8 \times 2 = 1.6$	1	
$0.6 \times 2 =$		

We see that after four multiplications in Example 4-6 we come back to the decimal 0.6. Repeated multiplications can only result in a repetition of the decimal sequence 0.6, 0.2, 0.4, 0.8, and the binary sequence 1, 0, 0, 1. We have a *repeating binary* 0.1001 .... We can continue this process as far as we want, getting closer and closer to the exact result but never reaching it. This also occurs of course in decimal. For example, $\frac{1}{3} = 0.3333 \ldots$.

We can then write

$$0.6_{10} \approx 0.1001 \ldots$$

EXAMPLE 4-7   Find the decimal equivalent of 0.1001_2.

Solution

$$0.1001 = 0.5 + 0 + 0 + 0.0625 = 0.5625$$

The solution to Example 4-7 is not a very good approximation to 0.6_{10}, but if we were to go one term further in the approximation, we could write

$$0.6 \approx 0.10011\ldots$$
$$= 0.5625 + 0.03125 = 0.59375$$

and this is quite close to 0.6. Additional terms will give an even closer approximation.

For a combined number such as 47.6, we would split it into the whole part and the fractional part and write

$$47.6 = 101111.10011$$

4-6 DECIMAL-TO-BINARY CONVERSION: THE SUBTRACTIVE METHOD

Another approach to decimal-to-binary conversion is a comparison subtraction method and the use of Table 4-2. Consider 47_{10}. The highest power of 2 that is less that 47 is $32 = 2^5$.

EXAMPLE 4-8   Convert 47_{10} to binary using the subtractive method.

Solution

$$47 - 32 = 15 \qquad 32 = 1 \times 2^5$$
$$15 - 8 = 7 \qquad 8 = 1 \times 2^3$$
$$7 - 4 = 3 \qquad 4 = 1 \times 2^2$$
$$3 - 2 = 1 \qquad 2 = 1 \times 2^1$$
$$1 - 1 = 0 \qquad 1 = 1 \times 2^0$$
$$47 = 1 \times 2^5 + 0 \times 2^4 + 1 \times 2^3 + 1 \times 2^2 + 1 \times 2^1 + 1 \times 2^0$$
$$= 101111$$

Note that this procedure is exactly the reverse of the binary-to-decimal conversion.

EXAMPLE 4-9   Convert 0.6_{10} to binary using the subtractive method.

Solution

$$0.6 - 0.5 = 0.1 \qquad\qquad 0.5 = 1 \times 2^{-1}$$
$$0.1 - 0.0625 = 0.0375 \qquad\qquad 0.0625 = 1 \times 2^{-4}$$
$$0.0375 - 0.03125 = 0.00625 \qquad\qquad 0.03125 = 1 \times 2^{-5}$$
$$0.6 = 1 \times 2^{-1} + 0 \times 2^{-2} + 0 \times 2^{-3} + 1 \times 2^{-4} + 1 \times 2^{-5} + \ldots$$
$$= 0.10011\ldots$$

4-7 BINARY ADDITION

Before discussing binary addition, let us review decimal addition. When we learned decimal addition, we did so by setting up an addition table. For example, $2 + 3 = 5$ is a decimal addition for which the sum does not equal or exceed the base 10. What about $6 + 4 = 10$? In this case the sum term $= 0$, and there is a carry of 1 to the next significant column to the left. Table 4-3 is the table for decimal addition of two digits, $X + Y$.

Table 4-3 Decimal addition : $X + Y$

X \ Y	0	1	2	3	4	5	6	7	8	9
0	0	1	2	3	4	5	6	7	8	9
1	1	2	3	4	5	6	7	8	9	10
2	2	3	4	5	6	7	8	9	10	11
3	3	4	5	6	7	8	9	10	11	12
4	4	5	6	7	8	9	10	11	12	13
5	5	6	7	8	9	10	11	12	13	14
6	6	7	8	9	10	11	12	13	14	15
7	7	8	9	10	11	12	13	14	15	16
8	8	9	10	11	12	13	14	15	16	17
9	9	10	11	12	13	14	15	16	17	18

Binary addition follows the same rules as decimal addition except that it is much simpler. The addition table can be written as follows:

1 Addition of

$$0 + 0 = 0$$
$$0 + 1 = 1$$
$$1 + 0 = 1$$
$$1 + 1 = 10 \quad \text{(sum of 0 and carry of 1)}$$

The 2-bit addition table is given in Table 4-4.

94

Table 4-4 Two-bit binary
addition: $X + Y$

X ⟍ Y	0	1
0	0	1
1	1	10

2 Addition of 3 bits, required when a carry is generated in the previous column:

$$0 + 0 + 0 = 0$$
$$0 + 0 + 1 = 1$$
$$0 + 1 + 0 = 1$$
$$0 + 1 + 1 = 10 \quad \text{(sum of 0 and carry of 1)}$$
$$1 + 0 + 0 = 1$$
$$1 + 0 + 1 = 10 \quad \text{(sum of 0 and carry of 1)}$$
$$1 + 1 + 0 = 10 \quad \text{(sum of 0 and carry of 1)}$$
$$1 + 1 + 1 = 11 \quad \text{(sum of 1 and carry of 1)}$$

Using the above rules, we can proceed to add two binary numbers.

EXAMPLE 4-10   Add the binary number 11 to 10.

Solution

$$
\begin{array}{cc}
11 & 3 \\
\text{carry} \rightarrow +_1 10 & +2 \\
\hline
101 & 5
\end{array}
$$

EXAMPLE 4-11   Add the binary number 11011 to 10011.

Solution

$$
\begin{array}{cc}
1\ 1\ 0\ 1\ 1 & 27 \\
\text{carry} \rightarrow {}_1 1\ 0_1 0_1 1\ 1 & 19 \\
\hline
1\ 0\ 1\ 1\ 1\ 0 & 46
\end{array}
$$

As can be seen, binary addition follows the same methods as decimal addition.

EXAMPLE 4-12   Add:

$$
\begin{array}{cc}
\begin{array}{r}
1\ 1\ 1 \\
1\ 0\ 1 \\
1\ 0\ 1 \\
{}_{101}1{}_0 0\ 1 \\
\hline
1\ 0\ 1\ 1\ 0
\end{array}
&
\begin{array}{r}
7 \\
5 \\
5 \\
5 \\
\hline
22
\end{array}
\end{array}
$$

Solution

Note that in the first column of Example 4-12 there are four 1s. In binary addition this is 100: a sum of 0, a carry of 0 to the next column, and a carry of 1 to the second column to the left. In the third column we now have five 1s, four from the original problem and a 1 carry from the first column. The addition of five 1s is 101, a sum of 1 and a 10 carry.

Of course we could have added the numbers successively:

$$111 + 101 = 1100 \qquad 7 + 5 = 12$$
$$1100 + 101 = 10001 \qquad 12 + 5 = 17$$
$$10001 + 101 = 10110 \qquad 17 + 5 = 22$$

This is how the computer does it.

4-8   BINARY SUBTRACTION

Let us review decimal subtraction. As an example,

$$
\begin{array}{r}
7 \\
-5 \\
\hline
2
\end{array}
$$

and the difference is a positive number, since 7 is greater than 2. What about $5 - 7$?

$$
\begin{array}{r}
5 \\
-7 \\
\hline
-2
\end{array}
$$

and the difference is negative since 7 is greater than 5. Let us try $55 - 27$.

$$
\begin{array}{r}
55 \\
-27 \\
\text{Borrow} \quad -1 \\
\hline
28
\end{array}
$$

In this case we subtract 7 from 5, obtain a difference of 8, and generate a borrow of 1 from the next column. This 1 and 2 are now combined and subtracted from 5 to give a difference of 2. The final result is 28.

Binary subtraction follows the same principles.

1   Subtraction of 2 bits:

$$0 - 0 = 0$$
$$0 - 1 = -1 \qquad \text{(or 1 and borrow 1 from the next column)}$$
$$1 - 0 = 1$$
$$1 - 1 = 0$$

2   Subtraction of 3 bits, which occurs when a borrow is required from the next column:

$$0 - 0 - 0 = 0$$
$$0 - 0 - 1 = -1 \qquad \text{(or 1 with a borrow of 1)}$$
$$0 - 1 - 0 = -1 \qquad \text{(or 1 with a borrow of 1)}$$
$$0 - 1 - 1 = -10 \qquad \text{(or 0 with a borrow of 1)}$$
$$1 - 0 - 0 = 1$$
$$1 - 0 - 1 = 0$$
$$1 - 1 - 0 = 0$$
$$1 - 1 - 1 = -1 \qquad \text{(or 1 with a borrow of 1)}$$

Using these rules, let us perform some problems in subtraction.

EXAMPLE 4-13   Subtract 101 from 111.

Solution

$$
\begin{array}{r}
111 \\
-101 \\
\hline
010
\end{array}
\qquad
\begin{array}{r}
7 \\
-5 \\
\hline
2
\end{array}
$$

EXAMPLE 4-14   Subtract 1110 from 1010.

Solution

$$
\begin{array}{r}
1010 \\
-1110
\end{array}
\qquad
\begin{array}{r}
10 \\
-14
\end{array}
$$

By inspection we can see that 1110 is greater than 1010; therefore, the result is a negative number. We can then subtract the smaller from the larger and write that the answer is negative.

$$
\begin{array}{r}
1110 \\
-1010 \\
\hline
100 \rightarrow \quad -100
\end{array}
$$

The correct answer is $1010 - 1110 = -100$.

Let us try another example for which borrows are required.

EXAMPLE 4-15   Subtract 011 from 100.

Solution

$$
\begin{array}{r} 1\ 0\ 0 \\ -\ _{-1}0_1 1\ 1 \\ \hline 0\ 0\ 1 \end{array}
\qquad\qquad
\begin{array}{r} 4 \\ -3 \\ \hline 1 \end{array}
$$

In the right column, $0 - 1$ is 1 borrow 1. In the second column we have $0 - 1 - 1$. Following the rules, this is equal to a 0 with a borrow into the third column. In the third column we have $1 - 0 - 1 = 0$.

4-9 COMPLEMENT SUBTRACTION:
SUBTRACTION BY COMPLEMENT ADDITION

As we shall see in Chapter 5, subtraction and addition require different electronic equipment. In this section we shall show how subtraction can be performed by an addition technique called complement subtraction. This will simplify the electronics needed to perform arithmetic in computers and calculators. Let us perform a problem that has a positive difference in decimal subtraction, in three different ways: (a) conventional subtraction, (b) nines (9s) complement subtraction, and (c) tens (10s) complement subtraction.

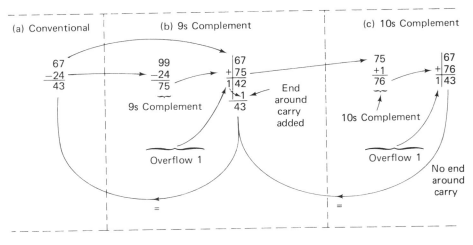

In decimal subtraction, consider the problem $67 - 24 = 43$.

a Conventional: $67 - 24 = 43$
b 9s complement subtraction:

1 Obtain the 9s complement of the subtrahend by subtracting each digit from 9 (the largest digit): $99 - 24 = 75$

2 Add the 9s complement to the minuend: $67 + 75 = 142$

3 In step 2, there is an overflow 1. This indicates a positive answer for the difference.

4 Perform an end-around carry with the 1 and add it to the least significant bit. However, discard the overflow 1 in the final result. This yields the same answer as in method a: $42 + 1 = 43$
$+43$

c 10s complement subtraction:

1 Obtain the 10s complement of the subtrahend by adding 1 to the 9s complement: $75 + 1 = 76$

2 Add the 10s complement to the minuend: $67 + 76 = 143$

3 In step 2, the overflow 1 indicates a positive difference.

4 Discard the overflow 1 to give a final difference of 43; the same result as in methods a and b. Practically, the end-around carry was performed in step 1: $+43$

What really has been done in methods b and c is to perform a problem in subtraction by means of an addition.

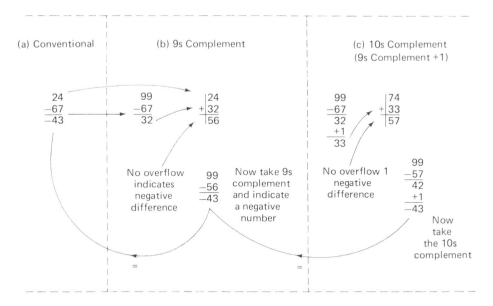

Let us now do a problem in subtraction with a negative difference. In decimal subtraction, consider the problem $24 - 67 = -43$.

a Conventional:

$$24 - 67 = -43$$

b 9s complement subtraction:

  1 Obtain the 9s complement of the subtrahend:

$$99 - 67 = 32$$

  2 Add the 9s complement to the minuend:

$$24 + 32 = 56$$

  3 In step 2 there is no overflow 1. This indicates a negative result for the difference.

  4 Now take the 9s complement of step 2 and indicate a negative difference. This is the same answer as in method a:

$$99 - 56 = 43$$
$$\equiv -43$$

c 10s complement subtraction:

  1 Obtain the 10s complement of the subtrahend:

$$32 + 1 = 33$$

  2 Add the 10s complement to the minuend:

$$24 + 33 = 57$$

  3 In step 2 there is no overflow 1. This indicates a negative result for the difference.

  4 Since we are working with the 10s complement, we have to take the 10s complement of step 2 and indicate a negative difference. This gives the same answer as in methods a and b:

$$99 - 57 = 42$$
$$42 + 1 = 43$$

$$-43$$

We see from this that the no overflow 1 in the next column tells us that the result is negative. With a negative difference, we have to take the complement. To obtain the correct answer, we have to be consistent and take the 9s complement if we are using the 9s complement method, or the 10s complement if the 10s complement is being used.

The computer follows the same procedure, but with binary numbers. The complements are called 1s or 2s complements. The computer will sense the presence or absence of the overflow 1 and make the appropriate decision.

Let us now find the 1s complement of a binary number. As in decimal, the binary number will be subtracted from a number of all 1s to obtain the 1s complement.

EXAMPLE 4-16 Find the 1s complement of 110011.

Solution

$$
\begin{array}{r}
111111 \\
-110011 \\
\hline
001100
\end{array}
$$

We see from Example 4-16 that the 1s complement is the inverse of the number. To obtain the complement we simply replace 1s with 0s and 0s with 1s. In the same way as in decimal, the 2s complement is the 1s complement plus 1. For Example 4-16, the 2s complement is

$$001100$$
$$+ \qquad 1$$
$$\overline{001101}$$

The fact that the 1s complement is the inverse of the number is quite fortunate. Many electronic circuits automatically provide the complement of a bit. If it is not available, an inverter will generate it.

Following are examples of binary complement subtraction.

EXAMPLE 4-17 For a positive difference: 1101 − 110.

Solution Binary complement subtraction follows the same principles as decimal complement subtraction, using the 1s or 2s complements.

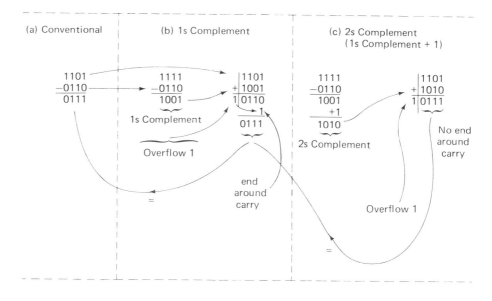

a Conventional binary subtraction: $\qquad$ $1101 - 110 = 0111$

b 1s complement subtraction:

$\quad$ **1** Obtain the 1s complement of the subtrahend. Interchange 1s and 0s (*Note:* the subtrahend must have the same number of bits as the minuend): $\qquad$ $0110 \rightarrow 1001$

$\quad$ **2** Add the 1s complement to the minuend: $\qquad$ $1101 + 1001 = 10110$

$\quad$ **3** In step 2, the overflow 1 indicates a positive difference.

$\quad$ **4** Perform an end-around carry with the overflow 1. Discard the overflow 1 in the final

result and indicate a positive answer. The
result is the same as in method a:

$$0110 + 1 = 0111$$
$$+0111$$

c 2s complement subtraction:

1 Obtain the 2s complement of the subtrahend by adding 1 to the 1s complement:

$$1001 + 1 = 1010$$

2 Add the 2s complement to the minuend:

$$1101 + 1010 = 10111$$

3 The difference is positive because of the overflow 1. Discard the overflow 1 in the final answer. This is the same as in methods a and b.

$$+0111$$

EXAMPLE 4-18   For a negative difference: $110 - 1101$.

Solution

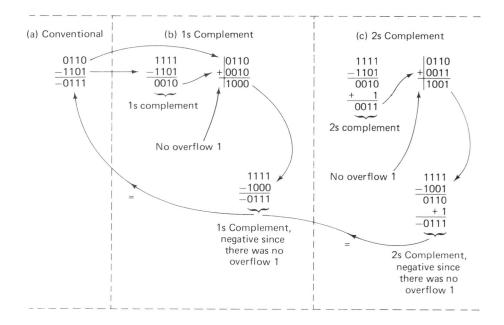

a Conventional binary subtraction:

$$0110 - 1101 = -0111$$

b 1s complement subtraction:

1 Obtain the 1s complement of the subtrahend by interchanging the 1s and 0s:

$$1101 \rightarrow 0010$$

2 Add the 1s complement to the minuend:

$$0110 + 0010 = 1000$$

3 No overflow 1 indicates a negative difference.

4 Obtain the 1s complement of 2 and indicate a negative difference. The answer is the same as in method a:

$$1000 \rightarrow 0111$$
$$-0111$$

c   2s complement subtraction:

1   Obtain the 2s complement of the subtrahend:

$0010 + 1 = 0011$

2   Add the 2s complement to the minuend:

$0110 + 0011 = 1001$

3   In step 2 there is no overflow 1. This indicates a negative difference.

4   Since we are working with the 2s complement, we have to take the 2s complement of step 2 and indicate a negative difference. This gives the same difference as in methods a and b:

$1001 \rightarrow 0110$

$0110 + 1 = 0111$

-0111

Note that in Example 4-18 the number 110 was written as 0110, with the same number of bits as there are in 1101.

There are complex ICs, such as the TTL arithmetic logic unit (ALU) 54/74181,† that can be programmed to do addition or subtraction as well as many other functions on two 4-bit numbers. The time of operation is very short, approximately 24 ns. Subtraction is performed by the 1s complement addition method, with the 1s complement generated internally. If we were performing the previous problem, $1101 - 110$, which has a positive difference, this ALU would provide the output 10110 exactly as in the illustration. Additional circuitry would then sense the presence of the leftmost 1, which we have called the overflow 1, do the end-around (also called forced) carry to give 0111, and assign positive (+) to the answer. For the subtraction, $110 - 1101$, the ALU unit will provide an output 01000. Additional circuitry would then sense the leftmost 0. This 0 causes additional circuitry to generate the complement, which is equal to 0111, and assign negative (−) to the result of the subtraction.

Examples 4-17 and 4-18 showed that subtraction can be performed using either the 1s or the 2s complement method. How do computers do it? Both methods are used but when signed arithmetic is being performed 2s complement arithmetic has an advantage. This is discussed further in Sec. 15-4.

4-10 BINARY MULTIPLICATION

The process of multiplication in binary is performed exactly as in decimal. The multiplication rules are as follows:

$$0 \times 0 = 0$$

$$0 \times 1 = 0$$

$$1 \times 0 = 0$$

$$1 \times 1 = 1$$

†Texas Instruments, SN54/74181; Motorola, MC74181; Fairchild, 9341/74181; National Semiconductor, DM 54/74181.

The multiplication table is given in Table 4-5.

Table 4-5 Binary multiplication: $X \cdot Y$

X \ Y	0	1
0	0	0
1	0	1

EXAMPLE 4-19   Multiply 11001 by 1101.

Solution

$$
\begin{array}{rr}
11001 & 25 \\
\times\ 1101 & \times 13 \\
\hline
11001 & \\
00000 & \\
11001\ \ & \\
11001\ \ \ & \\
\hline
101000101 & 325 \\
\end{array}
$$

In the first line of Example 4-19 we have multiplication of 11001 by 1. In the second line, 11001 is multiplied by 0 and shifted one column to the left. The third line is 11001 multiplied by 1 and shifted over an additional column. This shifting operation is performed by an electronic circuit called a shift register, which will be studied in Chapter 8. The fourth line is 11001 multiplied by 1 and again shifted one column to the left. The final result is obtained by binary addition of all columns.

EXAMPLE 4-20   Multiply 101.11×1.01.

Solution

$$
\begin{array}{r}
101.11 \\
\times\ \ \ 1.01 \\
\hline
10111 \\
00000\ \ \\
10111\ \ \ \\
\hline
111.0011 \\
\end{array}
$$

4-11   BINARY DIVISION

Binary division is performed in binary arithmetic by repeated division and subtraction in the same manner as in decimal arithmetic.

EXAMPLE 4-21 $1111101_2 \div 101_2$.

Solution

$$
\begin{array}{r}
11001 \\
101\overline{)1111101} \\
101 \\
\overline{101} \\
101 \\
\overline{} \\
101 \\
101 \\
\overline{0}
\end{array}
\qquad
\frac{1111101}{101} = 11001
$$

$$
\frac{125_{10}}{5_{10}} = 25_{10}
$$

EXAMPLE 4-22 $1111101_2 \div 110_2$; $125 \div 6$.

Solution

$$
\begin{array}{r}
10100.1 \\
110\overline{)1111101.0} \\
110 \\
\overline{111} \\
110 \\
\overline{1010} \\
110 \\
\overline{100}
\end{array}
\qquad
\frac{1111101}{110} = 10100.1 + \text{remainder}
$$

4-12 OCTAL NUMBER SYSTEM

The *octal* number system is used in some computers and is closely related to the binary number system. It has eight symbols: 0, 1, 2, 3, 4, 5, 6, and 7. Table 4-6 gives the first 12 octal numbers and their decimal and binary equivalents.

Table 4-6 Decimal–octal–binary equivalents

Decimal	Octal	Binary
0	0	0
1	1	1
2	2	10
3	3	11
4	4	100
5	5	101
6	6	110
7	7	111
8	10	1000
9	11	1001
10	12	1010
11	13	1011
12	14	1100

Note that, when the numbers become greater than 7, we again use the position method to indicate this, with the 1 in the second column representing 1×8^1. A number like 467.32 has a decimal equivalent of

$$467.32_8 = 4 \times 8^2 + 6 \times 8^1 + 7 \times 8^0 + 3 \times 8^{-1} + 2 \times 8^{-2}$$

$$\underset{\text{octal point}}{\uparrow} = 4 \times 64 + 6 \times 8 + 7 \times 1 + \tfrac{3}{8} + \tfrac{2}{64}$$

$$= 256 + 48 + 7 + \tfrac{24}{64} + \tfrac{2}{64}$$

$$= (311\tfrac{26}{64})_{10}$$

The addition and multiplication tables for octal numbers are given in Tables 4-7 and 4-8, which can be used to perform arithmetic in the octal system.

EXAMPLE 4-23 $36_8 \times 24_8$.

Solution

$$
\begin{array}{r}
36 \\
\times 24 \\
\hline
170 \\
{}_174 \\
\hline
1130_8
\end{array}
$$

$4_8 \times 36_8 = 140 + 30 = 170_8$
$2_8 \times 36_8 = 60 + 14 = 74_8$

Check

$$36_8 = 3 \times 8^1 + 6 \times 8^0 = 24 + 6 = 30_{10}$$

$$24_8 = 2 \times 8^1 + 4 \times 8^0 = 16 + 4 = 20_{10}$$

$$1130_8 = 1 \times 8^3 + 1 \times 8^2 + 3 \times 8^1 + 0 \times 8^0$$

$$= 512 + 64 + 24 = 600_{10} = 30_{10} \times 20_{10}$$

To convert a decimal number to an octal number, we proceed as with binary by repeated division by 8, and write the remainders in order reading up.

EXAMPLE 4-24   Convert $217_{10} = X_8$.

Solution

$$
\begin{array}{r}
8)\overline{217} \\
\hline
27
\end{array}
\qquad +1 \uparrow
$$

$$
\begin{array}{r}
8)\overline{27} \\
\hline
3
\end{array}
\qquad +3
$$

$$
\begin{array}{r}
8)\overline{3} \\
\hline
0
\end{array}
\qquad +3 \quad \Big| \quad \text{Read up}
$$

$$217_{10} = 331_8$$

Table 4-7 Octal addition: $X + Y$

X \ Y	0	1	2	3	4	5	6	7
0	0	1	2	3	4	5	6	7
1	1	2	3	4	5	6	7	10
2	2	3	4	5	6	7	10	11
3	3	4	5	6	7	10	11	12
4	4	5	6	7	10	11	12	13
5	5	6	7	10	11	12	13	14
6	6	7	10	11	12	13	14	15
7	7	10	11	12	13	14	15	16

Table 4-8 Octal multiplication: $X \cdot Y$

X \ Y	0	1	2	3	4	5	6	7
0	0	0	0	0	0	0	0	0
1	0	1	2	3	4	5	6	7
2	0	2	4	6	10	12	14	16
3	0	3	6	11	14	17	22	25
4	0	4	10	14	20	24	30	34
5	0	5	12	17	24	31	36	43
6	0	6	14	22	30	36	44	52
7	0	7	16	25	34	43	52	61

Check

$$331_8 = 3 \times 8^2 + 3 \times 8^1 + 1 \times 8^0$$
$$= 3 \times 64 + 3 \times 8 + 1$$
$$= 192 + 24 + 1 = 217$$

The relationship between octal and binary is of greater interest. Consider the binary number

$$111011011.11010101$$

Starting from the binary point in both directions, break up the binary number into groups of three and express each group with its octal symbol from Table 4-6.

$$\underset{7}{\underbrace{111}} \ \underset{3}{\underbrace{011}} \ \underset{3}{\underbrace{011}} \cdot \underset{6}{\underbrace{110}} \ \underset{5}{\underbrace{101}} \ \underset{2}{\underbrace{010}} \qquad \text{add extra 0 to complete group of 3}$$

The octal equivalent is 733.652.

What is the advantage of the octal system? In the above illustration, the expression in octal requires six characters instead of 17 in binary. This results in lower printout time and less space requirements for printouts. Moreover, in the illustrative example $217_{10} = 331_8$, three divisions by 8 were required, with a subsequent simple conversion to get to binary. To have performed this conversion by repeated division by 2 would have required eight divisions.

4-13 HEXADECIMAL NUMBER SYSTEM

Table 4-9   Decimal–hexadecimal–binary equivalents

Decimal	Hexadecimal	Binary
0	0	0000
1	1	0001
2	2	0010
3	3	0011
4	4	0100
5	5	0101
6	6	0110
7	7	0111
8	8	1000
9	9	1001
10	A	1010
11	B	1011
12	C	1100
13	D	1101
14	E	1110
15	F	1111

The *hexadecimal* number system uses base 16. It is even more economical than the octal system and is used quite extensively in computers. The 16 symbols are given in Table 4-9, together with the binary and decimal equivalents.

Tables can be written for hexadecimal addition and multiplication similar to Tables 4-7 and 4-8. The conversions from decimal to hexadecimal are done in the same manner as in other base systems.

EXAMPLE 4-25   Convert $4C5_{16}$ to decimal.

Solution

$$4C5 = 4 \times 16^2 + C \times 16^1 + 5 \times 16^0$$
$$= 4 \times 16^2 + 12 \times 16 + 5 \times 1$$
$$= 1024 + 192 + 5$$
$$= 1221_{10}$$

EXAMPLE 4-26   Convert $4C5_{16}$ to binary.

Solution   Replace each hexadecimal character by its binary equivalent from Table 4-9.

$$\underbrace{0100}_{4} \quad \underbrace{1100}_{C} \quad \underbrace{0101}_{5} = 010011000101_2$$

EXAMPLE 4-27   Convert 10101101100.010111_2 to hexadecimal.

Solution   Starting at the binary point, break up the binary number into groups of four and express each group with its hexadecimal equivalent from Table 4-9. Incomplete groups are to be completed with additional zeros.

$$\underbrace{0101}_{5} \quad \underbrace{0110}_{6} \quad \underbrace{1100}_{C} \; . \; \underbrace{0011}_{3} \quad \underbrace{1101}_{D} = 56C.3D$$

Problems

Unless otherwise indicated, base 2 is to be used.

4-1. Express the decimal number 4296.32 in the form of Eq. (4-1).

4-2. Write the decimal numbers from 10 to 34 in binary notation.

4-3. Convert the following binary numbers to decimal: (a) $1101_2 = ?_{10}$; (b) $1100111_2 = ?_{10}$; (c) $10101_2 = ?_{10}$.

4-4. Convert the following binary numbers to decimal: (a) $11101_2 = ?_{10}$; (b) $111000_2 = ?_{10}$; (c) $1011011_2 = ?_{10}$.

4-5. Convert the following binary numbers to decimal; express in fractional and decimal form: (a) $0.111_2 = ?_{10}$; (b) $0.0011_2 = ?_{10}$.

4-6. Convert the following binary numbers to decimal; express in decimal form: (a) $1101.101_2 = ?_{10}$; (b) $110011.011_2 = ?_{10}$.

4-7. Convert the following binary numbers to decimal; express in decimal form: (a) $111001.01 = ?_{10}$; (b) $1010101.0101 = ?_{10}$.

4-8. What are the values of 2^{11}, 2^{12}, 2^{-11}, and 2^{-12}?

4-9. Convert the following decimal numbers to binary: (a) $29_{10} = ?_2$; (b) $61_{10} = ?_2$; (c) $46_{10} = ?_2$; (d) $156_{10} = ?_2$.

4-10. Convert the following decimal numbers to binary: (a) $23_{10} = ?_2$; (b) $74_{10} = ?_2$; (c) $83_{10} = ?_2$.

4-11. Convert the following decimal numbers to binary: (a) $0.625_{10} = ?_2$; (b) $0.7_{10} = ?_2$; (c) $0.21875_{10} = ?_2$.

4-12. Convert the following decimal numbers to binary: (a) $0.875_{10} = ?_2$; (b) $0.9_{10} = ?_2$; (c) $0.59375_{10} = ?_2$.

4-13. Convert the following decimal numbers to binary: (a) $57.25_{10} = ?_2$; (b) $29.375_{10} = ?_2$.

4-14. $0.4_{10} = ?_2$. Continue until the approximation is accurate to within 1 percent.

4-15. (a) $110011 + 110111$; (b) $1011101 + 1011111$; (c) $1011 + 1011 + 1001 + 1001$. Check your answers.

4-16. $1101 + 1010 + 10110 + 1111$.

4-17. Use 9s and 10s complement: (a) $68 - 32$; (b) $657 - 225$; (c) $823 - 128$; (d) $32 - 68$; (e) $225 - 657$.

4-18. Use 9s and 10s complement: (a) $48 - 26$; (b) $627 - 306$; (c) $269 - 483$.

4-19. Subtract directly and also using 1s and 2s complements. Check your answer by decimal conversion: (a) $11011 - 10101$; (b) $110101 - 101100$; (c) $11011 - 1101$; (d) $1011 - 10111$.

4-20. Repeat Prob. 4-19: (a) $101101 - 10111$; (b) $11001 - 1011$; (c) $10111 - 1001$; (d) $100 - 1111$.

4-21. Use 1s complement subtraction: (a) $111001 - 11010$; (b) $101 - 10011$. Repeat using 2s complement subtraction.

4-22. Use 1s complement subtraction: (a) $1100 - 10101$; (b) $1100111 - 1011$. Repeat using 2s complement subtraction.

4-23. Multiply in binary and check your result by decimal conversion: (a) 1101×1011; (b) 11001×111.

4-24. Multiply in binary: (a) 10101×10101; (b) 1101×1011.

4-25. Multiply in binary: $101 \times 101 \times 101$.

4-26. Multiply in binary: $1011 \times 1011 \times 1011$.

4-27. Multiply in binary; check your answer by decimal conversion: (a) 10.1×10.1; (b) 11.11×1.1.

4-28. Multiply in binary: 101.11×1010.11.

4-29. Divide in binary (continue to 2 binary places): (a) $1111 \div 101$; (b) $11001100 \div 1001$; (c) $1100100 \div 1011$.

4-30. Write the decimal numbers 13 through 25 in base 8.

4-31. (a) $271_8 = ?_{10}$; (b) $174_8 = ?_{10}$.

4-32. (a) $0.32_8 = ?_{10}$; (b) $0.47_8 = ?_{10}$.

4-33. (a) $24_8 + 17_8 = ?_8$; (b) $42_8 \times 26_8 = ?_8$; (c) $275_{10} = ?_8$. Check your answer to parts (a) and (b) by conversion to decimal.

4-34. (a) $427_{10} = ?_8$; (b) $53_8 + 127_8 = ?_8$; (c) $27_8 \times 35_8 = ?_8$.

4-35. (a) $110111011110.011011 = ?_8$; (b) $42.37_8 = ?_2$.

4-36. (a) $625.43_8 = ?_2$; (b) $11011.11 = ?_8$.

4-37. (a) $18_{10} = ?_{16}$; (b) $38_{10} = ?_{16}$; (c) $143_{10} = ?_{16}$; (d) $CA_{16} = ?_{10}$.

4-38. (a) $42_{10} = ?_{16}$; (b) $83_{10} = ?_{16}$; (c) $127_{16} = ?_{10}$.

4-39. (a) $11001111_2 = ?_{16}$; (b) $AC47_{16} = ?_2$; (c) $101.111_2 = ?_{10}$.

4-40. (a) $1101011.01011_2 = ?_{16}$; (b) $FCB_{16} = ?_2$.

4-41. Write an addition table like Table 4-7 for base 3 addition.

4-42. Write an addition table like Table 4-7 for base 5 addition.

4-43. Write a multiplication table like Table 4-8 for base 3 multiplication.

4-44. Write a multiplication table like Table 4-8 for base 5 multiplication.

FIVE

EXCLUSIVE-OR and Arithmetic Operations

5-1 INTRODUCTION

In Chapter 4 we discussed arithmetic: how to add, subtract, multiply, and divide. Computers that perform arithmetic operations do so using the 1s and 0s of logic operations and this leads to the binary arithmetic system. The purpose of this chapter is to develop the logic equations and logic systems that will enable us to perform these arithmetic functions.

Consider the procedure involved in the addition of two binary numbers $X + Y$.

$$
\begin{array}{ll}
X & 1011 \quad \text{Least significant bit} \\
Y & +0101 \quad \text{(half-adder)} \\
\end{array}
$$

The first step in the addition is to add the least significant bits, in this example two 1s. Two-number addition in the least significant column involves only 2 bits. This addition results in a sum term S and a carry-out term C_o, which will be added in the next column. The truth table for the least significant bit is given in Table 5-1,

Table 5-1 Truth table: addition of least significant bit

X	Y	S	C_o
0	0	0	0
0	1	1	0
1	0	1	0
1	1	0	1

which is a statement of the following arithmetic process:

$$0 + 0 \longrightarrow \text{sum } S = 0, \quad \text{carry out } C_o = 0$$
$$0 + 1 \longrightarrow \text{sum } S = 1, \quad \text{carry out } C_o = 0$$
$$1 + 0 \longrightarrow \text{sum } S = 1, \quad \text{carry out } C_o = 0$$
$$1 + 1 \longrightarrow \text{sum } S = 0, \quad \text{carry out } C_o = 1$$

5-2 EXCLUSIVE-OR†

Consider the truth table for the S column alone. The results of this column are important enough and occur so frequently in digital logic that it has been assigned a separate term called the EXCLUSIVE-OR. The EXCLUSIVE-OR truth table is given in Table 5-2.

Note that in the EXCLUSIVE-OR the output EO is a 1 only when X or $Y = 1$ the output EO is a 0 when X and Y are alike. Frequently this is written as

$$Z = X \oplus Y \tag{5-1}$$

Table 5-2 EXCLUSIVE-OR truth table

X	Y	EO
0	0	0
0	1	1
1	0	1
1	1	0

†M. E. Levine, *Digital Theory and Experimentation Using Integrated Circuits.* Englewood Cliffs, N.J.: Prentice-Hall, Inc., 1974, Expt. 5, The "EXCLUSIVE-OR" and Its Applications.

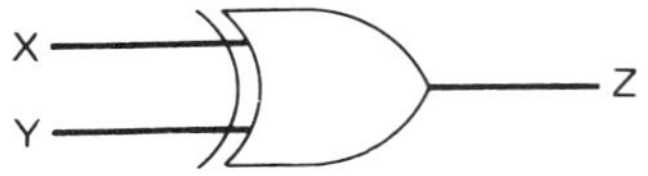

Figure 5-1 Logic symbol for EX-CLUSIVE-OR

In addition, the logic symbol shown in Fig. 5-1 has been assigned for the EXCLUSIVE-OR.

The Karnaugh map for the EXCLUSIVE-OR is given in Fig. 5-2a, which by extension is the equivalent of Fig. 5-2b. From the map it can be seen that there is no ready simplification of the EXCLUSIVE-OR.

	$\bar{X}$	X
$\bar{Y}$		1
Y	1	

(a)

	$\bar{X}$	X
$\bar{Y}$	0	
Y		0

(b)

Figure 5-2 Map of EXCLUSIVE-OR

The Boolean logic expression for the EXCLUSIVE-OR based upon truth table. Table 5-2, is given by

$$Z = \bar{X}Y + X\bar{Y} \tag{5-2a}$$

and from Fig. 5-2b this can also be written as

$$Z = \overline{\bar{X}\bar{Y} + XY} \tag{5-2b}$$

The complement $\bar{Z}$ of the EXCLUSIVE-OR is sometimes required and is called the EXCLUSIVE-NOR or EXCLUSIVE-AND.

$$\bar{Z} = \bar{X}\bar{Y} + XY \quad \text{or} \quad X \odot Y \tag{5-3a}$$

$$Z = X \odot Y \tag{5-3b}$$

Many different ways have been devised to generate the EXCLUSIVE-OR function, both with separate individual gates or within ICs that directly provide the EXCLUSIVE-OR function. They represent various ways of implementing Eq. (5-2).

The method of Fig. 5-3a is based directly upon the logic of Eq. (5-2a) and can be used if the inputs and their complements are available. When the complements are not available, inverters are required, as shown in Fig. 5-3b. The method of Fig. 5-3c makes use of wired-collector, wired-OR logic and can be used when the internal gate construction permits this. Since many ICs provide quad two-input NAND gates in a single package, the method of Fig. 5-3c generates the EXCLUSIVE-OR function in a single package and simultaneously generates the EXCLUSIVE-NOR function. (Its operation is left as a problem.) The other methods shown

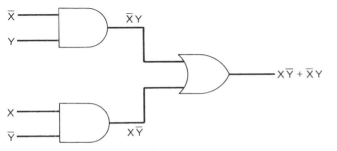

Figure 5-3a EXCLUSIVE-OR based on Eq. (5-2a) logic

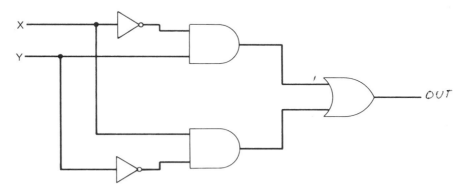

Figure 5-3b EXCLUSIVE-OR with inverters

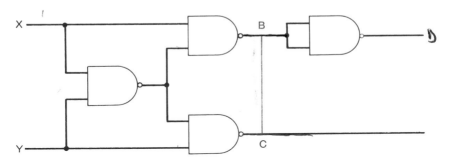

Figure 5-3c EXCLUSIVE-OR with wired-collector, wired-OR logic

Figure 5-3d EXCLUSIVE-OR based on Eq. (5-2b) logic

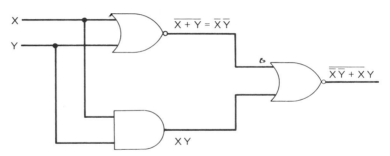

require both AND and OR functions, and these are normally not available in a single IC package. The method of Fig. 5-3d makes use of the logic of Eq. (5-2b).

As was discussed in the introduction to this chapter, the EXCLUSIVE-OR is used in binary addition. Since speed is an important consideration in the design of the modern computer, and because a limiting factor in logic-system speed is the delay through each logic gate, those designs of the EXCLUSIVE-OR which require a minimum number of propagation gates are frequently the best design. ICs are available that provide four EXCLUSIVE-ORs or four EXCLUSIVE-NORs in a single package.

Applications of the EXCLUSIVE-OR will be discussed in later sections of this chapter.

5-3 HALF-ADDER

The introduction to this chapter discussed the problem of adding the least significant bits of two binary numbers, which is a process of adding 2 bits. This step occurs in a circuit called a *half-adder*. The truth table, Table 5-1, is repeated as Table 5-3. The logic implementation of Table 5-3 is shown in Fig. 5-4. This also shows that a sum will be generated when either X or $Y = 1$, and the carry will be generated only when X and Y both have the value of 1. The Boolean equations for the HALF-ADDER are

$$S = X \oplus Y \tag{5-4a}$$

$$C_o = XY \tag{5-4b}$$

Table 5-3 HALF-ADDER : $X + Y$

X	Y	S	C_o
0	0	0	0
0	1	1	0
1	0	1	0
1	1	0	1

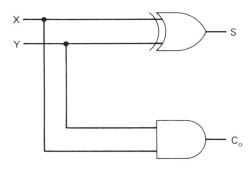

Figure 5-4 HALF-ADDER

5-4 HALF-SUBTRACTOR

The HALF-SUBTRACTOR occurs as the subtraction procedure in the least significant bit of binary subtraction.

$$
\begin{array}{rl}
X & 1011 \\
-Y & \underline{1001}
\end{array}
\quad \text{Half-subtractor column}
$$

As in the half-adder, only 2 bits are involved, one for X and one for Y. The subtraction process generates a difference term D and a borrow-out term B_o which is required for the next column. The arithmetic process is as follows:

$$0 - 0 \longrightarrow \text{difference } D = 0, \quad \text{borrow out } B_o = 0$$
$$0 - 1 \longrightarrow \text{difference } D = 1, \quad \text{borrow out } B_o = 1$$
$$1 - 0 \longrightarrow \text{difference } D = 1, \quad \text{borrow out } B_o = 0$$
$$1 - 1 \longrightarrow \text{difference } D = 0, \quad \text{borrow out } B_o = 0$$

The truth table for the HALF-SUBTRACTOR is given in Table 5-4. As can be seen, the difference D is the EXCLUSIVE-OR function, and the borrow out B_o occurs when $X = 0$ and $Y = 1$. The logic implementation of Table 5-4 is shown in Fig. 5-5.

Table 5-4 HALF-SUBTRACTOR: $X - Y$

X	Y	D	B_o
0	0	0	0
0	1	1	1
1	0	1	0
1	1	0	0

Figure 5-5 HALF-SUBTRACTOR: $X - Y$

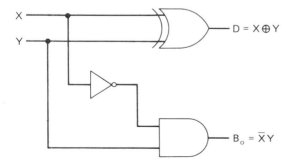

5-5 FULL-ADDER†

Let us reconsider the addition problem, but now let us direct our attention to the column to the left of the least significant column.

$$
\begin{array}{ccc}
 & B & A \\
X & 1 & 1 \\
Y & 1 & 1 \\
 & 1 & \\
\hline
 & 0 &
\end{array}
$$

$\longleftarrow$ Carry out C_o from column A becomes carry in C_i for column B

In this example, in column A, $1 + 1$ results in a sum $= 0$ and a carry-out term $C_o = 1$, which has to be added in the next column. This term is called the carry-in term C_i for column B. We can see that in all columns other than the A column we must deal with the addition of 3 bits, X, Y, and C_i. This will result again in a sum term S and a carry-out term C_o to the next column. This addition process,

$$ X + Y + C_i \longrightarrow \text{sum } S \text{ and } C_o $$

is called full addition. The addition process results in the following:

$$ 0 + 0 + 0 \longrightarrow \text{sum } S = 0, \quad \text{carry out } C_o = 0 $$
$$ 0 + 0 + 1 \longrightarrow \text{sum } S = 1, \quad \text{carry out } C_o = 0 $$
$$ 0 + 1 + 0 \longrightarrow \text{sum } S = 1, \quad \text{carry out } C_o = 0 $$
$$ 0 + 1 + 1 \longrightarrow \text{sum } S = 0, \quad \text{carry out } C_o = 1 $$
$$ 1 + 0 + 0 \longrightarrow \text{sum } S = 1, \quad \text{carry out } C_o = 0 $$
$$ 1 + 0 + 1 \longrightarrow \text{sum } S = 0, \quad \text{carry out } C_o = 1 $$
$$ 1 + 1 + 0 \longrightarrow \text{sum } S = 0, \quad \text{carry out } C_o = 1 $$
$$ 1 + 1 + 1 \longrightarrow \text{sum } S = 1, \quad \text{carry out } C_o = 1 $$

The truth table for the FULL-ADDER is given in Table 5-5.

The Boolean equations for the FULL-ADDER are

$$ S = \bar{X}\bar{Y}C_i + \bar{X}Y\bar{C}_i + X\bar{Y}\bar{C}_i + XYC_i \tag{5-5a} $$
$$ C_o = \bar{X}YC_i + X\bar{Y}C_i + XY\bar{C}_i + XYC_i \tag{5-5b} $$

Equation (5-5) shows, as does the truth table, that there are four possible combinations of X, Y, and C_i which will result in a sum $S = 1$, and similarly for the carry-out term C_o.

To see if any simplification is possible, let us plot S and C_o on a Karnaugh map. From Table 5-6 it can easily be seen that there is no ready simplification for

†M. E. Levine, *Digital Theory and Experimentation Using Integrated Circuits*. Englewood Cliffs, N.J.: Prentice-Hall, Inc., 1974, Expt. 6, Full-Adder/Full-Subtractor.

Table 5-5 FULL-ADDER: $X + Y + C_i$

X	Y	C_i	Sum S	Carry Out C_o
0	0	0	0	0
0	0	1	1	0
0	1	0	1	0
0	1	1	0	1
1	0	0	1	0
1	0	1	0	1
1	1	0	0	1
1	1	1	1	1

Table 5-6 FULL ADDER, $X + Y + C_i$: (a) sum S; (b) carry out C_o

	$\bar{X}\bar{Y}$	$\bar{X}Y$	XY	$X\bar{Y}$
$\bar{C}_i$		1		1
C_i	1		1	

(a)

	$\bar{X}\bar{Y}$	$\bar{X}Y$	XY	$X\bar{Y}$
$\bar{C}_i$			1	
C_i		1	1	1

(b)

Figure 5-6 Sum S in FULL-ADDER

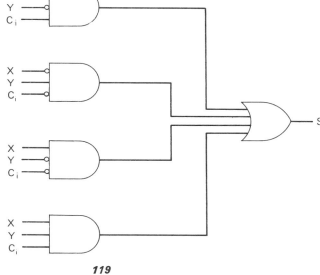

the sum term, but the adjacent box terms will permit simplification of the carry-out term.

Straightforward implementation of the sum function is given in Fig. 5-6. As can be seen both from Eq. (5-5a) and Fig. 5-6, implementation of the sum requires four AND and one OR gates. Since IC logic families are available as NAND or NOR families, it is advantageous to convert Eq. (5-5a) by De Morgan's theorem to uniform logic terms. We can write Eq. (5-5a) by De Morgan's theorem as

$$S = \overline{(\overline{\overline{X}\,\overline{Y}C_i})(\overline{\overline{X}\,Y\overline{C}_i})(\overline{X\overline{Y}\,\overline{C}_i})(\overline{XYC_i})} \tag{5-6}$$

Equation 5-6 can be implemented by four three-input NAND gates and one four-input NAND gate, which are available in DTL and TTL.

Suppose that we now convert Eq. (5-5a) to NOR logic. Converting each term individually results in

$$S = (\overline{X + Y + \overline{C}_i}) + (\overline{X + \overline{Y} + C_i}) + (\overline{\overline{X} + Y + C_i}) + (\overline{\overline{X} + \overline{Y} + \overline{C}_i}) \tag{5-7}$$

This can be implemented by four three-input NOR gates and one four-input OR gate. The OR gate is awkward, since four-input OR gates may not be available, whereas four-input NOR gates are readily available. To realize the OR function would require an additional inverter.

If use is now made of the missing boxes of Table 5-6a, the sum S also can be realized by

$$S = \overline{\overline{X}\,\overline{Y}\,\overline{C}_i + \overline{X}\,YC_i + XY\overline{C}_i + X\overline{Y}C_i} \tag{5-8}$$

which can be converted by De Morgan's theorem to

$$S = \overline{(\overline{X + Y + C_i}) + (\overline{X + \overline{Y} + \overline{C}_i}) + (\overline{\overline{X} + \overline{Y} + C_i}) + (\overline{\overline{X} + Y + \overline{C}_i})} \tag{5-9}$$

This can be implemented by four three-input NORs and one four-input NOR gate, and these are available in NOR logic such as RTL.

The Karnaugh map for the carry-out C_o term has adjacent boxes. Combining adjacent boxes, the carry-out term can be written as

$$C_o = YC_i + XY + XC_i \tag{5-10}$$

This represents three two-input AND gates and one three-input OR and is obviously simpler than Eq. (5-5b). Its logic diagram is shown in Fig. 5-7a. However, it is not in NAND or NOR logic and requires both AND and OR. Applying De Morgan's theorem,

$$C_o = \overline{(\overline{YC_i})(\overline{XY})(\overline{XC_i})} \tag{5-11}$$

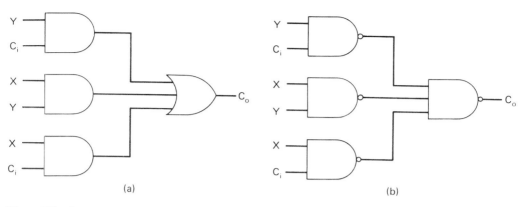

Figure 5-7 Carry out C_o in the FULL-ADDER: (a) AND-OR gates; (b) NAND gates only

This can be realized with three two-input NAND and one three-input NAND. In addition, it does not require the complemented values of the inputs. Figure 5-7b is the logic diagram corresponding to Eq. (5-11).

Equation (5-11) and Fig. 5-7 are a logical representation for the carry-out term. The carry-out term is a majority term function and indicates that there will be a carry-out term whenever any two of the inputs have the value 1.

Equation (5-5a) can be factored as follows:

$$S = \bar{C}_i(\bar{X}Y + X\bar{Y}) + C_i(\bar{X}\bar{Y} + XY)$$
$$= \bar{C}_i(X \oplus Y) + C_i(\overline{X \oplus Y}) \tag{5-12}$$

the second term being the missing boxes of the EXCLUSIVE-OR function or the EXCLUSIVE-NOR. This indicates that a simplification may be possible using the EXCLUSIVE-OR function.

Equation (5-6a) and Table 5-5 were obtained by the simultaneous addition of X, Y, and C_i. Suppose that we add X and Y separately and then combine with C_i. If we add $X + Y$, we obtain a carry-out C_o term if X and Y are both 1, or we obtain a carry-out C_o term if X or Y are 1 and $C_i = 1$. We obtain a sum S term if X or $Y = 1$ and $C_i = 0$, or if X and $Y = 0$ and $C_i = 1$, or if X or $Y = 1$ and $C_i = 1$. These results can be implemented by the logic blocks of Fig. 5-8. Note that it

Figure 5-8 FULL-ADDER with HALF-ADDERS

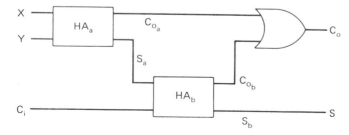

requires two HALF-ADDERS and one OR gate. Its advantage is that it enables the function of full addition to be performed with half-adders, and reduces the number of functions required to build a multibit adder. The same building blocks can now be used in the half-adder and full-adder positions.

5-6   FULL-SUBTRACTOR

As in the FULL-ADDER, a FULL-SUBTRACTOR is required for all bits other than the least significant bit.

$$
\begin{array}{ccc}
\text{Bit} & B & A \\
X & 1\ 0 & 0 \\
-Y & 1 & 1 \\
\hline
 & 1 & \\
\hline
 & 1 &
\end{array}
$$

1 ←——— Borrow-out B_o term from column A becomes borrow-in B_i term to column B

The FULL-SUBTRACTOR has to perform the following arithmetic process: $X - Y - B_i \rightarrow$ difference D and borrow out B_o. The B_o term becomes the B_i term to the next more significant column. The subtraction process results in the following:

$$0 - 0 - 0 \longrightarrow \text{difference } D = 0, \quad \text{borrow out } B_o = 0$$
$$0 - 0 - 1 \longrightarrow \text{difference } D = 1, \quad \text{borrow out } B_o = 1$$
$$0 - 1 - 0 \longrightarrow \text{difference } D = 1, \quad \text{borrow out } B_o = 1$$
$$0 - 1 - 1 \longrightarrow \text{difference } D = 0, \quad \text{borrow out } B_o = 1$$
$$1 - 0 - 0 \longrightarrow \text{difference } D = 1, \quad \text{borrow out } B_o = 0$$
$$1 - 0 - 1 \longrightarrow \text{difference } D = 0, \quad \text{borrow out } B_o = 0$$
$$1 - 1 - 0 \longrightarrow \text{difference } D = 0, \quad \text{borrow out } B_o = 0$$
$$1 - 1 - 1 \longrightarrow \text{difference } D = 1, \quad \text{borrow out } B_o = 1$$

The truth table for the FULL-SUBTRACTOR is given in Table 5-7.

The Boolean equations for the FULL-SUBTRACTOR based upon Table 5-7 are

$$D = \bar{X}\bar{Y}B_i + \bar{X}Y\bar{B}_i + X\bar{Y}\bar{B}_i + XYB_i \qquad (5\text{-}13a)$$
$$B_o = \bar{X}\bar{Y}B_i + \bar{X}Y\bar{B}_i + \bar{X}YB_i + XYB_i \qquad (5\text{-}13b)$$

The Karnaugh maps for Eqs. (5-13) are given in Table 5-8. The map for the difference D is the same as that for the FULL-ADDER. The same logic and components can be used. The map for B_o differs from that of C_o. This requires different circuitry.

Table 5-7   FULL-SUBTRACTOR: $X - Y - B_i$

X	Y	B_i	Difference D	Borrow Out B_o
0	0	0	0	0
0	0	1	1	1
0	1	0	1	1
0	1	1	0	1
1	0	0	1	0
1	0	1	0	0
1	1	0	0	0
1	1	1	1	1

Table 5-8   FULL-SUBTRACTOR, $X - Y - B_i$: (a) difference D; (b) borrow out B_o

	$\bar{X}$ $\bar{Y}$	$\bar{X}$ Y	X Y	X $\bar{Y}$
$\bar{B_i}$		1		1
B_i	1		1	

(a)

	$\bar{X}$ $\bar{Y}$	$\bar{X}$ Y	X Y	X $\bar{Y}$
$\bar{B_i}$		1		
B_i	1	1	1	

(b)

Using similar logic and reasoning as for the FULL-ADDER, the FULL-SUBTRACTOR can be implemented with two HALF-SUBTRACTOR blocks as shown in Fig. 5-9. Operation of the circuit of Fig. 5-9 can be demonstrated by means of a truth table (Prob. 5-13).

Figure 5-9   FULL-SUBTRACTOR with HALF-SUBTRACTORS

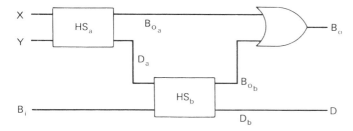

5-7 FULL-ADDER: SPEED AND ECONOMY CONSIDERATIONS

The FULL-ADDER of Figs. 5-6 and 5-7 requires two levels (AND gates, one level, and the OR gate, the second level) of logic and a total number of 25 gate inputs. In addition, it requires that the complements of X, Y, and C_i be available. If they are not available, a third level of logic is required. Since speed of operation is determined by the delay through each level of logic, designs that minimize the number of logic levels perform the addition in the least amount of time. FULL-ADDERs have been developed that reduce the number of gate inputs or do not require the complements, and hence are more economical; but such designs increase the number of logic levels, and hence the time required to complete the addition.

Consider the problem of adding two multibit binary numbers such as

$$\begin{array}{ll} X & 111100111 \\ Y & +101111011 \\ \hline & 1101100010 \end{array}$$

When we do this addition manually, we start by adding the two least significant bits, determining the sum and the carry. The carry is then added to the next column. We add one column at a time. This is serial addition. The same technique can be done electronically. If we convert X and Y from the parallel form shown above to a serial pulse form (Chapter 8), we obtain the result shown in Fig. 5-10.

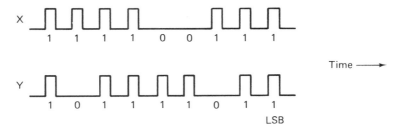

Figure 5-10 Serial pulse representation for X and Y

Let us now present these pulses to a serial full adder as shown in Fig. 5-11. Each pair of pulses is presented separately. The sum appears as a series of pulses on the S output. The carry pulses C_o are delayed by one pulse time and are combined with the next pair of pulses. The circuit of Fig. 5-11 is economical. Only one FULL-ADDER is required. It is slow, and the addition time is equal to the pulse time multiplied by the number of bits.

Alternatively, a PARALLEL-ADDER, such as shown in Fig. 5-12, can be used. From the figure it can be seen that addition of the first bit requires a HALF-ADDER, and then FULL-ADDERs are required for all succeeding pairs of bits. Compared to the circuit of Fig. 5-11, this is considerably more costly.

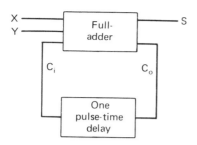

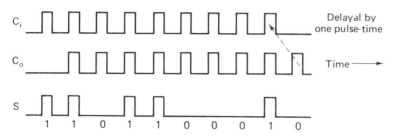

Figure 5-11 Serial FULL-ADDER

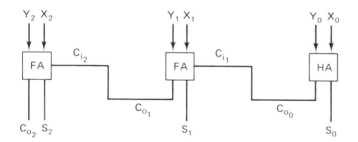

Figure 5-12 Three-bit PARALLEL-ADDER

At first glance it would appear that parallel addition would be quite fast. However, consider the following problem;

$$
\begin{array}{r}
11111111 \\
00000001 \\
\hline
100000000
\end{array}
$$

Addition of the first pair generates a carry. This carry must then "propagate down" the adder until the final carry is generated and after considerable delay. In any computer handling sums of many bits, such delay cannot be tolerated.

Just as logic equations were developed for the sum S and carry C_o of a FULL-ADDER, it is possible to develop logic equations for the addition of multibit numbers. This minimizes the "carry-propagate" delay. The greater the number of bits, the more complex the logic and the electronics. Practically, 4-bit FULL-ADDERs, such as the type 7483, are available. This incorporates a "carry-look-ahead" circuit, which minimizes the final carry delay. The 7483 is a TTL IC, and the carry-propagate delay is about 12 nanoseconds (ns). When adding more than 4-bit numbers, other ICs, called look-ahead-carry generators, incorporate the logic needed to minimize the carry-out delay.

5-8 APPLICATION OF THE EXCLUSIVE-OR

The EXCLUSIVE-OR is basically a comparison logic generator. It generates a 0 when both bits are alike and a 1 if the input bits are not alike. Many applications of the EXCLUSIVE-OR are based upon its use as a comparator.

Errors occasionally occur in the transmission of binary data. Error checking is an important aspect of digital logic. One way to error check is to compare the same binary number against itself. For example, we might transmit data twice and then compare the results of the two transmissions. If they are the same, the probabilities of an error are much more remote than if there were no check. Figure 5-13 is the logic diagram of a 4-bit comparator comparing the numbers $X_3 X_2 X_1 X_0$ against $Y_3 Y_2 Y_1 Y_0$. Any dissimilarity in any bits of the two numbers will generate a 1 in the output. Only if the two numbers are alike will the output be a 0. Similar matching

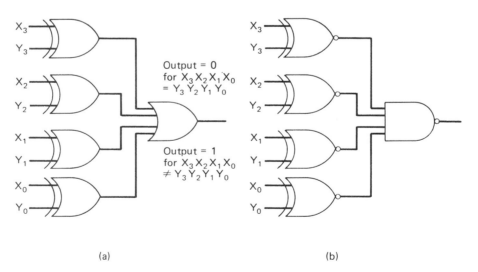

(a)   (b)

Figure 5-13 Four-bit comparator: (a) comparator;
(b) De Morgan equivalent of (a)

Table 5-9

Data Word	Even Parity Bit	Transmit
010011001	0	0100110010
010110000	1	0101100001
101000000	0	1010000000
111011000	1	1110110001

Data Word	Odd Parity Bit	Transmit
01011	0	010110
01111	1	011111
10111	1	101111
00010	0	001000

comparisons are required in computer programming and determine the next step in the program.

Another error test is the parity check. In using parity, all binary words are sent with an even number of 1s as even parity, or with an odd number of 1s as odd parity. Which one is used is the option of the equipment designer. Suppose that we are transmitting with even parity and the word has an odd number of 1s. We then transmit an extra bit, called a parity bit to make the number of 1s even. The extra bit does not affect the data; it is sent only for error checking. Determining whether the number of 1s in a binary word is odd or even is quite simple, as we shall see, using EXCLUSIVE-OR techniques. Table 5-9 shows examples of parity bits and parity transmission. The error testing comes in checking for parity. Suppose that we are using odd parity and an error occurs in one bit. This changes the number of 1s to an even number of 1s, and we know that there is an error. For example, suppose that we were sending the word 10110000 with odd parity, and a transmission error occurred in the third bit. The word would be transmitted as 10010000 with an even number of 1s, an indication of an error. The question then is what if simultaneous errors had occurred in 2 bits, say the first and third; we would then have the word 0010000, which is in odd parity with no indication of error. The possibility of two errors exists, but the odds against its occurring are considerably higher. For example, if the possibility of an error in 1 bit is 1 in 10,000, the possibility of errors in 2 bits is 1 in 100,000,000, quite small. With parity checking, all we know is that an error has occurred. How to locate the error is another matter. This subject is discussed further in Chapter 10.

Both odd and even parity are used. Odd parity has an advantage in that it is not possible to transmit all 0s. All 0s might be transmitted with a momentary power failure. Parity is discussed further in Sec. 10-5.

Parity checking and parity generation are readily performed using EXCLU-SIVE-OR gates. Figure 5-14 is the logic diagram of a 3-bit parity checker and

Table 5-10 Three-bit parity generator

X_0	X_1	Output EO_1	X_2	Output EO_2	No. of 1s in $X_2X_1X_0$
0	0	0	0	0	0 = even
0	0	0	1	1	1 = odd
0	1	1	0	1	1 = odd
0	1	1	1	0	2 = even
1	0	1	0	1	1 = odd
1	0	1	1	0	2 = even
1	1	0	0	0	2 = even
1	1	0	1	1	3 = odd

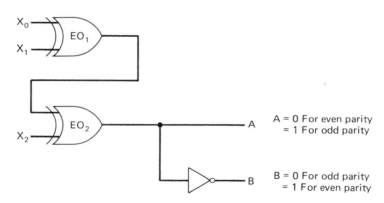

Figure 5-14 Three-bit parity generator–parity checker

parity generator, and Table 5-10 is the truth table for the diagram of Fig. 5-14. As can be seen from the truth table, the outputs A and B can be used for parity checking or for parity generation. To convert the word $X_2X_1X_0$ to even parity use output A. To convert to odd parity, use output B.

Another method of parity generation checking is the parity tree. An 8-bit parity tree is shown in Fig. 5-15. It uses EXCLUSIVE-ORs. EXCLUSIVE-NORs can be used in place of the EXCLUSIVE-ORs in Fig. 5-15 with resultant inverted output.

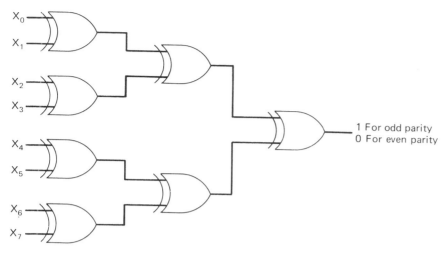

Figure 5-15   EXCLUSIVE-OR parity tree generator–checker

5-9   INTEGRATED-CIRCUIT LOGIC FUNCTIONS

All the logic functions discussed in this chapter are available in integrated circuits (ICs) in all the IC families. More and more complex functions in MSI and LSI form are being developed and are becoming commercially available to simplify logic implementation. Typical examples of these functions are given in the following paragraphs.

The EXCLUSIVE-OR function is generally available in quad form with four EO gates available in a single IC package. Some units provide the EXCLUSIVE-OR and some provide the EXCLUSIVE-NOR. Typical QUAD EXCLUSIVE-OR gates are the TTL SN54L86, the DTL MC1812P, the RTL MC971, and the CMOS CD4030A. The MC7242P is a quad EXCLUSIVE-NOR gate with open collector outputs.

The SN54180 is an EXCLUSIVE-NOR TTL parity-tree 8-bit odd–even parity generator–checker. The MC4008L contains 8 EXCLUSIVE-NOR functions. Seven are connected together to form an 8-bit parity tree providing a 1 output for even parity. The eighth EXCLUSIVE-NOR can be used as an inverter to invert the tree output to provide a 1 for odd parity, or it can be used to interconnect two 8-bit parity trees to form a 16-bit parity tree.

HALF-ADDERS are available in all logic families. Typical is the RTL MC775 dual HALF-ADDER.

Many types of FULL-ADDERS are available. The U6A938059X is a gated full adder. The U6A938259X is a 2-bit full adder, and the U6A938359X is a 4-bit full adder. These three units operate as serial-carry adders. For higher speed, adders with look-ahead circuits are used. The SN54181N is a high-speed arithmetic logic unit (ALU) function generator intended to perform many of the basic arithmetic functions, such as addition, complement subtraction, multiplication, and comparison, as well as a variety of logic functions. In performing addition it adds two 4-bit numbers in 24 ns.

The SN54182N is a look-ahead-carry generator, which provides look-ahead carry over four binary adders or four groups of adders. A single SN54182N combines with four SN54181N to provide second-level look-ahead circuitry for 16-bit addition. The SN54182N can perform this look-ahead function in 13 ns. These adder functions are in TTL logic. Similar functions are available in ECL.

Problems

5-1. The IC MC1231P EXCLUSIVE-NOR gate generates the function in the circuit shown. Explain its operation by expressing the logic function at each point on the logic diagram. [*Hint:* show that the output is given by either Eq. (5-2a) or (5-2b).] Draw a Karnaugh map.

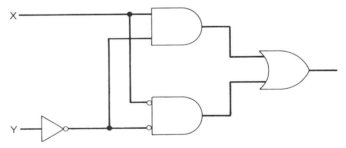

5-2. By means of a truth table for each point of the circuit diagram, show that the circuit of Fig. 5-3c generates the EO function. [*Hint:* write out truth tables for points B and C, assuming that the connection BC is not present. Then connect B to C and apply wired-OR logic at BC.]

5-3. The IC MC812G is the EXCLUSIVE-OR section of a half-adder. Its logic diagram is shown below Explain its operation by expressing the logic func-

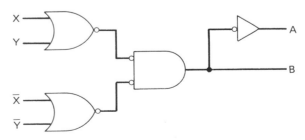

Figure P5-3 EXCLUSIVE-OR

tion at each point of the logic diagram. What are the functions at points A and B? [*Hint:* show that the output at B is given by either Eq. (5-2a) or (5-2b).] Draw Karnaugh maps.

5-4. Show how the IC MC746P whose logic diagram is given below generates the EXCLUSIVE-OR–NOR function.

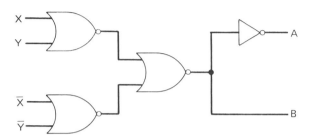

Figure P5-4   EXCLUSIVE-OR–NOR gate

5-5. The IC MC808G is a half-adder generating both the sum S and carry out C_o. Its logic diagram is shown below. Explain its operation.

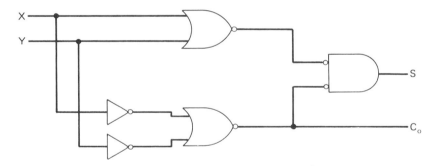

Figure P5-5   HALF-ADDER

5-6. In a parallel adder for adding two 16-bit numbers, how many HALF-ADDERs and FULL ADDERs are required?

5-7. In a parallel adder using the HALF-ADDER method of Fig. 5-8, how many HALF-ADDERs and OR gates are needed to add two 16-bit numbers?

5-8. (a) Draw the logic diagram for the difference D of the full subtractor of Eq. (5-13a).
(b) How many gates and gate inputs does it require and what types?

5-9. (a) Express the difference D of the full subtractor in the missing box (complement) form.
(b) Draw the logic diagram.
(c) How many gates does it require and what are the types?

5-10. Using Table 5-8b, simplify the expression for the borrow out B_o term of the full subtractor: (a) draw the logic diagram; (b) How many gates are required? (c) what are their types?

5-11. Express the borrow out B_o of the full subtractor in the missing box form: (a) simplify and draw a logic diagram; (b) how many gates are required and what are the types?

5-12. Complete the following truth table for the logic levels at each point of Fig. 5-8 to show that the circuit acts as a FULL-ADDER.

X	Y	C_i	C_{o_a}	S_a	C_{o_b}	$S_b = S$	C_o
0	0	0					
0	0	1					
0	1	0					
0	1	1					
1	0	0					
1	0	1					
1	1	0					
1	1	1					

5-13. Repeat Prob. 5-12, but for the FULL-SUBTRACTOR of Fig. 5-9.

5-14. The following waveforms are applied to inputs X, Y, and C_i of a FULL-ADDER. What are the S and C_o waveforms?

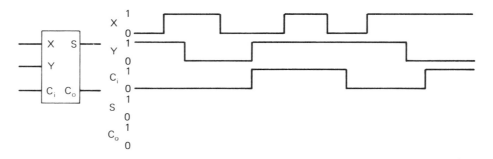

5-15. The following waveforms are applied to inputs X, Y, and B_i of a FULL-SUBTRACTOR. What are the D and B_o waveforms?

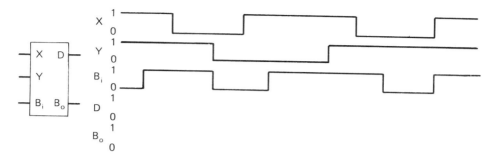

5-16. (a) Draw the logic diagram for a 2-bit comparator in which two 2-bit words are compared.

    (b) Write out a truth table.

5-17. (a) Draw the logic diagram for a 4-bit parity generator using the method of Fig. 5-14.

    (b) Write out a truth table to show that the output A is a 1 for odd parity and 0 for even parity.

5-18. Draw the logic diagram for a 4-bit parity tree using EXCLUSIVE-OR gates, and determine the output for odd and even parity.

5-19. Repeat Prob. 5-18 using EXCLUSIVE-NOR gates.

SIX

The Bistable

6-1 INTRODUCTION†

6-1.1 Basic Two-Inverter Bistable

The bistable or flip-flop (FF) is a circuit with two (and only two) stable states. It is a fundamental building block in digital circuitry and serves as the basis for memories, counters, frequency dividers, and shift registers. Since there are two stable states involved, means are provided for changing from one state to the other. We shall be concerned with symmetrical circuits in which equal amounts of energy are required to change from one state to the other. A second class of circuits, such as those based upon four-layer devices (the SCR is a typical example), are also bistables, but the energy and methods required to change from one stable state to the other are considerably different.

A quite familiar mechanical bistable is the common wall toggle switch used in the home to turn lights on and off. Push it up, it stays in that position and the lights are turned on; push it down, it stays in that position and the lights are turned off:

†M. E. Levine, *Digital Theory and Experimentation Using Integrated Circuits.* Englewood Cliffs, N.J.: Prentice-Hall, Inc., 1974, Expt. 7, Bistable or Flip-Flop (FF).

134

The basic electronic bistable consists of two dc-coupled inverters as shown in Fig. 6-1. Let $A = 1$; then $Q = 0$. This makes $B = 0$ and $\bar{Q} = 1$. But $\bar{Q}$ is connected to A, and we therefore have a stable configuration. Similarly, if $\bar{Q}\ (= A)$ was equal to 0, Q and B would be equal to 1, and we would again have a second stable state. The truth table is given in Table 6-1. Note particularly that the truth

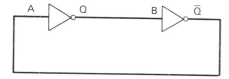

Figure 6-1 Basic two-inverter bistable

Table 6-1   Two-inverter bistable

State	Q	$\bar{Q}$
1	0	1
2	1	0

table and the circuit configuration allow only two possible sets of stable conditions. A set of forced conditions that might make both Q and $\bar{Q}$ alike is unnatural, does not satisfy allowed logic levels, and represents a not-allowed state.

The logic symbol for a bistable is shown in Fig. 6-2. Note that these are two output terminals, Q and $\bar{Q}$. (Although it is more common to use the letters Q and $\bar{Q}$ for the output terminals, the outputs are also frequently indicated by the levels 1 and 0.) Although Table 6-1 shows that Q can be either at a 1 or a 0, depending upon which state the bistable is in, it is also customary to consider the Q output as the 1 output terminal and the $\bar{Q}$ output as the 0 output terminal. As can be seen from the logic symbol and from Fig. 6-1, both the logic variable and its complement are available simultaneously.

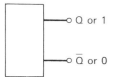

Q or 1

$\bar{Q}$ or 0

Figure 6-2   Bistable logic symbol

6-1.2   Basic Two-Transistor Bistable

A simple configuration of a two-transistor bistable in the basic manner of Fig. 6-1 is shown in Figs. 6-3a and 6-3b, where the circuit is designed and resistors chosen so that one transistor is heavily in saturation and the other cut off. This makes the 1 level (positive logic) $\approx V_{cc}$ and the 0 level ≈ 0. Fortunately, in a transistor $V_{CE\ \text{sat}} < V_{BE\ \text{sat}}$. With direct coupling from the collector of a transistor heavily in saturation to the base of a second transistor, as shown in Fig. 6-3, the low value of $V_{CE\,sat}$ is enough to cut off the second transistor. Practical bistables have been built using this principle.

A brief comment about the use of the letter Q in Fig. 6-3. It is common to use it both for transistors in circuit diagrams and for the output terminals of bistables. In this text, wherever transistors are used they will be indicated with a subscript such as in Fig. 6-3. Output terminals such as in Fig. 6-3 will not have a subscript.

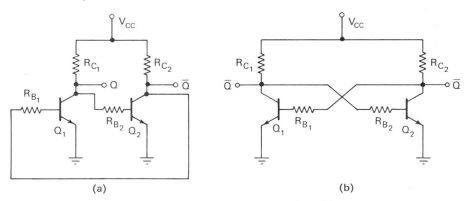

Figure 6-3 Two-transistor bistable: (a) cascade representation; (b) conventional representation

We have not discussed how to make the circuitry of Fig. 6-3 change from one state to the other. Suppose that transistor Q_1 is cut off ($Q = 1$) and Q_2 is in saturation ($\bar{Q} = 0$). Let us now physically short the collector of Q_1 to ground with a wire. No base current can flow into Q_2 and it cuts off. This now allows base current to flow into the base of Q_1 so that, when the short is removed, this base current will keep Q_1 in saturation, and Q_1 being in saturation will keep Q_2 cut off. This has changed the state of the bistable. This shorting technique incidentally is a very practical way of checking transistor bistables for basic operation when possible malfunction exists in an electronic system.

Of course it is not very practical to apply such a short circuit to a bistable for control purposes, but an electronic short circuit can be applied.

In Fig. 6-4, transistor Q_3 has been placed with its collector in parallel with Q_1, but with a separate base input, and similarly for transistors Q_2 and Q_4. Let inputs R and S be at ground potential. Q_3 and Q_4 do not conduct. If Q_1 is cut off (Q_2 in saturation), a sufficient positive potential at input R can drive Q_3 into saturation, putting an effective short at the collector of Q_1, and similarly for input S, thus providing independent control of outputs Q and $\bar{Q}$. We shall define this type of bistable later as an RS (set–reset) flip-flop. Its practical implementation is in the RTL IC type MC-802.

In the bistable of Fig. 6-4, what would happen if simultaneous positive potentials were applied to both S and R? This would drive both Q_3 and Q_4 into saturation, making both the Q and $\bar{Q}$ outputs at logic 0 level. But a bistable is defined and must operate so that only one of the two states of Table 6-1 can occur. Q and $\bar{Q}$ at logic level 0 is a not-allowed condition, and the system designer must take care that such an occurrence does not happen. For example, $Q = 0$ might control the clockwise direction of a motor, and $\bar{Q} = 0$ might control its counterclockwise direction. If both are at 0, the motor direction is indeterminate.

But suppose it does happen that S and R are made momentarily positive. What will be the state of the bistable when S and R are both returned to a ground-level input? Essentially, we are discussing a *race condition*. Either S or R must return to ground level first. If input R gets to ground first, the S input will keep Q_4 in saturation, the final FF state will be determined by the Q_4 saturation, and $Q = 1$, $\bar{Q} = 0$. In the same way, if S gets to ground first, Q_3 is in saturation when Q_4 is not, and the final FF state is $Q = 0$, $\bar{Q} = 1$. The stable state that finally results when both R and S are at ground is therefore determined by which of the S and R inputs first results in an allowed condition.

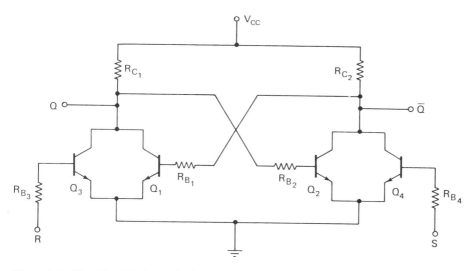

Figure 6-4   Bistable with electronic short circuit

6-1.3   Noise in Bistables

The noise immunity of the bistable of Fig. 6-3 is poor (as is typical of RTL logic). The base input terminal of the cutoff transistor is a high-impedance point and is therefore susceptible to *impulse noise* interference. Let Q_1 of Fig. 6-3 be the transistor in saturation. The base terminal of Q_2 is separated from ground ($V_{CE \text{ sat}}$ of Q_1) by resistor R_{B_2}. Q_2 is cut off; since no base current flows, its input impedance is high and susceptible to noise. Noise in a bistable is much more serious than in a gate. In a gate, a noise pulse comes and goes, leaving the final logic level unchanged. But in a bistable the noise pulse can result in a change of state, which will remain after the noise pulse has disappeared.

6-2 INTEGRATED-CIRCUIT BISTABLES

The previous discussion of the discrete transistor FF provided basic background material. The remainder of this chapter will be devoted to the development of various types of IC FFs in use today in digital systems. For each FF a truth table will be developed to explain its operation. In using FFs, it is of extreme importance that the user know *exactly when* FFs are enabled and data transferred into the FF. To illustrate the operation of FFs and when they are enabled, timing diagrams will be used. These diagrams are typical of those which occur in the practical usage of FFs in digital systems.

6-2.1 *RS* Flip-Flops, NOR Gate and NAND Gate

The *RS* FF is the simplest of the IC FFs. It is used for temporary storage of data, and frequently in control applications where an operation can be put into effect with a command and halted with another command.

NOR Gate *RS* FF

To begin the discussion of the IC bistable, let us look again at Fig. 6-4. Transistors Q_1 and Q_3 form a basic RTL NOR gate, as do transistors Q_2 and Q_4. The FF of Fig. 6-4 can now be redrawn using logic symbols as shown in Fig. 6-5. In Fig. 6-5a the diagram is drawn in series representation, as in Fig. 6-1. Figure 6-5b is the conventional way of drawing this FF, and Fig. 6-5c is its logic symbol. The logic symbol has two input enable terminals S and R and two output terminals Q and $\bar{Q}$.

Figure 6-5 IC NOR gate flip-flop: (a) series representation; (b) conventional representation; (c) logic symbol; (d) timing diagram

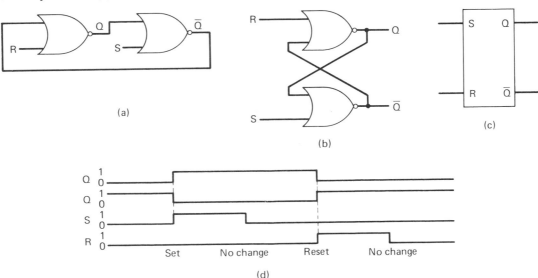

Table 6-2

	(a)			(b)	
	NOR: $C = \overline{A + B}$			NOR: $C = \overline{A + B}$; ×, don't care	

A	B	C
0	0	1
0	1	0
1	0	0
1	1	0

A	B	C
0	0	1
1	×	0
×	1	0

Since the performance of this FF is based on NOR gate operation, to aid in the explanation of the FF operation, NOR truth tables are given in Tables 6-2a and 6-2b. Table 6-2a is the NOR table as developed in Chapter 1. Table 6-2b is a truth table using *don't care* terminology. It follows from Table 6-2a that, by noting if any input is made a 1, we *don't care* whether the other input is a 1 or a .0, since the output C is a 0. Using these truth tables, we can develop Table 6-3, the

Table 6-3 NOR RS flip-flop

R	S	Q	$\bar{Q}$	
1	0	0	1	← Reset
0	1	1	0	← Set
0	0	No change, NC		← Previous state is maintained
1	1	Not allowed, NA		← Q and $\bar{Q} = 0$

truth table for the NOR gate RS FF. We can interpret this table, based upon the logic diagram and NOR truth tables, in the following manner:

1 If $S = 0$ and $R = 1$, the $R = 1$ makes Q go to a 0. Line 1 of the table.
2 Q and S are both equal to 0. This makes $\bar{Q} = 1$. Line 1 of the table.
3 $\bar{Q}$ and R are both 1. This make $Q = 0$.
4 $R = 1$ is the activating or *reset–R* condition that forces the FF into the condition $Q = 0$ and $\bar{Q} = 1$. Line 1 of the table.
5 If R is returned to 0 (S still at 0), $\bar{Q} = 1$ will keep Q at 0, and there is no change in the output level. Line 3 of the table.
6 Since the FF is symmetrical, if we make $S = 1$ and $R = 0$, the $S = 1$ will make $\bar{Q}$ go to a 0. $S = 1$ is the activating or *set–S* condition that forces the FF into the condition $Q = 1$, $\bar{Q} = 0$. Line 2 of the table.
7 If both S and R are made equal to 1 simultaneously, both Q and $\bar{Q}$ will go to a 0. This is a not-allowed condition.

139

This flip-flop is known as an *RS* Set-reset flip-flop. The FF can be set into either condition by connecting either *S* to 1 and returning *S* to 0, or by connecting *R* to 1 and returning it to 0. It is common practice to define the set condition as $Q = 1$, $\bar{Q} = 0$, and the reset condition as $Q = 0$, $\bar{Q} = 1$. An *RS* RTL FF which operates in the above manner is commercially available as the RTL *RS* FF type MC802.

Figure 6-5d is a timing diagram for this NOR gate *RS* FF. This FF is activated as soon as the *R* and *S* inputs are brought to the logic 1 level, as shown in the timing diagram.

In Figure 6-5, *R* is in the same gate as *Q*, and *S* is in the same gate as $\bar{Q}$. If we look at the first line of Table 6-3, we see that, when $R = 1$, $Q = 0$, and when $S = 1$, $\bar{Q} = 0$. From the second line, when $R = 0$, $Q = 1$, and when $S = 0$, $\bar{Q} = 1$. These have the appearance of inverter action.

EXAMPLE 6-1   The NOR gate bistable of Fig. 6-5 has the waveforms shown in Fig. 6-6 applied to its *R* and *S* terminals. What are the *Q* and $\bar{Q}$ waveforms?

Solution

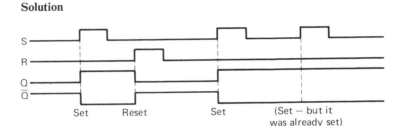

Figure 6-6   Solution to Example 6-1

NAND Gate *RS* FF

NAND gates can also be used as the basis for *RS* FFs. Figure 6-7a shows a series representation, and Fig. 6-7b shows the conventional way of drawing this

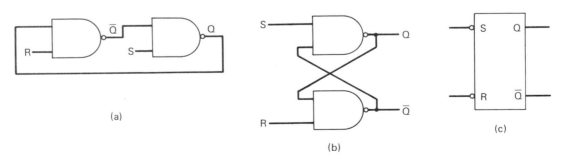

Figure 6-7   IC NAND *RS* flip-flop: (a) series representation; (b) conventional representation; (c) logic symbol; (d) timing diagram

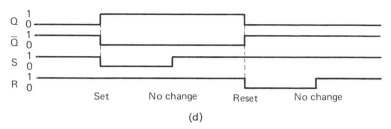

Set   No change   Reset   No change

(d)

Figure 6-7 (Continued)

FF. Figure 6-7c gives its logic symbol. This FF is sometimes called a *latch*. Since the performance of this FF is based upon NAND gate operation, NAND gate truth tables are given in Tables 6-4a and 6-4b. Table 6-4a is the truth table of Chapter 1, whereas Table 6-4b uses the $\times = don't\ care$ terminology. This follows from Table 6-4a by noting that if any input is made a 0 the gate output is a 1, and we don't care whether the other input is a 1 or a 0. With these truth tables we can

Table 6-4

(a)
NAND : $C = \overline{AB}$

A	B	C
0	0	1
0	1	1
1	0	1
1	1	0

(b)
NAND : $C = \overline{AB}$; $\times$, don't care

A	B	C
0	$\times$	1
$\times$	0	1
1	1	0

develop the NAND *RS* FF truth table, Table 6-5. We interpret this table, based upon the logic diagram and NAND truth tables, in the following manner:

Table 6-5   NAND *RS* flip-flop

S	R	Q	$\bar{Q}$	
1	0	0	1	Reset
0	1	1	0	Set
1	1	No change, NC		Previous state is maintained
0	0	Not allowed, NA		Q and $\bar{Q} = 1$

1   If $R = 0$ and $S = 1$, the $R = 0$ makes $\bar{Q}$ go to a 1. Line 1 of the table.

2   $\bar{Q} = 1$ and $S = 1$; together they make $Q = 0$. Line 1 of the table.

3   Q and R are both 0, making $\bar{Q} = 1$.

4   $R = 0$ is the activating or *reset–R* condition that forces the FF into the condition $Q = 0$ and $\bar{Q} = 1$. Line 1 of the table.

5   If R is returned to 1 (S still at 1), $Q = 0$ keeps $\bar{Q}$ at 1. $\bar{Q} = 1$ and $S = 1$ keep Q at 0, and there is no change in the output level. Line 3 of the table.

6   Since the FF is symmetrical, if we make $S = 0$ and $R = 1$, the $S = 0$ will make Q go to a 1 and $\bar{Q}$ go to a 0. $S = 0$ is the activating or *set–S* condition that forces the FF into the condition $Q = 1$ and $\bar{Q} = 0$. Line 2 of the table.

7   If both S and R are made equal to 0 simultaneously, both Q and $\bar{Q}$ will go to a 1. This is a not-allowed condition.

Figure 6-7d is a timing diagram for this NAND gate *RS* FF. This FF is activated as soon as the R or S inputs are brought to logic 0, as shown in the timing diagram. This is shown on the logic symbol by means of circles on the input R and S lines.

In this FF, S is in the same gate as Q, and R is in the same gate as $\bar{Q}$. The truth table again gives the appearance of inverter action. Both the NOR and NAND gate *RS* FFs are modifications of the basic inverter FF of Fig. 6-1.

EXAMPLE 6-2   The NAND gate *RS* FF of Fig. 6-7 has the waveforms shown in Fig. 6-8 applied to its R and S inputs. What are the Q and $\bar{Q}$ waveforms?

Solution

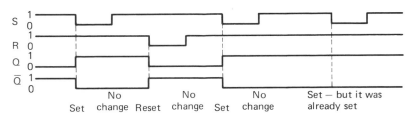

Figure 6-8   Solution to Example 6-2

Logic card manufacturers have built *RS* flip-flops from two NAND gates using the circuit shown in Fig. 6-7. Figure 6-9† shows such a logic card with 10 *RS* FFs. This card has 5 HTL MC668L integrated circuits. Each IC has four two-input NAND gates, and two *RS* FFs are made from each IC.

†Datascan Division of Anadex Instruments: logic card 413.

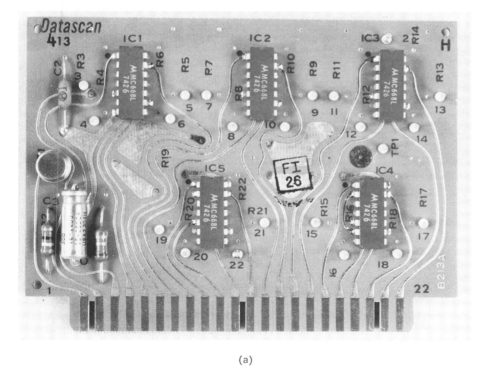

(a)

Figure 6-9 Integrated-circuit logic card with 10 *RS* flip-flops:
(a) front side; (b) back side

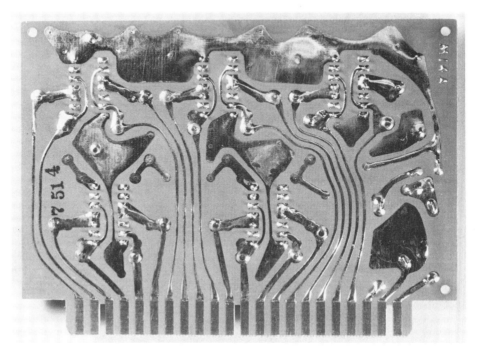

(b)

6-2.2 Clocked *RS* Flip-Flop (NAND)

The clocked *RS* flip-flop is the basic memory element in semiconductor memories. In many applications a stream of data is being generated during a process, but it is desired to store only the final result of the data process and to ignore, from a storage standpoint, the intermediate data. The storage element has to be isolated from the data stream and finally enabled by a clock (gate, strobe) pulse. Typical examples are computer multiplication from which only the final result is to be stored, and electronic counters and digital voltmeters for which the intermediate counts and counter display during counting are both annoying and irrelevant.

Figure 6-10a is the logic diagram for a clocked NAND gate *RS* FF, and Fig. 6-10b is its logic symbol. It is constructed with four two-input NAND gates. Gates *W* and *X* act as gating elements. Gates *Y* and *Z* are the *RS* bistable of Fig. 6-7. Its truth table is given in Table 6-6.

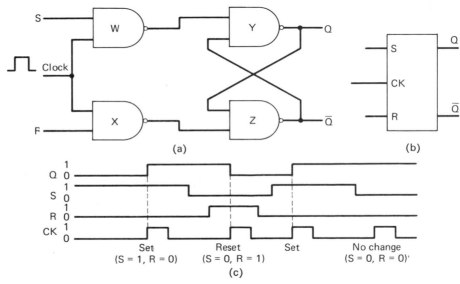

Figure 6-10 Clocked NAND gate *RS* flip-flop: (a) logic diagram; (b) logic symbol; (c) timing diagram

Table 6-6 Clocked NAND *RS* flip-flop (after clock pulse)

S	R	Q	$\bar{Q}$
1	0	1	0
0	1	0	1
0	0	NC	NC
1	1	NA	NA

In the clocked NAND gate FF,

1 If the clock is at a 0 level:
    a The outputs of gates W and X must be at the 1 level.
    b Referring to Table 6-5, FF YZ is in the no-change condition. The 0 clock level isolates the bistable from the data lines.
2 Let the clock go to the 1 level:
    a Gates W and Y act as inverters $(\overline{1 \cdot A} = \bar{A})$.
    b Table 6-5 is applicable. Table 6-6 is the same as Table 6-5 except for the inverter action of gates W and X, which makes the SR data line up with the $Q\bar{Q}$ lines.
    c $S = 0$ and $R = 0$ converts to the stable state of Table 6-5. The FF does not change with the clock pulse.
    d $S = 1$ and $R = 1$ converts to the NA condition of Table 6-5.

The effect of the clock pulse is to transfer the data at the S and R lines into the FF. S and R should not change when the clock is at 1. If they change during the time the clock is high, the data stored are that of the final S and R values.

Figure 6-10c shows the timing diagrams. Note that information appears at the Q output when the clock goes to the 1 level. This timing should be compared with the RS FF of Fig. 6-7. Only the Q output need be given in the timing diagram since $\bar{Q}$ is the invert of Q.

Other terms used in place of clocked are "gated" and "strobe." Another term used for the data storage block of Fig. 6-10 is the word "latch."

The FF of Fig. 6-10 required two data input lines, S and R. Common terminology is to refer to data on these two data lines as *double-rail* data.

Example 6-3 The waveforms in Fig. 6-11 are applied to the S, R, and clock (CK) inputs of the FF of Fig. 6-10. What is the Q waveform?

Solution

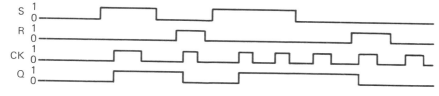

Figure 6-11 Solution to Example 6-3

6-2.3 The *D* Latch (NAND)

The circuit of Fig. 6-10 required double-rail input data. If S and R should both be 1, we have a NA condition. However, if an inverter is placed between S and R, as shown in Fig. 6-12a, S and R can never be 1 at the same time, and the NA condition cannot occur. Only a single data line is needed, and we say we have *single-rail* data. The FF of Fig. 6-12a is called a *D* or *data* FF. Figure 6-12b gives the logic symbol, and Fig. 6-12c is the timing diagram. This FF again is enabled at the leading edge of the clock pulse, and the D line data appear at the output as soon as the FF is enabled.

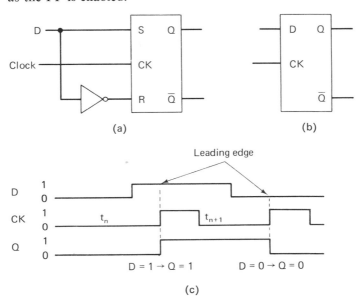

(a) (b)

(c)

Figure 6-12 The *D* latch: (a) logic diagram; (b) logic symbol; (c) timing diagram.

Table 6-7 *D* Latch

t_n		t_{n+1}
D	Q_n	Q_{n+1}
0	0	0
0	1	0
1	0	1
1	1	1

There are many different ways of writing FF truth tables. Table 6-7 illustrates one way of doing this with a table defining the operation of the *D* latch.

1 The conditions at *D* and *Q before* the clock pulse are given as Q_n at time t_n.

2   The conditions at *D* and *Q* *after* the clock pulse are given as Q_{n+1} at time t_{n+1}.

3   There is no need to give $\bar{Q}$. It is the invert of *Q*.

4   The FF transfers the data at their input to their output after the clock pulse. The *Q* data after the clock pulse equal the *D* data before the clock pulse.

The levels at *D* arrive at *Q* delayed by one clock pulse time; thus the latch of Fig. 6-12 is also known as a *D* or delay latch. (Unfortunately, data and delay both start with the letter *D* so that there may be some confusion as to terminology.) Figure 6-12c shows the timing diagram, with *D* data at *Q* at the conclusion of the timing pulse.

Gated latches are common in all the logic families. Typical examples are the TTL SN54/7475 and the DTL MC1813P and MC1814P. These units are quad latches and can provide temporary storage of 4 bits.

6-2.4   Master–Slave *RS* Flip-Flop

In the *RS* flip-flop, there is inadequate isolation between the input and output. To provide isolation, the master–slave FF has been developed; it is shown in Fig. 6-13. It consists of two gated *RS* latches in cascade (a master and a slave), with an

Figure 6-13   Master-slave *RS* flip-flop: (a) diagram; (b) master-slave timing; (c) logic symbol; (d) timing diagram

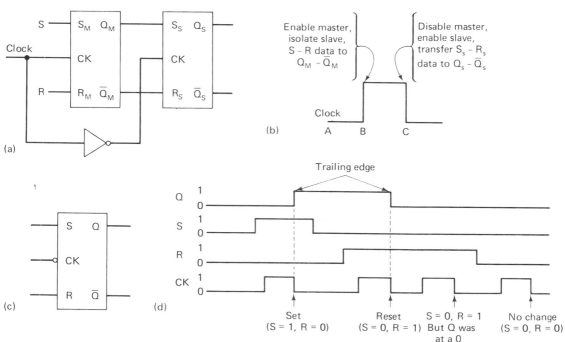

inverted clock connected to the second (or slave) unit. *RS* input data are transferred to the output in the following manner:

1 If the clock is low, the master is disabled (properties of clocked *RS* FF), *A* to *B* (Fig. 6-13b).
2 Let the clock pulse arrive and go high; *B* in Fig. 6-13b. The master is enabled.
3 *S–R* data are transferred to Q_M–$\bar{Q}_M$ as in a clocked *RS* FF.
4 The inverter between the high clock and the slave input makes the slave input low. The slave is disabled and the Q_M–$\bar{Q}_M$, S_S–R_S data are isolated from Q_S–$\bar{Q}_S$. Also *S* and *R* can change, but the isolation of the slave prevents it from appearing at the output.
5 Now make the clock pulse go low (Fig. 6-13b). The slave is enabled and the S_S–R_S data get to Q_S–$\bar{Q}_S$.
6 The *S–R* data have appeared at the output when the clock pulse arrived at 0, at the trailing edge of the clock pulse (Fig. 6-13b). This allows for accurate timing of data arrival.
7 The master is disabled, and *S–R* input is isolated from the Q_S–$\bar{Q}_S$ output.

The truth table for the master–slave *RS* FF is given by Table 6-8. Comparing this with Table 6-6, it can be seen that they are the same. Table 6-8 is Table 6-6 through two clocked FFs. In this case Q_n is the logic level before the clock pulse, and Q_{n+1} is the logic level after the clock pulse. This is another way of indicating the before and after clock pulse level, rather than using t_n and t_{n+1}. In Table 6-8, *S* and *R* both equal to 1 is not allowed (the state of *Q* and $\bar{Q}$ are indeterminate), and when *S* and *R* are not alike, the final result is determined by the levels of S and *R*.

Table 6-8 Master–slave *RS* flip-flop

S	R	Q_n	Q_{n+1}	
0	0	0	$0 \equiv$ NC	No change
0	0	1	$1 \equiv$ NC	
0	1	0	0	
0	1	1	0	
1	0	0	1	
1	0	1	1	
1	1	0	U	Final state is undetermined
1	1	1	U	

Figure 6-13c gives the logic symbol for this FF. Note the inverting circle at the clock input. This circle indicates that the data appear at the output of the FF at the *trailing* edge of the clock pulse when it returns to logic 0. This is shown in the timing diagram of Fig. 6-13d. This differs from the clocked FF, in which data appear at the output at the leading edge of the clock pulse.

Connecting an inverter between S and R and calling the single-rail data input the D or data input does exactly the same thing as in the RS FF, and converts the master–slave FF into a D (data or delay) FF.

6-2.5 Toggle or *T* Flip-Flop

Let us take the master–slave FF and cross-connect the outputs back to the input as shown in Fig. 6-14. Note that $\bar{Q}_S$ goes to S_M. The operation is as follows:

1 $\bar{Q}_S$ goes to S_M, and Q_S goes to R_M.
2 Assume that $Q_S = 1$ and $\bar{Q}_S = 0$. This makes $S_M = 0$ and $R_M = 1$.
3 Apply a clock pulse. After the clock pulse, what will be the levels of Q_S and $\bar{Q}_S$?
4 $S_M = 0$ and $R_M = 1$ make $Q_S = 0$ and $\bar{Q}_S = 1$. But this is a reversal of state or a *toggle*.
5 Each clock pulse toggles the FF at the trailing edge of the clock pulse.
6 This FF is called a T FF, and the clock input is called the T input.

Figure 6-14 Toggle (T) flip-flop: (a) T FF conversion; (b) logic symbol; (c) timing diagram

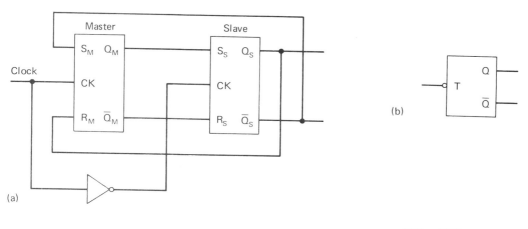

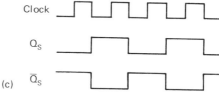

A most important factor is the timing. Since data are transferred from the master to the slave when the clock pulse goes to 0, the toggled output appears at the FF output terminals when the clock pulse goes to 0, or at the negative-going transition. This is shown in the timing diagram of Fig. 6-14c.

Note that this bistable is direct coupled. It toggles at the negative-going transition or trailing edge of the clock pulse. Theoretically, even a very slow† waveform may be applied to the bistable and the bistable will toggle.

The bistable of Fig. 6-14 is known by many titles, among which are the following:

1. Toggle or T flip-flop: state change for each incoming pulse.
2. Binary: it will be shown later that it counts in binary.
3. Divide by 2: the output frequency is half that of the input frequency.
4. Scaler: from its use as a pulse counter.
5. Eccles–Jordan multivibrator: after the inventors of the vacuum-tube equivalent.

6-2.6   The *JK* Flip-Flop

Let us now modify the master inputs of the NAND gate master–slave flip-flop by making the clocking input gates three-input NANDs, as shown in Fig. 6-15. This FF is more universal than the RS FF and, as we shall see, the NA condition of the RS FF is replaced by a toggle.

Note that the cross-connection from $\bar{Q}_S$ and Q_S to the master inputs has been retained. Figure 6-15a has apparently omitted the inverter to the slave. However, the standard inverting logic circle has been drawn at the slave input. This indicates inverter action. The inverter is in the slave part of the FF, and the slave is enabled when the clock pulse has returned to 0. The logic symbol of Fig. 6-15b shows the clock input with an inverting circle. This is to indicate that data are transferred into the slave and appear at the FF output when the clock pulse has returned to 0.

The operation of the JK FF is as follows. There are two additional inputs J and K:

1. Let J and K both be equal to 1. Since in a NAND gate $\overline{1 \cdot A} = \bar{A}$, making J and $K = 1$ makes the FF of Fig. 6-15 identical to that of the T FF (Fig.

---

†Unfortunately, the real world is different from the theoretical world. Slow waveforms can be used with the slower forms of IC logic, such as RTL, DTL, HTL, and CMOS. A long-rise-time waveform will keep the transistors in the bistable in the active region for some time. With fast logic such as TTL, this time in the active region, together with the high f_T transistors, permits the circuit to break out into a parasitic oscillation, and the resultant state of the bistable becomes indeterminate. Circuits such as the Schmitt trigger (Chapter 9) must be used to sharpen the waveform and reduce the time in the active region. Even with slower logic, if a long-rise-time waveform is applied to the bistable, it would be well to check for correct toggle operation.

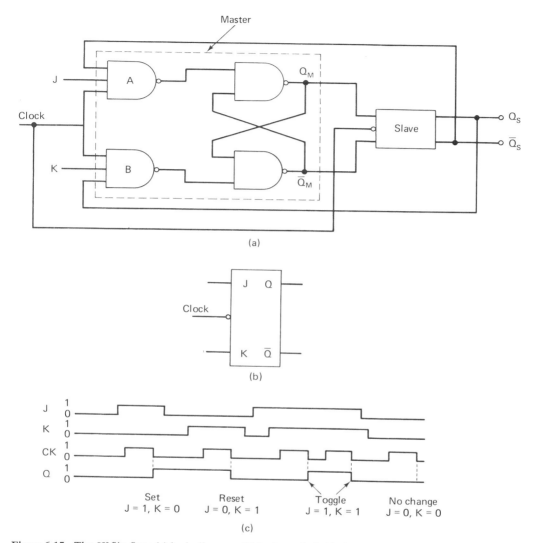

Figure 6-15 The *JK* flip-flop: (a) logic diagram; (b) logic symbol; (c) timing diagram

6-14). Hence J and $K = 1$ results in a toggle condition, and the NA condition of the RS master–slave FF has been eliminated.

2 Let $J = 1$ and $K = 0$:

   a Assume that $Q_S = 1$ and $\bar{Q}_S = 0$. Let the clock go to a 1. The inputs to gates A and B are

        Gate A: $J = 1$, $\bar{Q}_S = 0$, clock $= 1$

        Gate B: $K = 0$, $Q_S = 1$, clock $= 1$

But a single 0 into a NAND gate makes its output equal 1. The outputs of gates A and B are 1, and this is a stable no-change condition. The FF does not change its levels with the clocking pulse. *Note that* $J = 1$ *and* $K = 0$ *lines up with* $Q_S = 1$ *and* $Q_S = 0$.

    b Assume that $Q_S = 0$ and $\bar{Q}_S = 1$. Let the clock go to a 1. The inputs to gates A and B are

          Gate A: $J = 1$, $Q_S = 1$, clock $= 1$
          Gate B: $K = 0$, $\bar{Q}_S = 0$, clock $= 1$

This makes the output of gate $A = 0$ and gate $B = 1$. But from Table 6-5 we know that this will make $Q_M = 1$ and $\bar{Q}_M = 0$. At the conclusion of the clock pulse, this also will make $Q_S = 1$ and $\bar{Q}_S = 0$. *Note that again* $J = 1$ *and* $K = 0$ *lines up with* $Q_S = 1$ *and* $Q_S = 0$ *after the clock pulse.*

  3 Let $J = 0$ and $K = 1$. Since the FF is symmetrical, after the clock pulse Q_S will be equal to 0 and $\bar{Q}_S = 1$. This follows the same reasoning as for $J = 1$ and $K = 0$.

  4 Let $J = 0$ and $K = 0$. The outputs of gates A and B are always equal to 1. This is the stable state, and there is no change in the outputs as the FF is clocked.

The JK truth table is given in Table 6-9. The notation again is slightly different and represents another way of indicating the truth table of bistables. Q_{n+1} is the output of the FF at time t_{n+1} or after the clock pulse. Q_n is the output of the bistable at time t_n or before the clocking pulse. The first two lines say that, if J and K are not alike, the final state of the FF is determined by the J and K values, with the Q output lining up with the J values. The third line says that if $J = K = 0$, $Q_{n+1} = Q_n$, a no-change condition. The final line says that if $J = K = 1$, the FF toggles and $Q_{n+1} = \bar{Q}_n$.

Note that the JK bistable is more universal than any of the previous bistables. The J and K inputs, if not alike, make the bistables have the properties of an RS FF. When $J = K = 0$, the no-change condition of the RS FF is maintained. But when $J = K = 1$, the not-allowed condition of the RS FF is changed to a toggle result.

Table 6-9 The JK flip-flop

J	K	Q_{n+1}	
1	0	1	
0	1	0	
0	0	Q_n	No change
1	1	$\bar{Q}_n$	Toggle

JK flip-flops are available in all the logic families. One additional feature is required to make a practical *JK* FF. When power is turned on, a FF goes into one of the two stable states. Which state is determined by the component values and transistor parameters. However, in many applications it is necessary to start with the FF in a particular desired state. Provisions have to be made for a direct set S_D or direct reset R_D (the term clear C_D is used interchangeably with R_D), which forces the FF into the desired initial condition. Preset (PR) and clear (CLR) also are terms used for this function and these inputs. These inputs are also called the asynchronous inputs.

To implement the S_D and C_D inputs, let us now modify the slave section of the *JK* FF, which we have previously shown to be the clocked NAND GATE *RS* FF of Fig. 6-10, by converting the *YZ* gate section to three-input NAND gates, as shown in Fig. 6-16. This modification operates in the following manner.

1   Let S_D and C_D both equal 1. Levels of 1 leave NAND gates unchanged and the FF operation is unchanged.

2   Let S_D be made equal to a 0. This 0 into a NAND gate forces its output Q to a 1. The output from gate *W* has no effect.

3   Let C_D be made equal to 0. This forces $\bar{Q}$ to a 1.

4   Let S_D and C_D both be at 0. Both Q and $\bar{Q}$ go to a 1, a not-allowed state.

Figure 6-16   *RS* master-slave clocked flip-flop modified for S_D and C_D inputs: (a) NAND gate modification (slave); (b) logic symbol; (c) timing diagram

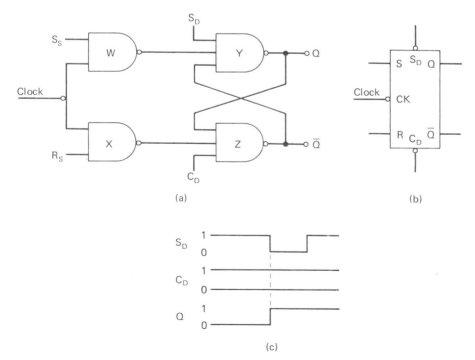

(This illustrates the basic S_D and C_D function. Within the IC, additional interconnections may be needed to ensure that the S_D and C_D functions override all others.) To provide the S_D and C_D performance, the S_D and C_D inputs have to go to the 0 or ground level. This is shown on the logic symbol with inverting circles (Fig. 6-16b). Since the S_D and C_D inputs override all other inputs, they activate the output as soon as they are applied, as shown in the timing diagram. In this respect they act exactly like the basic *RS* FF of Fig. 6-7. There is no clock pulse delay until the negative-going transition.

6-2.7   Universal Flip-Flop

An additional modification of the *RS* master–slave clocked FF is to modify the *R* and *S* inputs and make them two-input AND inputs, as shown in Fig. 6-17. We shall see in Chapter 7 that this is of great usefulness in building electronic counters. The letter *C* is used in place of *R* in Fig. 6-17. To activate the *S* input, both S_1 and S_2 have to be made equal to a 1, and similarly for the *C* input. We can then write out the truth table for the universal FF as in Table 6-10. Note in this table that both *S* inputs have to be 1 to get a 1 output, and similarly with C_1 and C_2. Once an input such as S_1 is made a 0, the internal input *S* becomes a 0. The other input, such as S_2, can be a 1 or a 0 and is represented by an × (don't care) in Table 6-10. This follows standard AND action. If both AND gates are disabled (0 into S_1 or S_2, and 0 into C_1 or C_2), both the *S* and *C* inputs are equal to 0. After the clock pulse, there is no change shown by Q_n in lines 5–8. If all four inputs are made 1, we are in a not-allowed condition, and the final state is determined by internal transistor parameters and not by the inputs.

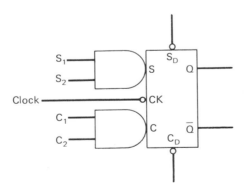

Figure 6-17   Universal flip-flop

In the FF of Fig. 6-17, if we connect $\bar{Q}$ to S_2 and Q to C_2, the FF becomes a *JK* flip-flop with the *J* input the S_1 input and the *K* input the C_2 input. This demonstrates some of the versatility of the universal FF.

Figure 6-18 is the symbol for a *JK* FF having two-input AND gates for its data input lines. Table 6-11 gives its truth table.

Table 6-10 Universal flip-flop

S_1	S_2	C_1	C_2	Q_{n+1}	
1	1	0	×	1	
i	1	×	0	1	
0	×	1	1	0	
×	0	1	1	0	
0	×	×	0	Q_n	
0	×	0	×	Q_n	
×	0	×	0	Q_n	
×	0	0	×	Q_n	
1	1	1	1	U	Output is undefined

× = don't care

Table 6-11 Two-input AND JK flip-flop

J_1	J_2	K_1	K_2	Q_{n+1}	
1	1	0	×	1	
1	1	×	0	1	
0	×	1	1	0	
×	0	1	1	0	
0	×	×	0	Q_n	
0	×	0	×	Q_n	
×	0	×	0	Q_n	
×	0	0	×	Q_n	
1	1	1	1	$\overline{Q_n}$	toggle

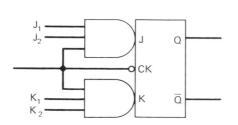

Figure 6-18 Two-input AND JK flip-flop

The type 7472 is a three-input AND *JK* FF with *PR* and *CLR* inputs. Figure 6-19 gives its logic symbol and Table 6-12 gives its complete operation. Note that the *PR* and *CLR* inputs do not require a clock ($\times$ in table), whereas utilization of the *JK* data inputs requires a clock pulse.

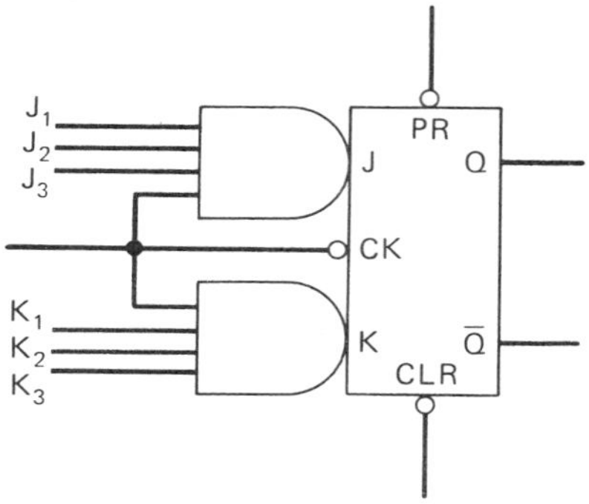

Table 6-12[a] Type 7472 function table

Inputs					Outputs	
Preset	Clear	Clock	*J*	*K*	*Q*	$\bar{Q}$
L	*H*	$\times$	$\times$	$\times$	*H*	*L*
H	*L*	$\times$	$\times$	$\times$	*L*	*H*
L	*L*	$\times$	$\times$	$\times$	*H**	*H**
H	*H*	⊓	*L*	*L*	Q_o	$\bar{Q}_o$
H	*H*	⊓	*H*	*L*	*H*	*L*
H	*H*	⊓	*L*	*H*	*L*	*H*
H	*H*	⊓	*H*	*H*	Toggle	

Positive logic: $J = J_1 \cdot J_2 \cdot J_3$, $K = K_1 \cdot K_2 \cdot K_3$.
[a]Texas Instruments, Inc.

Figure 6-19 Type 7412

6-3 GENERAL DESIGN CONDITIONS

Let us return to Fig. 6-14, the *T* flip-flop. Q_S is connected to R_M and $\bar{Q}_S$ to S_M. But the master and the slave FFs are identical, and in the truth tables *S* lines up with *Q* and *R* with $\bar{Q}$. Why not connect Q_S to R_S and $\bar{Q}_S$ to S_S? It eliminates one clocked bistable. Suppose that we do this and enable the slave using the clock. If $Q_S = 1$ and $\bar{Q}_S = 0$, the cross-connection and the truth table say that $Q_S = 0$ and $\bar{Q}_S = 1$. This is an apparent inconsistency, but it is and it isn't. If we were to suddenly apply $S_S = 1$ and $R_S = 0$ to the bistable of Fig. 6-14, it takes some finite time for the 1 and 0 to appear at Q_S and $\bar{Q}_S$. With the cross-connection, a short time later Q_S and $\bar{Q}_S$ will be 0 and 1. We see that the bistable outputs continually keep changing from 0 to 1 and back again, uncontrolled by the clock pulse, which is allowing this to occur only as a gate. We call this a *race-around condition*; it is to be avoided in a bistable since the final state of the bistable is determined by the state existing when the clock goes to a 0 and disables the circuit. (In Chapter 9 we shall make use of this race-around condition to develop the pulse generator known as the *ring oscillator*.) This modification of Fig. 6-14 fails because the input and output are not isolated from each other. The master–slave principle of Fig. 6-13 does isolate the input and output and therefore is very successful. This brings up another point. We must be very careful in the master–slave bistable to keep the master and slave from being activated at the same time; otherwise, the race-around condition occurs. The design of integrated circuits must take this into account, and the sequence of events that occur in a *JK* flip-flop is shown in Fig. 6-20:

Figure 6-20 Clock waveform

1 Isolate slave from master.
2 Enter information from *J* and *K* input to master.
3 Disable *J* and *K* inputs.
4 Transfer information from master to slave.

All these considerations make the IC master–slave FF a complex device. Figure 6-21 shows the circuit diagram of the type 945/845 DTL master–slave FF and its truth tables. Transistors Q_{10} to Q_{13} are the master FF, Q_4 to Q_9 the slave FF, and Q_1 to Q_3 the voltage level inverter transistors.

1 In the master S_1, S_2, C_1, and C_2 are typical DTL inputs. The cross-coupling occurs between the collector of Q_{12} and the anode of D_1 and between the collector of Q_{11} and the anode of D_2.
2 In the slave the cross-coupling is more complex. Q_4 and Q_5 are an output pair with the collectors of Q_5 and Q_6 tied together. If Q_4 and Q_5 are not conducting, Q_6 can conduct, and its emitter current drives Q_8 and Q_9. Similarly for the other side.
3 Now ground the C_D input. This puts the collector of Q_1 and the base of Q_6 one diode drop (0.7 V) above ground. In order for Q_6 to conduct, its base must be at 2.1 V (3 V_{BE} drops, Q_6, Q_8, and Q_9) above ground. Hence Q_6, Q_8, and Q_9 do not conduct and $\bar{Q} = 1$. Similarly for the S_D input.
4 Q_1, Q_2, and Q_3 are the clock pulse isolating transistors. When the clock pulse goes high, it enables the master by means of the bottom diodes to the bases of Q_{19} and Q_{13}. When the clock pulse goes high, the emitter of Q_3 goes high, cutting Q_3 off. Since the collector of Q_3 cannot conduct, neither can Q_1 and Q_2, which couples Q_{11} and Q_{12} data into the slave. The master is enabled and the slave isolated. When the clock goes low, Q_3 can conduct and allow either Q_1 or Q_2 to conduct and transmit Q_{11} or Q_{12} data to the slave.

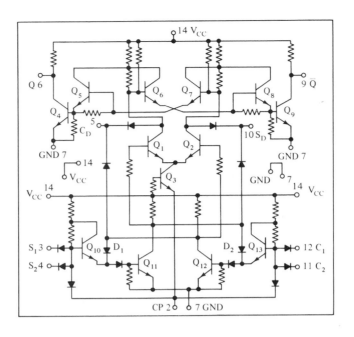

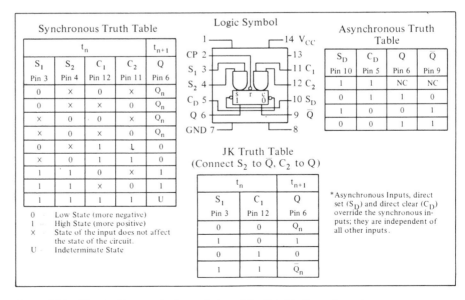

Synchronous Truth Table

t_n				t_{n+1}
S_1	S_2	C_1	C_2	Q
Pin 3	Pin 4	Pin 12	Pin 11	Pin 6
0	X	0	X	Q_n
0	X	X	0	Q_n
X	0	0	X	Q_n
X	0	X	0	Q_n
0	X	1	L	0
X	0	1	1	0
1	1	0	X	1
1	1	X	0	1
1	1	1	1	U

0	Low State (more negative)
1	High State (more positive)
X	State of the input does not affect the state of the circuit.
U	Indeterminate State

Logic Symbol

Asynchronous Truth Table

S_D	C_D	Q	$\overline{Q}$
Pin 10	Pin 5	Pin 6	Pin 9
1	1	NC	NC
0	1	1	0
1	0	0	1
0	0	1	1

JK Truth Table
(Connect S_2 to $\overline{Q}$, C_2 to Q)

t_n		t_{n+1}
S_1	C_1	Q
Pin 3	Pin 12	Pin 6
0	0	Q_n
1	0	1
0	1	0
1	1	Q_n

*Asynchronous Inputs, direct set (S_D) and direct clear (C_D) override the synchronous inputs; they are independent of all other inputs.

Figure 6-21 Circuit diagram, logic symbol, and truth table for universal master-slave flip-flop: type 945/845 (Courtesy Motorola Semiconductor Products, Inc.)

6-4 INTEGRATED-CIRCUIT FLIP-FLOPS: ADDITIONAL CONSIDERATIONS AND LIMITATIONS

The basic principles of the bistable have been discussed. The JK FF is the most versatile of the FFs discussed and is available in all the IC logic families. To provide greater facility in circuit design, multiple JK inputs are available in many ICs. The numbers of inputs and functions that can be provided are limited in almost all cases by the number of available pin connections from the IC package.

The SN54L72 is a TTL typical JK master–slave FF with three J inputs, $J_1, J_2,$ and J_3, and three K inputs, $K_1, K_2,$ and K_3. The application of multiple J and K inputs to circuit design will be shown in Chapter 7.

Dual JK bistables are available. Not enough pins are available, so some functions must be omitted. In DTL, the MC853 is a dual unit with separate clocks to each FF and separate S_D inputs to each FF, but no C_D function. The MC855 has a common clock and C_D and separate S_D inputs.

D flip-flops are available in most logic families. The SN5474/7474 are TTL dual D FFs.

RS universal clocked FFs are available. The MC845 has been previously discussed. The RS inputs are two-input gates. In TTL, the SN54L71/SN74L71 have three-input AND gates for the RS inputs. In DTL, the MC1815 is a parallel gated clocked FF. Each S and $C(R)$ input is a two-wide, two-input AND–OR gate (two two-input AND gates driving an OR gate).

Latches such as the DTL MC1813, DTL MC1814, and TTL SN5475/SN7475 are made in quad units and are used for temporary bit storage.

The direct-coupled master–slave ICs have had wide success. In addition, the transition on the trailing edge of the clock pulse (trailing-edge logic) is used to provide a uniform time of transition of data to all the FF output terminals in a system using a common clock pulse.

Toggling speed is a major factor in system design. Table 6-13 lists the toggling speeds of the common IC logic families. Much development work is currently under way to improve toggling rates. As a rough rule, increases in speed are accompanied by increases in current and dissipation. However, improvements in the technology are continually modifying this tendency.

Table 6-13 Logic family clock rates (MHz)

RTL	8
DTL	12–30
HTL	4
TTL[a]	15–60
ECL	60–400
MOS	2
CMOS	5

[a]Schottky-clamped TTL is a modified version of TTL with toggling rates slightly lower than ECL.

6-5 SUMMARY OF FLIP-FLOP TYPES USING NAND GATES

1 The two-inverter (basic) FF

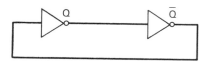

State	Q	$\bar{Q}$
Set	1	0
Reset	0	1

2 *RS* FF (latch)

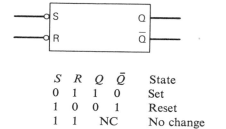

S	R	Q	$\bar{Q}$	State
0	1	1	0	Set
1	0	0	1	Reset
1	1	NC		No change
0	0	1	1	Not allowed

Activated as soon as Set–Reset conditions are activated.

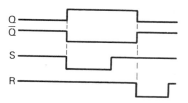

3 Clocked *RS* FF

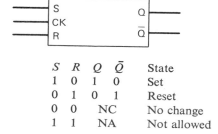

S	R	Q	$\bar{Q}$	State
1	0	1	0	Set
0	1	0	1	Reset
0	0	NC		No change
1	1	NA		Not allowed

Two data inputs S and R (double rail). Enabled as soon as clock level (leading edge) goes to logic 1. Output corresponds to the S and R inputs when clock level returns to 0.

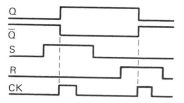

4 *D* (Data) latch

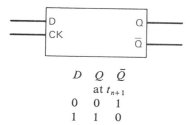

D	Q	$\bar{Q}$
	at t_{n+1}	
0	0	1
1	1	0

Single data (*D*) line (single rail). Same as clocked *RS* FF but inverter between S and R prevents not-allowed condition. Enabled as soon as clock level goes to logic 1. Data stored corresponds to D level when clock returns to logic 0.

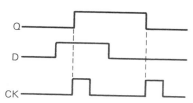

5 Master–slave *RS* FF

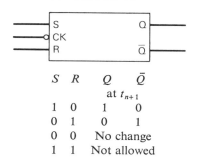

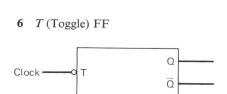

S	R	Q	$\bar{Q}$
		at t_{n+1}	
1	0	1	0
0	1	0	1
0	0	No change	
1	1	Not allowed	

Two clocked *RS* FFs in cascade with inverted clock to second (slave). When clock pulse goes to logic 1:

    Master enabled⎱ Data transferred
    Slave disabled ⎰to master.

When clock pulse returns to logic 0:

    Master disabled⎱ Data transferred
    Slave enabled ⎰as output to slave.

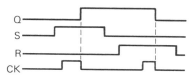

6 *T* (Toggle) FF

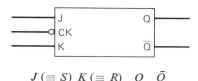

Toggles as each clock pulse is applied. Can be made from master–slave *RS* FF by cross-connecting *S* to $\bar{Q}$ and *R* to *Q*. Toggles at trailing edge of clock pulse if slave is enabled at trailing edge of clock pulse.

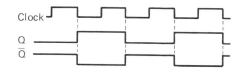

7 *JK* FF

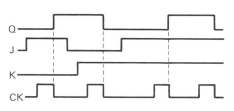

Combination of *RS* and *T*. Eliminates not-allowed conditions by toggling.

$J (\equiv S)$	$K (\equiv R)$	Q	$\bar{Q}$	
1	0	1	0	
0	1	0	1	
0	0	NC		No change
1	1	$\bar{Q}$	Q	Toggle

8 Asynchronous FF

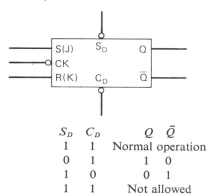

The direct set (S_D) and direct clear (C_D) override all other inputs. Outputs appear as soon as S_D and C_D are activated.

S_D	C_D	Q	$\bar{Q}$	
1	1	Normal operation		
0	1	1	0	
1	0	0	1	
1	1	Not allowed		

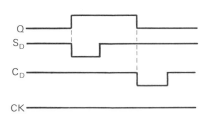

Problems

6-1. Define (a) NOR *RS* FF, and (b) NAND *RS* FF.

6-2. The NOR *RS* bistable of Fig. 6-5 has the following waveform applied to the *RS* terminals. What are the waveforms of the Q and $\bar{Q}$ outputs?

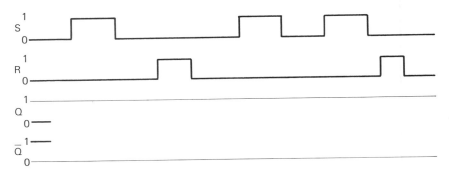

6-3. The NOR *RS* bistable of Fig. 6-5 has the following waveform applied to the *RS* terminals. What are the waveforms of the Q and $\bar{Q}$ outputs?

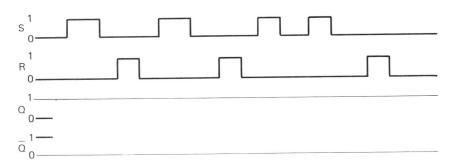

6-4. The NAND *RS* bistable of Fig. 6-7 has the following waveform applied to its *RS* terminals. What are the output waveforms of the Q and $\bar{Q}$ outputs?

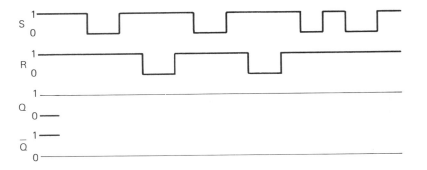

6-5. The NAND *RS* bistable of Fig. 6-7 has the following waveforms applied to its *RS* terminals. What are the output waveforms at the Q and $\bar{Q}$ outputs?

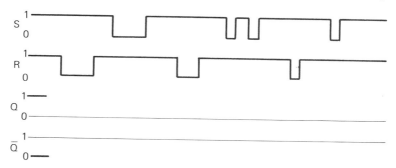

6-6. The gated latch of Fig. 6-10 has the following waveforms applied to its S, R, and clock terminals. What is the Q output waveform?

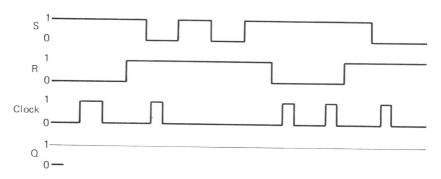

6-7. The gated latch of Fig. 6-10 has the following waveforms applied to its S, R, and clock terminals. What is the Q and $\bar{Q}$ output waveform?

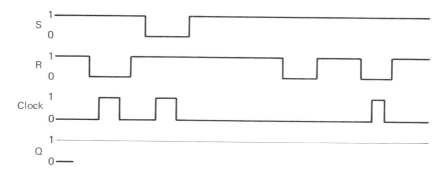

6-8. Draw a logic diagram for a clocked *RS* FF using NOR gates similar to Fig. 6-10. Write out and explain the truth table.

6-9. The following waveform data are presented to the *D* latch of Fig. 6-12. What is the *Q* output?

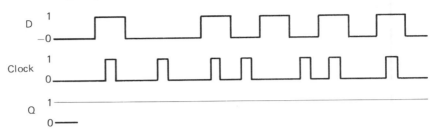

6-10. Give the logic symbols for the following: (a) bistable; (b) *RS* FF; (c) latch; (d) *D* FF; (e) *JK* FF.

6-11. The toggle FF of Fig. 6-14 has the following clock waveform applied to it. What is the *Q* output waveform? What is the clock frequency? What is the period of the *Q* output waveform, and what is the frequency of the *Q* output?

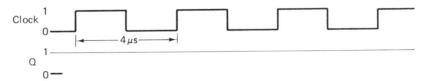

6-12. The toggle FF of Fig. 6-14 has the following waveform applied to its clock input. What is the *Q* output waveform?

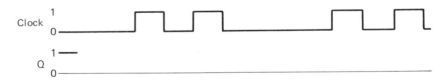

6.13.

(a) What is the waveform frequency at Q_A?
(b) What is the waveform frequency at Q_B?

6-14. The following inputs are applied to a *JK* FF. What is the *Q* output waveform?

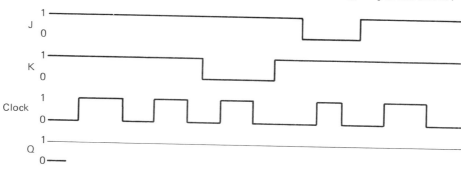

6-15. The following inputs are applied to the *S*, *R*, S_D, and C_D inputs of the FF of Fig. 6-16. What is the *Q* output waveform? The S_D and C_D inputs override all other inputs.

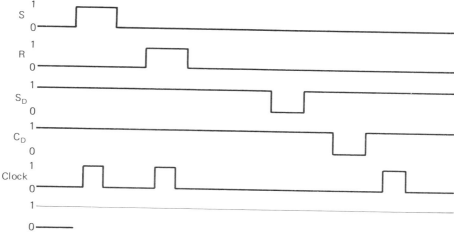

6-16. A *D* FF has the $\bar{Q}$ output terminal connected to the *D* input terminal. Show that this converts it into a *T* FF.

6-17. Explain the meaning of the inverting circles at the clock, C_D, and S_D inputs of the universal FF of Fig. 6-17.

6-18. Table 6-10 has the entries × and *U*. What do we mean by these two entries?

SEVEN

Series Counters

7-1 INTRODUCTION

A major application of the T flip-flop is its use as a basic building block in circuits that are used to count pulses. Electronic counters have many applications in electronic equipment. Counters are used for simple pulse counting, for frequency division, for the measurement of time, period, and frequency, in digital voltmeters, and for count control purposes in industrial systems. Electronic counters have so many applications that quite frequently they become a subsystem block in a much larger system. This chapter discusses how counters are built from the basic flip-flops and some of their applications.

7-2 BINARY RIPPLE COUNTERS†

A T flip-flop or binary has the basic property (discussed in Chapter 6) of changing state with each incoming pulse. After two pulses it is back to its original logic level, but there is nothing to indicate how many pulses have been applied. If the

†M. E. Levine, *Digital Theory and Experimentation Using Integrated Circuits*. Englewood Cliffs, N.J.: Prentice-Hall, Inc., 1974, Expt. 8, Binary Counters/The Binary Number System.

output of a binary is now connected to a second binary, the second binary changes state each time the first binary returns to its original state, and indicates by its own level when the first binary has returned to its original state.

Let us connect three *T* FFs in series as shown in Fig. 7-1a to form a three-stage binary ripple counter.

1 The counter consists of three *T* FFs in series.
2 The *Q* output of each FF is connected to the clock (*T*) input of the following stage. Hence the clock for any stage is the *Q* output of the preceding stage.
3 Each FF has a C_D (direct clear) input.
4 The circle at the *T* input to each FF tells us that the FF toggles at a negative transition when the clock input returns to level 0.

Figure 7-1 Three-stage binary ripple counter: (a) circuit diagram; (b) waveforms

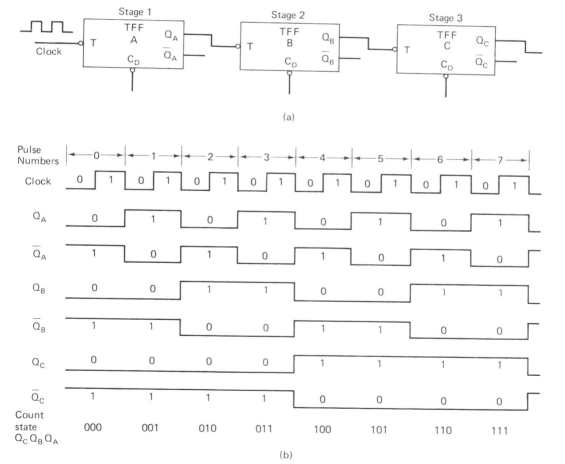

5 Using the C_D inputs, clear all the FFs so that the Q levels are all equal to 0.

6 Now apply a periodic square-wave clock-pulse waveform to the input of FF A as shown in Fig. 7-1b.

7 At the time when the Q output of FF A (Q_A) has a negative transition, FF B toggles. This is shown in Fig. 7-1b.

8 At the time when the Q output of FF B (Q_B) has a negative transition, FF C toggles. This is shown in Fig. 7-1b.

It can readily be seen from Fig. 7-1 that each time the signal goes through one of the FFs, commonly called a *stage*, the period is doubled and the frequency is halved. We can therefore write for the counter of Fig. 7-1

$$\text{At } Q_A \quad \text{Period} = 2 \times \text{clock period}$$

$$\text{Frequency} = \frac{1}{2} \times \text{clock frequency}$$

$$\text{At } Q_B \quad P_B = 2^2 \quad \text{or} \quad 4 \times P_{\text{clock}}$$

$$f_B = \frac{1}{2^2} \quad \text{or} \quad \frac{1}{4} \times f_{\text{clock}}$$

$$\text{At } Q_C \quad P_C = 2^3 \quad \text{or} \quad 8 \times P_{\text{clock}}$$

$$f_C = \frac{1}{2^3} \quad \text{or} \quad \frac{1}{8} \times f_{\text{clock}}$$

In general, therefore, for an *n*-stage binary counter, we can write

$$P = 2^n \times \text{clock period} \qquad (7\text{-}1a)$$

$$f = \frac{1}{2^n} \times \text{clock frequency} \qquad (7\text{-}1b)$$

Let us now examine the Q and $\bar{Q}$ levels at each of the clock pulses, following Fig. 7-1b.

1 Let pulse 0 be the first clock pulse. The reason for this will become apparent in a short time.

2 Tabulate the logic levels for the Q and $\bar{Q}$ outputs of each stage at each clock pulse. These are given in Table 7-1 as Q_A, $\bar{Q}_A$, Q_B, $\bar{Q}_B$, Q_C and $\bar{Q}_C$.

3 The Q_A outputs toggle at each clock pulse.

4 Every time Q_A goes from 1 to 0 (negative transition), Q_B toggles.

5 Every time Q_B goes from 1 to 0, Q_C toggles.

6 Express the Q_C, Q_B, and Q_A states as a binary number $Q_C Q_B Q_A$. Call this the *count state*. This is shown in column E of Table 7-1.

7 Column F is the decimal equivalent of column E. We see that the decimal numbers of column F start at 0, progress up to 7, and begin again at 0.

If we call the first pulse number 0, the pulse numbers agree with column E in binary and column F in decimal.

8 The count of column E progresses in binary. Hence we call the counter of Fig. 7-1 a *binary counter*. The count progresses up to 7, reverts to 0, and begins again.

9 In obtaining the decimal equivalent, the Q_C column has a weight of 4, the Q_B a weight of 2, and the Q_A a weight of 1.

Table 7-1 Logic levels: three-stage binary counter

Clock Pulse	Q_C	Q_B	Q_A	Count State $Q_C Q_B Q_A$ E	Decimal Equiv. of Column E F	$\bar{Q}_C$	$\bar{Q}_B$	$\bar{Q}_A$	$\bar{Q}_C \bar{Q}_B \bar{Q}_A$ G	Decimal Equiv. of Column G H
	Weighting						Weighting			
	4	2	1			4	2	1		
0	0	0	0	000	0	1	1	1	111	7
1	0	0	1	001	1	1	1	0	110	6
2	0	1	0	010	2	1	0	1	101	5
3	0	1	1	011	3	1	0	0	100	4
4	1	0	0	100	4	0	1	1	011	3
5	1	0	1	101	5	0	1	0	010	2
6	1	1	0	110	6	0	0	1	001	1
7	1	1	1	111	7	0	0	0	000	0
8	0	0	0	000	0	1	1	1	111	7
					etc.					etc.

10 In Table 7-1, the $\bar{Q}_C$, $\bar{Q}_B$, and $\bar{Q}_A$ columns are the inverse of the Q_C, Q_B, and Q_A columns. $\bar{Q}_C \bar{Q}_B \bar{Q}_A$, column G, is the complement of column E. Column G starts at a count of 7, counts down to 0, and begins again at a count of 7.

11 Column F counts up. Column H counts down. At each count, columns F and H add up to 7, the maximum count capability.

There is one possible point of confusion. The count states column and binary number notation are $Q_C Q_B Q_A$, whereas in the diagram of Fig. 7-1 the counter stages are drawn in the reverse direction. Unfortunately, the electronics and the way in which circuit diagrams are normally drawn, with signal flow from left to right, are in pictorial opposition. Many textbooks, therefore, draw Fig. 7-1 in the reverse direction, with signal input on the right and signal flow from right to left.

The largest count capability of a three-stage binary counter is 111 in binary, or $2^3 - 1 = 7$ in decimal. In general, the largest decimal count capability of an n-stage binary counter is

$$\text{Maximum count for } n \text{ stages} = n \text{ ls} \equiv 111 \ldots 1 \qquad (7\text{-}2)$$
$$= 2^n - 1$$

For intermediate counts we can write

$$\text{Count} = B_n \times 2^{n-1} + B_{n-1} \times 2^{n-2} + B_{n-3} \times 2^{n-3} + \ldots B_1 \times 2^0 \quad (7\text{-}3)$$

where B is the binary level (1 or 0) at each stage.

In the counter of Fig. 7-1, the first stage generates a weighting of the least significant bit (LSB) $\equiv 1$, and the final stage generates a weighting of the most significant bit (MSB) $\equiv 4$. For example, in our three-stage counter, at the count of 6 or 110, $B_3 = 1$, $B_2 = 1$, and $B_1 = 0$.

$$\text{Count} = 1 \times 2^{3-1} + 1 \times 2^{3-2} + 0 \times 2^0$$
$$= 4 + 2 + 0$$
$$= 6$$

Suppose that we are at the count of 011 and the next pulse is applied. The next count changes to 100. All three stages change. Do they change instantly and simultaneously? They change very fast with integrated circuits but not exactly simultaneously. For example, in the SN7472 *JK* FF, the delay time to logic level 1 from the clock to output is approximately 16 ns, and to logic level 0 from clock to output is approximately 25 ns. In the 011 to 100 transition, we therefore have a delay in time after the clock has gone to its 0 level before the final 1 is reached of

Stage 1	1 to 0	25 ns
Stage 2	1 to 0	25 ns
Stage 3	0 to 1	16 ns
		66 ns

Now, 66 ns is quite fast, but by modern standards it is measurable, noticeable, and at times bothersome. What happens is that each stage change follows the previous one. The change propagates down or ripples through the counter. It is common practice to call the series counter of Fig. 7-1 a *ripple counter*.

What kind of troubles can occur? If we just simply want to count pulses there is no problem. But suppose that we have a 10-stage counter with SN7472 *JK* FFs which are capable of toggling at 15 megahertz (MHz) and are counting at a

10-MHz rate. We wish to perform an operation at a time when the count is in the state 1000000000 (this will be more fully discussed under decoding, Sec. 10-9). For now let us see if the 1 in stage 10 and the 0 in stage 0 can exist within 100 ns, the 10-MHz rate. Following our previous reasoning, the previous count was 0111111111. After stage 1 has gone to a 0, eight more stages have to go to 0, and stage 10 to a 1, a total time of $1 \times 16 + 8 \times 25 = 216$ ns. But this is more than twice the period of the 100-ns clock rate. By the time the 1 has rippled to the final stage, two extra clock pulses have entered the system and we will fail in our attempt to perform the required operation. A synchronous or clocked counter (discussed in the next section) is used to solve the problem.

Of fundamental importance in understanding counters is the concept of *COUNT STATE*. In series counters the count state is the representation as a binary number of the Q output logic levels of each stage, starting with the final stage first. For example, in the counter of Fig. 7-1, at count 6 the logic levels are $Q_C = 1$, $Q_B = 1$ and $Q_A = 0$. The count state is

Final stage Q_C	Intermediate stage(s) Q_B	First stage Q_A	
1	1	0	(7-4)

In this case, at a decimal count of 6, the count state is 110.

The counter of Fig. 7-1 has eight different count states. For this counter, the count states progress in binary through eight distinct and different count states before returning to the original count state. This is an equivalent way of saying that this counter divides by 8, that it will generate one output count for every 8 input counts, or that its output frequency is $\frac{1}{8} \times f_{in}$. The count state concept is particularly applicable when we modify binary counters to count in nonbinary modes.

7-3 SYNCHRONOUS COUNTER

As previously discussed, ripple counters run into problems when count coincidences are required, and the time separation *between* pulses is comparable to the time it takes the count to ripple through the counter. What is needed in such cases is a system in which all stages change state at a time controlled by the clock pulse. Let us now modify the counter of Fig. 7-1 to do this.

Figure 7-2a is the circuit diagram of a three-stage synchronous counter. It has three T FFs, but it also has three additional gates X, Y, and Z.

1   Reset all the FFs so that Q_A, Q_B, and Q_C are initially in the 0 state.
2   Apply the first clock pulse (CP). It can go through gate X to toggle FF A, but it cannot go through gates Y and Z because of the 0s of Q_A and Q_B. Q_A is at a 1, Q_B at 0, and Q_C at 0.

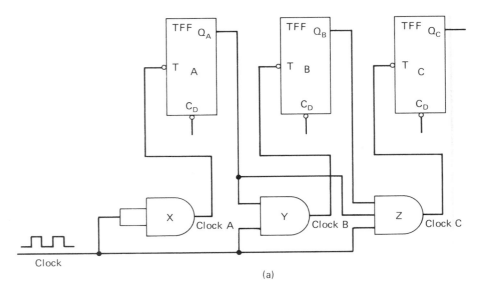

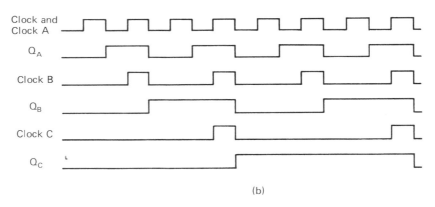

Figure 7-2 Three-stage synchronous counter and waveforms: (a) circuit diagram; (b) waveform

3 Apply a second CP. It goes through gate X to toggle FF A again. However Q_A is now at a 1. Hence gate Y is enabled, and the clock pulse is transmitted through gate Y to the clock input of FF B. The negative-going transition of the CP is applied at the *same* time to the T inputs of FF A and FF B, and both toggle at a time controlled by the CP. Gate Z is not enabled because $Q_B = 0$. $Q_A = 0$, $Q_B = 1$, $Q_C = 0$.

4 The third CP toggles FF A. $Q_A = 1$, $Q_B = 1$, $Q_C = 0$.

5 At the fourth CP, gate Y is enabled since $Q_A = 1$. Gate Z is also enabled, since Q_A and Q_B, which control gate Z, are both at 1. Hence the fourth

CP will toggle all three FFs. Again the toggling time is controlled by the negative-going CP transition, and all FFs toggle synchronously. There is no ripple-through problem.

6 Figure 7-2b shows the waveforms. It can be seen that each FF is controlled by the clock, and the final waveforms are the same as in the ripple counter, Fig. 7-1b.

What have we done to solve the required coincidence problem? To do this requires a more complex circuit. Moreover, in Fig. 7-1, only the first FF has to operate at the highest frequency; all the others can operate at progressively lower frequencies so that possible economies in FF cost and power consumption exist. In Fig. 7-2, besides the additional gates, all FFs must be capable of operating at the clock frequency. In general, the propagation delay through the AND gates X, Y, and Z is very small compared to the system requirements so that gate X, which was shown in Fig. 7-2 for gate propagation equalization, is frequently omitted. MSI synchronous binary counters are available in TTL. The SN54193/ 74193 is a 4-bit (stage) synchronous up–down counter in which the direction of counting can be revised.

7-4 PRINCIPLE OF DIVIDE-BY-N, MOD-N COUNTERS

The counters of Figs. 7-1 and 7-2 divide frequencies only by factors of 2^n and count in binary. Such counters have limited usefulness. Most counters divide by factors other than 2^n. For example, the divide-by-10 or *decade counter* is very useful and common.

What do we mean by divide by N? This means primarily that the counter generates an output frequency $1/N$ of the original frequency. The output waveshape, not necessarily square or rectangular, repeats itself at a rate $1/N$ of the original frequency and has a period N times the clock frequency. Whereas the output waveshapes of the counters of Figs. 7-1 and 7-2 had equal times for 1 and 0 levels (square-wave output), we shall see that many of the divide-by-N counters later discussed do not generate square waves. In discussing the design and operation of divide-by-N counters, the count-state concept is of major importance. Suppose that we have such a counter with no signal applied. We can measure the Q output levels at each stage and express the Q output levels in the count-state manner of Eq. (7-4). Now apply a single pulse. At least one, and possibly more than one, of the stages will change its state. Now again express the new count state. This can be repeated pulse by pulse. After N counts, the count state will return to the original count state. In a divide-by-N counter there are N different count states; after N counts, the count state has returned to its original count state. Such a counter with its N different count states is frequently designated as a *modulus-N* or *mod-N* counter.

Now consider the problem of a *mod-3 counter*. A single-stage binary divides by 2. This is not enough. A two-stage binary counter is capable of dividing by 4. If we were to take such a two-stage counter and omit one of the count states, we would then have a mod-3 counter. For example, we could count in the following sequence of count states: 00, 01, 10, and return to 00. This omits or disallows the count state 11. There are three different count states. This is in accordance with our definition of a mod-3 counter. However, this is not the only possibility. We can omit any one of the count states. As a matter of fact, there is no reason why the count states have to progress in a binary sequential order.

In Table 7-2, we tabulate all the possibilities for a mod-3 counter. We can see that even a simple mod-3 has many design possibilities. Which one to use is determined by the factors of cost, usage, ease of implementation and possibility of malfunction. In general, designs where the count state progresses up (as though it were in a binary counter sequence) are easiest to implement. In the listing of Table 7-2, counters A, B, C, and G follow this pattern and are easiest to implement. Designs that skip back and forth such as E (0, 3 2 in decimal equivalent) require more complex circuitry, but can be made if required.

Table 7-2   Mod-3 counter count states

Possibility	A	B	C	D	E	F	G	H
Count States	00	00	00	00	00	00	01	01
	01	01	10	10	11	11	10	11
	10	11	11	01	01	10	11	10

To go back to the counter of Fig. 7-1, it counted in a normal binary counting sequence up to the count of 111 and then back to zero. This counter has eight count states and is a mod-8 counter. It is apparent that by merely interchanging count states an enormous number of possibilities exist. One possibility could have interchanged only the 000 and the 001 states and counted in the following sequence: 001, 000, 010, 011, 100, 101, 110, and 111. The counter of Fig. 7-1 is quite simple to build. A counter with this apparently simple interchange becomes considerably more complex.

Mod-N counters are all based on a principle of omission of count states. For example, to build a mod-6 counter the design must start with a three-stage counter that is capable of dividing by 8. A mod-6 counter is made by omitting any two of the count states. We can see from this example that a three-stage counter is required for mod-5, mod-6, and mod-7 counters.

To design a mod-N counter, we must first determine the number n of stages required. A series connection of n T FFs, as we have seen, has 2^n count states. By omitting some of the count states, one can design a mod-N counter where N is less than 2^n. If we have one stage less, or $n - 1$ stages, the number of count states is 2^{n-1}. With n stages, it is practical to omit states between the values 2^n and 2^{n-1}. We can therefore write that, for a mod-N counter, N must fall between the limits

$$2^n > N > 2^{n-1} \tag{7-5}$$

and the counter will require n FFs.

Another method of determining the number of stages required for a mod-N counter is to express N as a binary number. The number of bits needed to do this gives the number of stages needed in the counter.

EXAMPLE 7-1 How many stages are needed for a mod-13 counter?

Solution

    a. $16 > 13 > 8$. $2^4 > 13 > 2^3$. Needs four stages.
    b. $13_{10} = 1101_2$. Four bits are needed in the binary number. Four stages are required.
    A four-stage counter is capable of generating $2^4 = 16$ count states. A mod-13 counter omits three count states.

EXAMPLE 7-2 How many stages are needed for a mod-151 counter?

Solution

    a. $256 > 151 > 128$.
    b. $2^8 > 151 > 2^7$. Needs eight stages.
    c. $151_{10} = 10010111_2$. Eight bits are needed for the binary number. Needs eight stages.

7-5 MOD-N COUNTERS: FEEDBACK TECHNIQUE

This section develops the use of the *feedback* principle. It is the technique used when counters were made with discrete components, that is, transistors, diodes, resistors, and capacitors. Although this method can also be used with ICs, the most popular IC mod-N counters use different techniques; the material in this section is presented as background information.

Figure 7-3a shows the diagram of a four-stage counter using feedback. If this were a four-stage ripple counter, it would count in binary from 0000 to 1111, return to zero, and divide by 16. The counter of Fig. 7-3a is modified so that when the $\bar{Q}_D$ output changes from a 1 to 0 (Q_D goes from 0 to 1), the negative-going transition generates a short-duration negative-going pulse that is applied to the S_D inputs of stages 2 and 3 of the counter; this sets them into the 1 state. This counter counts in the normal binary progression, as seen in Table 7-3, up to the count where Q_D changes from 0 to 1, count 8'. At this time, the negative-going transition from $\bar{Q}_D$ is used to generate a very short duration negative-going pulse, which is

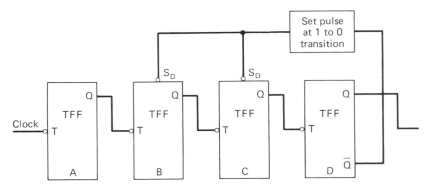

Figure 7-3a Mod-10 (decade) feedback counter

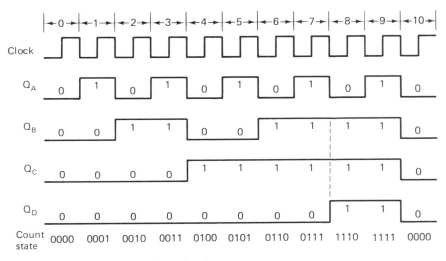

Figure 7-3b Wave patterns: 2′421 decade counter

applied to the S_D terminals of stages 2 and 3. Stages 2 and 3 had changed to 0s at count 8, but this pulse changes them back to 1s, at the count of 8 and a count state 1110. At the next count, the counter goes to the count state 1111, and at the tenth pulse all four stages toggle, and the counter is in the state 0000. Count state 8′ only exists for a very short time, since the set pulse is designed to be just long enough to perform the set operation and to be very short compared to the pulse period.

What has happened? If we count the number of count states we see that there are ten such states. This, therefore, is a mod-10 counter. If we examine the count state column as binary numbers, we see that we have counted up to the count of 7, then jumped to a count of 14, then 15, and back to 0. We have skipped counts 8–13, a total of 6, which is to be expected since the binary count capability of a four-stage counter is 16.

Table 7-3 Decade 2'421 feedback counter

Count Pulse	Q_D	Q_C	Q_B	Q_A	Count State	Decimal Count
		Weighting				
	2'	4	2	1		
0	0	0	0	0	0000	0
1	0	0	0	1	0001	1
2	0	0	1	0	0010	2
3	0	0	1	1	0011	3
4	0	1	0	0	0100	4
5	0	1	0	1	0101	5
6	0	1	1	0	0110	6
7	0	1	1	1	0111	7
8'	1	0	0	0		
8	1	1	1	0	1110	8
9	1	1	1	1	1111	9
10	0	0	0	0	0000	

Had the counter of Fig. 7-3 not been modified by feedback (Prob. 7-5), the count weightings of Q_D, Q_C, Q_B, and Q_A would have been 8, 4, 2, and 1, and these add up to the equivalent decimal count. How about the counter of Fig. 7-3? Can we devise a weighting scheme such that the weights add up to the decimal value? It is obvious that the binary weightings will not do. For example, at count 9 the binary weightings add up to $8 + 4 + 2 + 1 = 15$, not 9. However, if we assign a weight of 2 to the Q_D output, and 4, 2, and 1 to the Q_C, Q_B, and Q_A outputs, the weighting now comes out to the decimal value. At the eighth count, $2 + 4 + 2 + 0 = 8$, and at the ninth count $2 + 4 + 2 + 1 = 9$. This counter is therefore a weighted decade counter, known as the 2'421 counter in accordance with the columnar weights. In this counter, each decimal value can be expressed by an equivalent count state that is weighted by the 2'421 scheme. We call this counter a *binary-coded-decimal* (BCD) counter. More will be said about BCD counters in Sec. 7-7.

What about the feedback scheme of Fig. 7-3, which really should be called a skip-forward scheme? The feedback to Q_C advances (skips) four counts, and to Q_B, two counts. We see from this that we can readily control the modulus of the counting process by the way in which the feedback is performed. Some counters have the feedback set so that instead of a single skip of six counts, such as in the 2'421, partial skips are performed more than once to arrive at an equivalent result.

Suppose that to the counter of Fig. 7-3 we were to apply a periodic wave, for example a square wave, and view it on a cathode-ray oscilloscope (CRO). What waveshapes would we obtain? The waveshapes must be related to the count progression of Table 7-3. The individual-stage logic levels must progress with time in accordance with each column as we go down. If we follow this procedure, we obtain and would see the wave patterns of Fig. 7-3b. For example, reading down we see that Q_A alternates between 0 and 1, Q_B alternates between two 0s and two 1s, Q_C has four 0s and six 1s, and Q_D has eight 0s and two 1s. After 10 input counts (0 through 9), the viewed pattern returns to that where all four FFs are at a 0 level. Examine carefully the Q_D output waveshape. Note that it repeats itself after 10 input pulses, but its waveshape is far from being symmetrical or square. For 80 percent of the time it is in the 0 state, and is in the 1 state for only 20 percent of the time. Nevertheless, its output frequency is the input frequency divided by 10.

7-6 INTEGRATED-CIRCUIT MOD-3 SERIES COUNTERS†

Although the feedback technique can be used with integrated circuits, IC mod-N counters utilize the RS and JK properties of the master–slave flip-flop to skip the required number of unwanted states. Truth tables for these FFs were developed in Chapter 6. This provides simpler solutions. Both ripple and synchronous (clocked) counters can be built. Figure 7-4 is an example of a mod-3 counter in which the JK properties of the clocked universal FF are used.

In the circuit of Fig. 7-4,

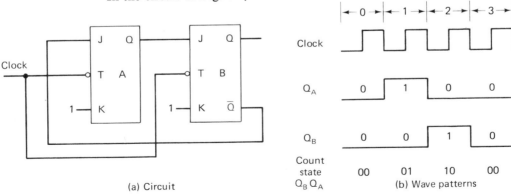

Figure 7-4 Mod-3 JK counter: (a) circuit; (b) wave patterns

†M. E. Levine, *Digital Theory and Experimentation Using Integrated Circuits.* Englewood Cliffs, N.J.: Prentice-Hall, Inc., 1974, Expt. 9, Divide-by-N Counters: Decade Counters.

1 J_A is connected to $\bar{Q}_B$, and $K_A = 1$. The state of FF A is controlled by the state of FF B.

   a If $\bar{Q}_B = 1$, a clock pulse (CP) to A will toggle FF A ($J = K = 1$).
   b If $\bar{Q}_B = 0$, a CP to A will make $Q_A = 0$ ($J = 0$, $K = 1$).

2 J_B is connected to Q_A, and $K_B = 1$. The state of FF B is controlled by the state of FF A.

   a If $Q_A = 1$, a CP to B will toggle FF B ($J = K = 1$).
   b If $Q_A = 0$, a CP to B will make $Q_B = 0$ ($J = 0$, $K = 1$).

Assume that the initial state is the 00 ($Q_B Q_A$) state. Table 7-4 shows the count sequence, and how the states of each of the FFs direct the state of the other FFs because of the JK properties of the universal FF. After the third pulse, the state of the counter has returned to the initial state. There are three separate count states. The 11 state has been skipped.

Table 7-4 Mod-3 counter of Figure 7-4

Clock Pulse	J_A	K_A	Q_A	$\bar{Q}_A$	J_B	K_B	Q_B	$\bar{Q}_B$		Count State $Q_A Q_B$
0			0	1			0	1		00
	1	1			0	1				
1			1	0			0	1	FF A has toggled	01
	1	1			1	1			FF B forced to stay the same	
2			0	1			1	0	Both FFs have toggled since $J_A = K_A = 1$ and $J_B = K_B = 1$	10
	0	1			0	1				
3			0	1			0	1	FF A forced to stay the same FF B forced into the 0 state	00

The wave pattern for this counter is shown in Fig. 7-4b. Note how the wave pattern follows the Q_A and Q_B logic levels reading down in the chart of Table 7-4 in columns Q_A and Q_B. Note also how the count state can be read directly from the wave patterns. The output wave pattern is not square wave.

It is quite conceivable that when the counter is initially turned on, or even during operation because of noise, the counter may get into the 11 state. Then there are two possibilities. The next pulse may get the counter into one of the normal states or the counter may lock up in the 11 state. If $Q_A = 1$, $J_B = 1$, and $K_B = 1$,

FF B will toggle and go to the 0 state. If $\bar{Q}_B = 0$ $(Q_B = 1)$, $J_A = 0$, and $K_A = 1$, FF A goes to the 0 state. Hence the next state after the unwanted 11 state is a normal 00 state and the counter has suffered only a temporary malfunction.

Another way of building a mod-3 counter that makes use of the JK and RS properties of the clocked master–slave FF is shown in Fig. 7-5.

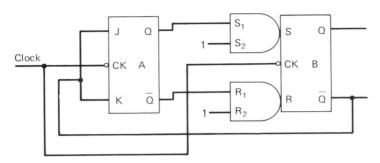

Figure 7-5   Mod-3 counter

In the circuit of Fig. 7-5,

1   Flip-flop A is a JK FF. Both J_A and K_A are connected to $\bar{Q}_B$.
   a   If $\bar{Q}_B = 1$, a CP will toggle FF A $(J = K = 1)$.
   b   If $\bar{Q}_B = 0$, there will be no change in FF A upon application of a CP $(J = K = 0)$.
2   Flip-flop B is an RS FF. Since S_2 and R_2 are equal to 1, $S = S_1$ and $R = R_1$. S_1 is connected to Q_A, and R_1 to $\bar{Q}_A$. The state of FF B is controlled by the state of FF A.
   a   If $Q_A = 1$ and $\bar{Q}_A = 0$, a CP applied to B will make $Q_B = 1$ and $\bar{Q}_B = 0$ $(S = 1, R = 0)$.
   b   If $Q_A = 0$ and $Q_B = 1$, a CP applied to B will make $Q_B = 0$ and $\bar{Q}_B = 1$ $(S = 0, R = 1)$.

Assume that the initial state is the 00 state. Table 7-5 shows the count sequence and how the states of each of the FFs direct the state of the other FF. As can be seen, there are three count states, with the 11 state skipped. This counter has the identical count state sequence as that of Fig. 7-4 and hence the same wave patterns.

Let us in this counter examine the unwanted 11 state. The 0 from $\bar{Q}_B$ again prevents FF A from changing its state. The 1 from Q_A to S_{1B} and the 0 from $\bar{Q}_A$ to R_{1B} makes FF B continue in the 1 state without a change. Hence the 11 state is continued and the counter is locked up in the 11 state. This shows that any mod-N counter must be carefully examined to prevent locking up in an unwanted sequence.

What to do about the counter of Fig. 7-5 and its locked-up 11 state? One way to overcome the problem is to sense when the 11 state does occur and then take action by means of feedback to prevent it. Figure 7-6 shows the counter of Fig. 7-5

Table 7-5 Mod-3 counter of Figure 7-5

Clock Pulse	J_A	K_A	Q_A	$\bar{Q}_A$	S_{1B}	R_{1B}	Q_B	$\bar{Q}_B$		Count State $Q_B Q_A$
0			0	1			0	1		00
	1	1			0	1				
1			1	0			0	1	FF A has toggled Since FF B was in the same state as FF A it did not change	01
	1	1			1	0				
2			0	1			1	0	FF A has toggled again FF B followed the state of FF A and changed	10
	0	0			0	1				
3			0	1			0	1	FF A does not change FF B follows FF A	00

with this modification. In Fig. 7-6, a NAND gate has been added to sense the 11 state. If a 11 state does occur, feedback to the C_D inputs of the two FFs will force them to the 00 state thereby effectively preventing the 11 state from occurring.

Although it is obvious that the mod-3 counter of Fig. 7-4 is preferable to that of Fig. 7-6, there are cases in counter design when a simpler solution, such as the one of Fig. 7-4, is not possible, and a method similar to that of Fig. 7-6 must be used to prevent lock out. The designer, however, must always strive to arrive at the simplest possible solution.

Figure 7-6 Mod-3 counter with lock-out inhibitor

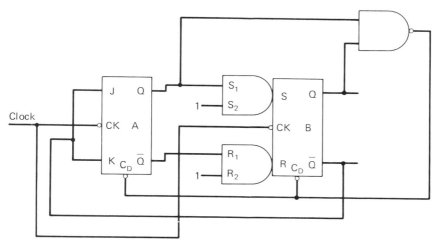

7-7 INTEGRATED-CIRCUIT MOD-6 COUNTERS

Suppose that we wish to build a mod-6 counter. The simplest way is to add a T FF that divides by 2 either before or after the counters of Fig. 7-4, 7-5, or 7-6. If we apply our input signal first to the T FF, the counter will divide the input by $2 \times 3 = 6$, and if we add the T FF after the mod-3 counter, we will divide by $3 \times 2 = 6$. Treating the mod-3 counter as a block, we can show this in Fig. 7-7. There are now three FFs, A, B, and C, in the circuit.

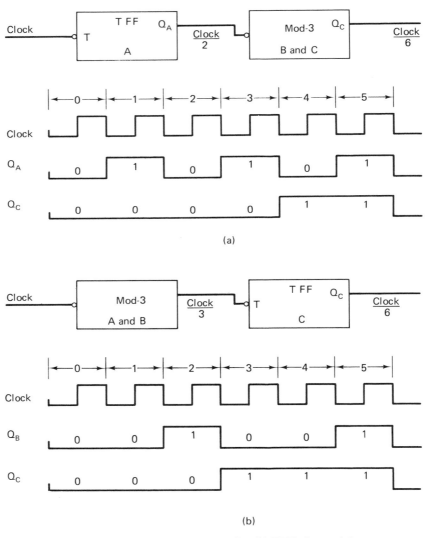

Figure 7-7 Mod-6 counters: (a) T FF before mod-3; (b) T FF after mod-3

In Fig. 7-7a, Q_A is the clock output divided by 2, and Q_C has the same wave pattern as that of Fig. 7-4b but is based upon the Q_B input waveform, which is the output of Q_A. In Fig. 7-7b, Q_B follows the wave pattern of Fig. 7-4b, and Q_C divides this frequency by 2 in the T FF. With the T FF following, the nonsquare output wave of Q_B is converted into a square wave. This is to be expected, since the negative Q_B transition occurs at equal time intervals. Some applications may require a non-square wave (see, for example, the NBCD counter of Sec. 7-8), and some may require a square-wave output such as that of Fig. 7-7b.

The type MC5492/7492 counter shown in Fig. 7-8 is a TTL MSI divide-by-12 counter. It has the mod-3 (outputs at pins 11 and 9) section of Fig. 7-4 followed by a T FF (output at pin 8) as in Fig. 7-7b. In addition, it has a separate (pin 12 output) T FF. By connecting the separate T FF in cascade with the mod-6 section, a divide-by-12 output is obtained. Depending upon the interconnection and the output terminal used, this IC then has the ability to divide by 2, 3, 6, and 12. Provisions are made to reset both sections to logic 0. Both R_0 inputs have to be at logic 1 to reset all the outputs to logic 0. This can be seen on the logic diagram of Fig. 7-8 as a two-input NAND.

The MC5492/7492 is particularly useful when converting 60-Hz pulses to 10-Hz pulses for timing systems and for electronic clocks in which the number of hours have to be counted.

COUNT SEQUENCE TRUTH TABLE

COUNT	OUTPUT			
	D	C	B	A
0	0	0	0	0
1	0	0	0	1
2	0	0	1	0
3	0	0	1	1
4	0	1	0	0
5	0	1	0	1
6	1	0	0	0
7	1	0	0	1
8	1	0	1	0
9	1	0	1	1
10	1	1	0	0
11	1	1	0	1

A connected to C2

V_{CC} = PIN 5
GND = PIN 10

Input Loading Factor:
R0 = 1
C1 = 2
C2 = 4
Output Loading Factor = 10
Total Power Dissipation = 160 mW typ/pkg
Propagation Delay Time = 60 ns typ

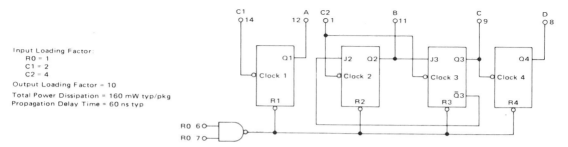

Figure 7-8   MC5492/7492 divide-by-2, divide-by-6 counter

By far the most important mod-N counter is the mod-10 or decade counter, which divides by 10 and has 10 count states. We have already discussed one such decade counter, the $2'421$ counter in Sec. 7-5, which used the feedback technique.

Many different decade-counter count-state sequences have been used. Outputs of counters are frequently required to interface with other equipment, such as numeric displays, printers, computers, and tape punches. Standards are important so that any particular piece of equipment can readily interface with any other. At the present time, one progression has virtually become the standard. This is the counter in which the count states progress in the natural binary progression up to 9 (0000 to 1001) and then reverts to 0. This skips six states, the decimal states 10 through 15. We shall see that it is a *weighted* counter and is known as the 8421 counter, corresponding to the weightings. The count states are expressed in binary notation and natural progression; it is commonly called the NBCD counter. However, the 8421 counter is sometimes called BCD, and this leads to confusion since the $2'421$ is also a BCD counter. When BCD is used, the probability is that the reference is to the 8421 counter, but one should check to be absolutely sure. At one time the $2'421$ and another weighted counter, the 4221, were quite popular, but the 8421 has become the most common of the decade cascade series counters.

The 8421 counter has to count up to 1001 and then go to the 0000 state. Thus at this time FF A must toggle, but this toggle must be prevented from making FF B toggle. In addition, FF D must toggle. Figure 7-9 is the logic diagram of an 8421-NBCD ripple decade counter. This counter makes use of JK and RS FFs. Flip-flops A, B, and C are JK FFs; FF D is an RS FF. Flip-flop A is a toggle FF; FF B is a JK FF under the control of FF D. If $\bar{Q}_D = 0$, FF B is at a 0 after the clock pulse. If $\bar{Q}_D = 1$, FF B toggles. Flip-flop C is a T FF, and FF D is an RS FF. The S input is an AND input under the control of Q_B and Q_C. The R input is controlled by the level of Q_D. Table 7-6 shows the count-state progression.

Examination of the count states shows that, if we assign a weight of 8 to the Q_D level, 4 to the Q_C level, 2 to Q_B, and 1 to Q_A, the sum of the weighted values adds up to the decimal values from 0 to 9. Hence this is a weighted 8421 counter, occasionally called a 1248 counter.

Figure 7-9 Integrated-circuit decade-8421-NBCD counter

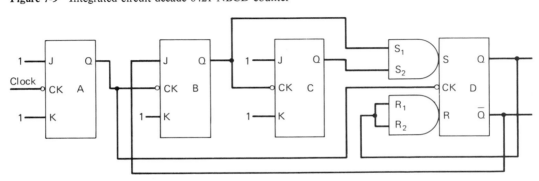

Table 7-6 Decade-8421-NBCD counter of Figure 7-9

Clock Pulse	Q_A	J_B	K_B	Q_B	Q_C	S_{1D}	S_{2D}	S_D	R_D	Q_D	$\bar{Q}_D$		Count State $Q_D Q_C Q_B Q_A$
0	0			0	0					0	1		0000
		1	1			0	0	0	0				
1	1			0	0					0	1	Only FF A toggles	0001
		1	1			0	0	0	0				
2	0			1	0					0	1	FF A toggles FF B toggles when FF A goes from 1 to 0	0010
		1	1			1	0	0	0				
3	1			1	0					0	1	Only FF A toggles	0011
		1	1			1	0	0	0				
4	0			0	1					0	1	FFs A, B, and C toggle because of the 1 to 0 change	0100
		1	1			0	1	0	0				
5	1			0	1					0	1	Only FF A toggles	0101
		1	1			0	1	0	0				
6	0			1	1					0	1	FFs A and B toggle	0110
		1	1			1	1	1	0				
7	1			1	1					0	1	FF A toggles FF D cannot change until FF A goes from 1 to 0	0111
		1	1			1	1	1	0				
8	0			0	0					1	0	FFs A, B, and C toggle; FF D is forced as an RS FF to change its state	1000
		0	1			0	0	0	1				
9	1			0	0					1	0	Only FF A toggles	1001
		0	1			0	0	0	1				
10	0			0	0					0	1	FF A toggles 1 to 0 FF B cannot toggle The $J_B = 0$, $K_B = 1$ keeps FF B in the 0 state FF D, whose clock comes from Q_A, is forced into the 0 state ($S = 0$, $R = 1$) The counter is back to the 0000 state	0000

The waveshapes that will be seen on a CRO if a periodic clock pulse is applied follow the downward columnar levels. They are shown in Fig. 7-10. By examining the levels at each pulse, it can be seen that the count state can readily be determined. For example, at pulse 7 and reading the levels beginning with Q_D, we can see that the count state is 0111.

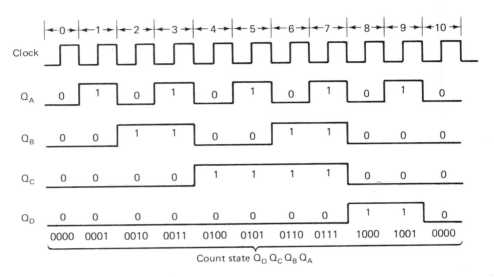

Figure 7-10   8421 counter waveforms

7-9 UNUSED STATES OF THE 8421 COUNTER

We must examine the counter of Fig. 7-9 for the unwanted states, 1010, 1011, 1100, 1101, 1110, and 1111. Let us now do this.

7-9.1 The 1010 State

$$J_B = 0 \ (\bar{Q}_D = 0), \qquad K_B = 1$$
$$\left. \begin{array}{l} S_{1D} = 1 \ (Q_B = 1) \\ S_{2D} = 0 \ (Q_C = 0) \end{array} \right\} \ S_D = 0 \quad \begin{array}{l} \text{this is } S \text{ of FF } D; \text{ do not confuse} \\ \text{with FF direct sets} \end{array}$$
$$R_D = 1 \ (Q_D = 1)$$

At the clock pulse, FF A toggles 0 to 1. This cannot affect either FF B or C. Hence the next state is 1011, an unwanted state.

7-9.2 The 1011 State

$$J_B = 0 \ (\bar{Q}_D = 0), \qquad K_B = 1$$
$$\left.\begin{array}{l} S_{1D} = 1 \ (Q_B = 1) \\ S_{2D} = 0 \ (Q_C = 0) \end{array}\right\} S_D = 0$$
$$R_D = 1 \ (Q_D = 1)$$

At the clock pulse, FF A toggles 1 to 0; FF B goes to the 0 state ($J_B = 0$, $K_B = 1$). This makes FF C toggle and go to the 1 state. Flip-flop D is forced ($S = 0$, $R = 1$) to the 0 state. The resultant state is the 0100 state, an allowed state.

This accounts for the 1010 and 1011 states. They return to a wanted state and the counter has suffered a temporary malfunction.

7-9.3 The 1110 State

$$J_B = 0 \ (\bar{Q}_D = 0), \qquad K_B = 1$$
$$\left.\begin{array}{l} S_{1D} = 1 \ (Q_B = 1) \\ S_{2D} = 1 \ (Q_C = 1) \end{array}\right\} S_D = 1$$
$$R_D = 1 \ (Q_D = 1)$$

At the clock pulse, FF A toggles 0 to 1. This cannot affect any other FF. Hence the 1110 state goes to the 1111 state, an unwanted state.

7-9.4 The 1111 State

$$J_B = 0 \ (\bar{Q}_D = 0), \qquad K_B = 1$$
$$\left.\begin{array}{l} S_{1D} = 1 \ (Q_B = 1) \\ S_{2D} = 1 \ (Q_C = 1) \end{array}\right\} S_D = 1$$
$$R_D = 1 \ (Q_D = 1)$$

At the clock pulse, FF A toggles 1 to 0. This makes FF B change its state 1 to 0, and this change makes FF C toggle 1 to 0. The S_D input is 1, and the R_D input is 1. This is an undefined state for FF D. It can go to either a 1 or a 0 state. After the clock pulse, the state of the counter is either 0000 or 1000, and both are allowed. Hence the counter has suffered a temporary malfunction.

7-9.5 The 1100 and 1101 States

These states also result in temporary malfunctions. Their evaluation is left as a problem (Prob. 7-21).

187

Hence all the unwanted states of the 8421 counter are temporary malfunctions and the counter progresses into the regular count sequence.

7-10 INTEGRATED-CIRCUIT 2'421 COUNTER

Figure 7-11 is the logic diagram of an IC 2'421 decade counter. Although it is also a 2'421 counter, its count-state progression is different from that of the 2'421 feedback counter of Sec. 7-5. It makes use of the *JK* and *RS* properties of clocked FFs. Table 7-7 gives the count states of this counter.

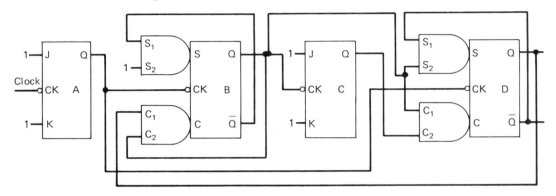

Figure 7-11 Logic diagram of IC 2'421 decade counter

Table 7-7 Count states IC 2'421 decade counter

D 2'	C 4	B 2	A 1	Decimal Equivalent
0	0	0	0	0
0	0	0	1	1
0	0	1	0	2
0	0	1	1	3
1	0	1	0	4
1	0	1	1	5
1	1	0	0	6
1	1	0	1	7
1	1	1	0	8
1	1	1	1	9

Although decoding will be discussed more completely in Sec. 10-8, it is appropriate to introduce some aspects of the decoding of decade counters here. *Decoding* is a process which selects a count state and generates an output at *that* count state and *no other*. Let us consider the two-stage binary counter whose logic diagrams, wave-shapes, and decoded outputs are given in Fig. 7-12. From the waveforms, to select the time of the first pulse corresponding to the count state 00, we connect $\bar{Q}_B$ and $\bar{Q}_A$ to the input of an AND gate. This AND gate is high only during the first pulse. This method is valid also for the other pulses. The basic procedure for decoding is to use an AND gate whose inputs are the count states. If the count-state level is at a 1, the input to the decoding AND is a direct input. If the count-state level is at a 0, connect the inverted input to the AND. It is because of the importance of decoding and its relationship to count states that so much emphasis has been placed on the count-state concept in this chapter.

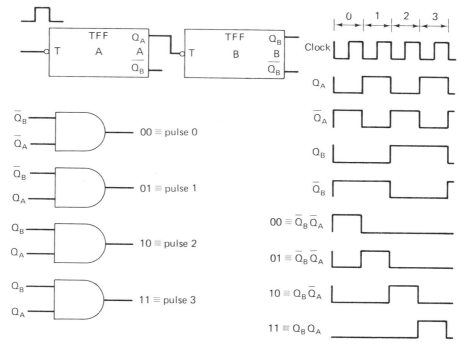

Figure 7-12 Decoding of two-stage binary counter

Since decade counters have four stages, it would appear that to decode the 10 count states would require 10 four-input AND gates. However, this is not quite correct. Some of the inputs are not always required. Why this is so will be discussed in Sec. 10-8. In the case of the 8421 counter, to decode all 10 states requires two four-input, six three-input, and two two-input AND gates. The 2'421 counter requires 10 three-input AND gates. These considerations sometimes determine the type of counter used.

The outputs of decade counters are frequently displayed on digital display devices, such as seven-segment or direct-reading numeric display devices. To do this, one must *decode* each count state.

7-12   TYPE 54/7490 IC 8421 COUNTER

The type 54/7490† is a TTL MSI decade counter. Its logic diagram, count sequence, truth tables, and waveshapes are shown in Fig. 7-13. It has a divide-by-2 section independent of a divide-by-5 section. If the output of the divide-by-2 section

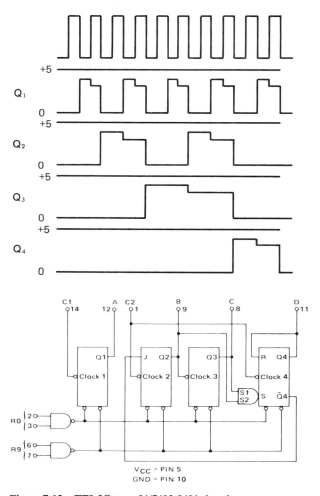

RESET/COUNT TRUTH TABLE

R0		R9		OUTPUT			
Pin 2	Pin 3	Pin 6	Pin 7	D	C	B	A
1	1	0	X	0	0	0	0
1	1	X	0	0	0	0	0
X	X	1	1	1	0	0	1
X	0	X	0	COUNT			
0	X	0	X	COUNT			
0	X	X	0	COUNT			
X	0	0	X	COUNT			

X = Don't care.

COUNT SEQUENCE TRUTH TABLE

COUNT	OUTPUT			
	D	C	B	A
0	0	0	0	0
1	0	0	0	1
2	0	0	1	0
3	0	0	1	1
4	0	1	0	0
5	0	1	0	1
6	0	1	1	0
7	0	1	1	1
8	1	0	0	0
9	1	0	0	1

A connected to C2.

Input Loading Factor:
R0, R9 = 1
C1 = 2
C2 = 4

Output Loading Factor = 10

Total Power Dissipation = 160 mW typ/pkg
Propagation Delay Time = 20 ns typ/bit

V_{CC} = PIN 5
GND = PIN 10

Figure 7-13   *TTL* IC type 54/7490 8421 decade counter

†Motorola Semiconductor Products, Inc.

(pin 12) is connected to the clock input of the divide-by-5 section 1 (pin 1), the resultant waveforms and the count-state progression make this an 8421 counter, whose operation was described in Sec. 7-8. Figure 7-13 shows the actual waveshapes as seen on a CRO for this counter. Note that the waveshapes are considerably different from the ideal waveshapes shown in Fig. 7-10. Although the ground levels are as expected, the high levels are not constant at the expected V_{cc} level of $+5$ V, but vary throughout the cycle. This is quite satisfactory, since the high level can vary arbitrarily throughout the cycle providing its level never drops below the threshold voltage after consideration of noise immunity.

Some applications such as *frequency synthesizers* require a square-wave output. If the input clock signal is applied to pin 1, and the pin 11 output connected to pin 12 (clock of FF 0), the resultant output is a divide-by-10 with square-wave output. Reset provisions are available for this counter. These are shown on the logic diagram. The counter can be reset to a count of 0 by means of the $R0$ input. Most counter applications require this. This is done by bringing both of the $R0$ inputs to logic 1 levels. The counter will remain at the 0 count (counting stopped) as long as both $R0$ inputs are at 1. When either (or both) of the $R0$ inputs is connected to logic 0, the counter can count. For 9s complement decimal applications, the counter has an independent set of reset inputs ($R9$) that reset the counter to a NBCD count of 9 (1001).

The type 54/7490 counter can be used to divide by 2, 5, or 10.

7-13 MSI IC COUNTERS

Other interesting IC counters are available. The 9305 is a TTL variable-mod counter. It has a separate divide-by-2 section, and a three-stage section that can be programmed to divide by 5, 6, 7, and 8. With this counter, one can divide by 2, 4, 5, 6, 7, 8, 10, 12, 14, and 16. If the output of the three-stage section drives the divide-by-2 section, division by 4, 6, 8, 10, 12, 14, and 16 with square-wave output is obtained.

With further increase in internal complexity ($\equiv$ 55 gates), MSI TTL up–down 25-MHz synchronous 4-bit binary (type 54/74193) and up–down synchronous decade (type 54/74192) counters are available. These counters have separate clock input terminals for up and down counting. An application of this type of counter to an analog-to-digital (A/D) conversion technique will be discussed in Chapter 13.

7-14 SERIES CONNECTION OF COUNTER BLOCKS

Let us connect a type 7490 decade and a type 7492 mod-12 counter in series. Suppose that we now apply a frequency of 240 kHz to the input, as shown in Fig. 7-14. We see from this that each block acts independently of the other. The 7490 divides the input frequency by 10, and the 7492 divides the output of the 7490 by 12. The

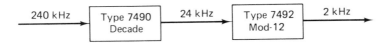

Figure 7-14 Counter blocks in series

overall frequency division is $10 \times 12 = 120$, and the output frequency is

$$\frac{240 \text{ kHz}}{120} = 2 \text{ kHz}$$

When large frequency division is required, such series cascading is almost always the only practical method available and is used extensively.

7-15 APPLICATION OF COUNTERS

In this section, we shall briefly indicate a few of the applications of counters.

7-15.1 Precision Time Intervals of 1 Second

Figure 7-15a One-MHz to 1-Hz counter

In this application, a precision stable 1 MHz frequency generator is divided by 10^6 by a series of six decade counters.

7-15.2 Frequency Measurement

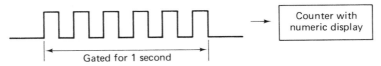

Figure 7-15b Frequency meter

Incoming pulses are counted for exactly 1 second(s) (or decimal multiple or submultiple). The number of counted pulses is the frequency.

7-15.3 Period Measurement

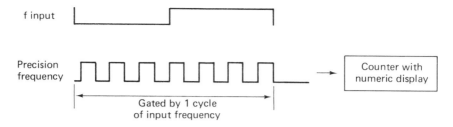

Figure 7-15c   Period counter

A precision stable frequency source is used to generate a series of accurately timed pulses. These pulses are gated by one cycle of the input frequency and counted. For example, if the precision frequency is 1 MHz, the resultant count gives the period of the wave in microseconds.

7-15.4 Television Horizontal-to-Vertical Generator Timing Pulses

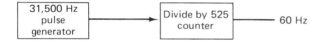

Figure 7-15d   Divide-by-525 counter

At the TV station a divide-by-525 counter accurately maintains the required relationship between the 31.5-kHz† horizontal timing and the 60-Hz vertical timing pulses.

7-15.5 Digital Voltmeter

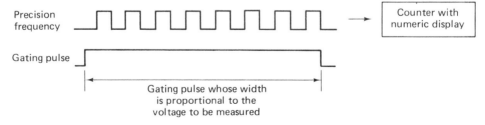

Figure 7-15e   Digital voltmeter

†Actual horizontal repetition rate is half this value (15.75 Hz).

193

A pulse is generated whose width is proportional to the analog voltage to be measured. This pulse is used to gate a pulse train, which is counted. The pulse count is displayed on a numeric display as the voltage.

7-15.6 Timing Source for Digital Wristwatch

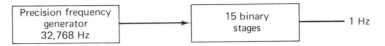

Figure 7-15f CMOS, 32,768-Hz frequency divider, 15 binary stages

A 15-stage binary counter divides a 32,768-Hz waveform down to 1-s pulses. This is used in a digital wristwatch to either drive a motor or to display the time through digital display devices.

Problems

7-1. The type 54/7493 is a four-stage binary ripple counter. A 40-kHz signal is applied to its input. Determine the frequency and period of the waveshapes at the outputs of each stage. What is the frequency division at each of the outputs?

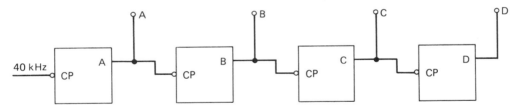

Figure P7-1 7493 Four-stage ripple counter

7-2. Sketch the waveshapes at outputs *A, B, C,* and *D* of Prob. 7-1. The clock input signal is a square wave, and the data are transferred to the FF output at the negative transition of the clock.

7-3. Repeat Prob. 7-2, but for outputs $\bar{A}, \bar{B}, \bar{C},$ and $\bar{D}$.

7-4. What is the count capability of the counter of Prob. 7-1? Express in binary and decimal.

7-5. Tabulate the count states at the *DCBA* terminals, starting from 0000, and give the decimal equivalents (Prob. 7-1).

7-6. Tabulate the count states at the $\bar{D}\bar{C}\bar{B}\bar{A}$ terminals, starting when *DCBA* = 0000, and give the decimal equivalents (Prob. 7-1).

7-7. Complete the following table for binary ripple counters.

Number of Stages	Frequency Division	Period Multiplication	Count Capability	
			Binary	Decimal
1	$\frac{1}{2}$	2	0–1	0–1
2	$\frac{1}{4}$	4	00–11	0–3
3				
4				
5				
6				
7				
8				

7-8. A *JK* FF has a propagation delay time of 10 ns to logic 1 from 0 level after the clock pulse, and a 16-ns delay time to logic 0 from the 1 level. A six-stage counter is in the 011111 count state. A clock pulse is applied. (a) What is the next count state? (b) Calculate the time delay before the final state is reached.

7-9. Repeat Prob. 7-8, but for the counter in the 010111 state.

7-10. Draw a logic diagram similar to that of Fig. 7-2 for a four-stage synchronous counter.

7-11. Repeat Prob. 7-10, but for a five-stage counter.

7-12. How many stages are required for a mod-12 counter? mod-23? mod-100?

7-13. Repeat Prob. 7-12, but for mod-31, mod-600, mod-6000.

7-14. Draw a logic diagram and show the feedback connections required to build a mod-11 counter similar to that of Fig. 7-3.

7-15. Repeat Prob. 7-14, but for a mod-21 counter.

7-16. Repeat Prob. 7-14, but for a mod-100 counter.

7-17. The following waveforms are viewed at outputs Q_A, Q_B, and Q_C of a three-stage counter. Determine the modulus of the counter and tabulate its count states.

Figure P7-17   Three-stage counter waveforms

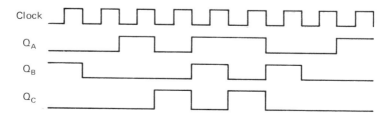

7-18. The following waveforms are viewed at the outputs Q_D, Q_C, Q_B, and Q_A of a four-stage counter. Determine the modulus of the counter and tabulate its count states.

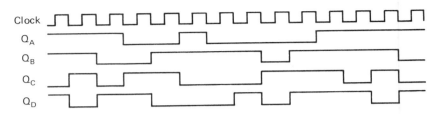

Figure P7-18 Four-stage counter waveforms

7-19. Figure 7-7a has omitted the Q_B waveshape. Draw it and tabulate the count states of the mod-6 counter.

7-20. Figure 7-7b has omitted the Q_A waveshape. Draw it and tabulate the count states of the mod-6 counter.

7-21. Following the procedure of Sec. 7-9, show that the unused state 1100 of the 8421 counter of Fig. 7-9 goes to the unused state 1101 upon the next count pulse. Show that the next pulse will bring it into an allowed state. What is the allowed state?

7-22. Draw the waveshapes for the 2'421 counter of Fig. 7-11.

7-23. Explain why the counter of Fig. 7-11 goes from the 0011 state to the 1010 state when the next pulse is applied. (*Hint:* tabulate all inputs to the FFs and apply *JK* and *RS* truth tables.)

7-24. Explain why the counter of Fig. 7-11 goes from the 1101 to the 1110 state when the next pulse is applied.

7-25. Explain why the IC mod-5 counter fails to count when it is in the unused state 111.

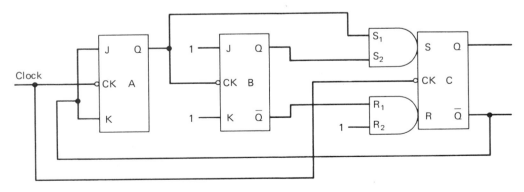

Figure P7-25 IC mod-5 counter

7-26. What are the frequencies and periods at U, V, W, and X?

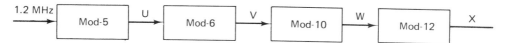

7-27. You have available type 7490 counters (mod-2, mod-4, mod-5, mod-10), type 7492 counters (mod-2, mod-3, mod-6, mod-12), and type 7493 counters (mod-2, mod-4, mod-8, mod-16). Show by means of block diagrams which counters you would use to divide a frequency by a factor of 9600.

7-28. Repeat Prob. 7-27, but for a factor of 72,000.

7-29. You have an oscillator operating at a frequency of 1.728 MHz. You wish to obtain a frequency of 1 Hz. Show how to use the counters of Prob. 7-27 to do this.

7-30. Show that the following counter will count up 00, 01, 10, 11 if $X = 1$, and count down 11, 10, 01, 00 if $X = 0$. This illustrates the technique of up–down counting.

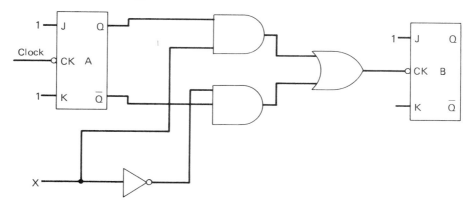

7-31. The TV horizontal-to-vertical generator of Fig. 7-15d converts a frequency of 31,500 Hz to 60 Hz, a division by 525. How many FFs are required to do this?

7-32. It is desired to obtain the TV frequency division of Prob. 7-31 by separately cascaded counters, with the limit of frequency division by each counter a maximum of 10. How many counters are required, and how many stages in each counter? Draw a block diagram showing the frequency division per counter and the frequency at each point in the counter chain. (*Hint:* determine the factors of 525.)

EIGHT

Shift Registers and Shift Counters[†]

8-1 INTRODUCTION

Consider the following problem in multiplication:

$$101 \times 111$$

$$
\begin{array}{r r}
101 & \text{(a)} \\
\times\ 111 & \\
\hline
101 & \text{(b)} \\
101 & \text{(c)} \\
\hline
1111 & \text{(d)} \\
101 & \text{(e)} \\
\hline
100011 & \text{(f)}
\end{array}
$$

In performing the multiplication,

1 The multiplicand (a) is stored in memory (b).
2 The multiplicand (a) is shifted one position to the left and stored in memory (c).

[†]M. E. Levine, *Digital Theory and Experimentation Using Integrated Circuits*, Englewood Cliffs, N.J.: Prentice-Hall, Inc., 1974, Expt. 10, Shift Registers and Ring Counters.

3 Then (b) and (c) are added to give a partial result (d).
4 The multiplicand (a) is shifted one more bit position to the left and stored in memory (e).
5 Then (d) and (e) are added to give the final result (f).

From this example we can see that to perform multiplication we need a memory in which data can be shifted one or more bit positions.

In computer terminology a memory is called a *register*, and a register in which the data can be shifted one bit position upon a command called a *shift pulse* is known as a *shift register*.

The shift register has important applications in

1 Shifting stored data bit positions, as shown in the previous example.
2 Data transmission:
    a Conversion of parallel data to serial data.
    b Conversion of serial data to parallel data.
    c Changing the rate of transmission of serial data (buffering).
    d Temporary data storage.
3 Delay of data by controllable times.
4 In a circulating shift register, to provide means for storage and recirculation of data. This is the electronic equivalent of the magnetic drum or disc.
5 Cathode-ray tube refresher memory for digital displays.
6 Pulse counter.
7 Act as a stack or push-down memory.

8-2 BASIC PRINCIPLES OF THE SHIFT REGISTER

The basic principle of operation of a shift register is shown in Fig. 8-1. Two *JK* or *RS* or *D* FFs are connected in cascade, with Q_A connected to $J_B(S_B)$ (D), and $\bar{Q}_A$ connected to $K_B(R_B)$. The levels that stage *B* assumes after a shift pulse depend upon the directed or steered properties of these FFs, which are given in Tables 8-1a, 8-1b, and 8-1c.

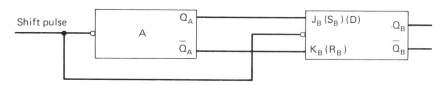

Figure 8-1   Basic shift register

From these truth tables we can see that, no matter what the previous state, after the shift pulse Q_B will follow Q_A, and $\bar{Q}_B$ will follow $\bar{Q}_A$, or a shift in data one position to the right will occur. We could just as easily have connected Q_B to J_A

Table 8-1

(a) Clocked *RS* FF	(b) Clocked *JK* FF	(c) Clocked *D* FF

		Before Clock Pulse		After Clock Pulse	
S	R	Q	$\bar{Q}$	Q	$\bar{Q}$
1	0	1	0	1	0
1	0	0	1	1	0
0	1	1	0	0	1
0	1	0	1	0	1

		Before Clock Pulse		After Clock Pulse	
J	K	Q	$\bar{Q}$	Q	$\bar{Q}$
1	0	1	0	1	0
1	0	0	1	1	0
0	1	1	0	0	1
0	1	0	1	0	1

	Before Clock Pulse		After Clock Pulse	
D	Q	$\bar{Q}$	Q	$\bar{Q}$
1	1	0	1	0
1	0	1	1	0
0	1	0	0	1
0	0	1	0	1

and $\bar{Q}_B$ to K_A. The shift pulse now would have made *A* follow *B* or a shift in data to the left. It is quite easy to arrange the Q to J and $\bar{Q}$ to K connections so that the connections are interchanged electronically and data are shifted either to the right or left as desired. In addition, the circuit of Fig. 8-1 can be expanded to include many FFs to form a multibit shift register.

8-3 SHIFT REGISTER FUNCTIONS AND APPLICATIONS

Figure 8-2 is the diagram of a 5-bit shift register that will shift data to the right. Each FF has a set (S) and a clear (CLR) input. In this illustration all the clear inputs are connected together.

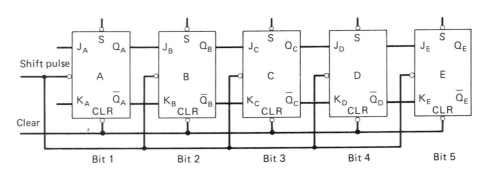

Figure 8-2   Five-bit shift-right shift register

In this function, data are converted from parallel data (available at the same time initially) to information in serial time form for transmission to a different location on a transmission line.

1 Connect J_A to level 0 and K_A to level 1. This will force FF A to take on a level equal to 0 after each shift pulse.
2 Clear all the FFs. This makes all the FFs have a level of 0.
3 Using the S inputs make the levels of FF A and FF C equal to 1. This makes the parallel data reading from left to right 10100 at pulse 0.
4 Now apply shift pulses to the register.

In Table 8-2 we show the states of the FFs as shift pulses are applied, and we can see the following:

Table 8-2 Parallel in–serial out

Shift Pulse	Q_A Bit 1	Q_B Bit 2	Q_C Bit 3	Q_D Bit 4	Q_E Bit 5
0	1	0	1	0	0
1	0	1	0	1	0
2	0	0	1	0	1
3	0	0	0	1	0
4	0	0	0	0	1

1 The data are shifted to the right with each shift pulse. 0s appear at the input since $J_A = 0$ and $K_A = 1$.
2 Compare the data at bit 5 reading down against the data at bit 0 reading right to left. They are the same.
3 Since the data at bit 5 appear there after a shift pulse or in a time sequence, bit 5 data are serial data in correct sequence.
4 The shift register has converted parallel data to serial data.

The data at bit 5 can now be sent down a single transmission line in time sequence rather than being sent from all 5 bits simultaneously down five transmission lines. Another application is parallel data conversion to serial form for use in a serial adder.

8-3.2 Serial In–Parallel Out

In this function data arrive at the serial inputs of the register, and after a number of shift pulses equal to the number of bits in the data stream, the data are shifted into the register and are available at the Q outputs in parallel form.

Table 8-3 Serial data to parallel data conversion

Shift Pulse	Serial Data	J_A	K_A	Q_A	Q_B	Q_C	Q_D	Q_E
0	—	—	—	0	0	0	0	0
1	1	1	0	1	0	0	0	0
2	0	0	1	0	1	0	0	0
3	1	1	0	1	0	1	0	0
4	1	1	0	1	1	0	1	0
5	0	0	1	0	1	1	0	1

Referring again to Fig. 8-2, the input as double-rail data (the input data is on two lines, a real input and an inverted input) is applied to the serial data inputs J_A and K_A and shifted into the register as shown in Table 8-3, where it can be seen that the data are appearing in time serial form as shown in the data column, and are shifted into the first stage at each shift pulse. Once in the shift register, data shift to the right bit by bit, and, finally, are available in parallel form. This can be seen if the data serial column reading down is compared with the Q outputs reading right to left (Q_E corresponds to the input data at shift pulse 1).

8-3.3 Buffering

In this function, serial data are shifted at one rate into the register. The data are then shifted out of the register at a different rate by shift pulses applied at a rate different from that used to shift data into the register. This is a serial-to-serial operation and enables one to change the rate at which data are being transmitted.

8-3.4 Delay

Since it takes one shift pulse to move data 1 bit, data arrive at the next bit position delayed by one pulse. By selecting the number of bits in the shift register and by controlling the shift pulse rate, desired delays can be imparted to the data stream.

EXAMPLE 8-1 Data are entering a 16-bit shift register. The shift pulse rate is 20 Kbit per second. By how much are the data delayed by the shift register?

Solution

$$T = \frac{1}{\text{PRR}} = \text{time of one shift pulse}$$

$$T = \frac{1}{20,000 \text{ pps}} = \frac{1}{20,000} \text{ s} = 50\mu s = \text{delay time per bit}$$

$$\text{Total time} = 16 \times 50\mu s = 800\mu s$$

8-3.5 Circulating Shift Register

Figure 8-3 shows the basic diagram of a circulating shift register. In a circulating shift register, Q of the final stage is connected to J of the first stage, and $\bar{Q}$ of the final stage is connected to K of the first stage. This makes any data appearing at stage N return to the initial stage and continue around the register. Figure 8-3 shows that the inputs to stage A have dual two-wide AND–OR gate arrangements, which serve as double-pole, double-throw switches. The recirculate/$\overline{\text{data}}$ operates in the following manner. To make a circulating shift register, this line has to be at logic 1, and to get serial data into the register this line must be at logic 0. For shift registers with a few bits and enough available pins, it is possible to have parallel data entry, but in multibit registers shift serial data is the only way to get data into the register.

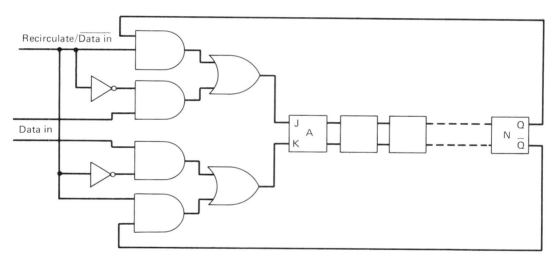

Figure 8-3 Circulating shift register

8-4 RING COUNTER

8-4.1 Decade Ring Counter

Consider a 10-stage circulating shift register, but with the limitation that one and only one stage is at logic 1 and all other stages are at logic 0. Each shift pulse moves the 1 one bit position. (An alternative version is to have one stage at a 0 and all other stages at a 1. This counter will move the 0.) If we begin with the 1 in the first stage and call this stage bit 0, each shift pulse shifts the 1 around the register until the 1 returns to bit 0 after 10 pulses. If the shift pulses are now renamed "count pulses," and we number each stage successively from 0 to 9, we have a decade counter. This counter is synchronous and has the property that one and only one stage has a 1 and all other stages are at 0. Hence each count state is automatically and simply defined; the counter is self-decoded. Figure 8-4 shows a diagram of a decade ring counter and its waveforms. Table 8-4 gives the count states for each of the counts. This is a weighted counter as is shown in this table. As seen in the waveforms of Fig. 8-4b, one pulse is developed at each output for every 10 input pulses. Hence this counter is a divide-by-10 counter.

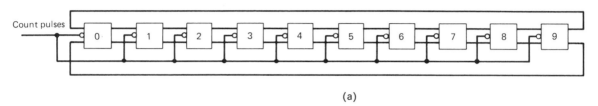

(a)

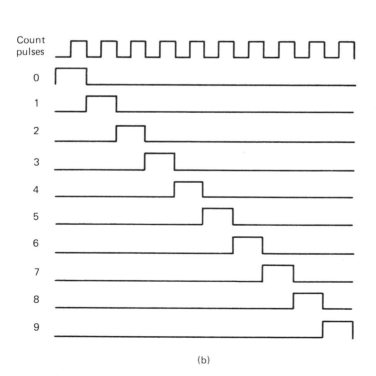

(b)

Figure 8-4   Decade ring counter: (a) block diagram; (b) waveforms

Table 8-4   Decade ring counter

Count	Count State
	Weight 9876543210
0	0000000001
1	0000000010
2	0000000100
3	0000001000
4	0000010000
5	0000100000
6	0001000000
7	0010000000
8	0100000000
9	1000000000

8-4.2 Quinary Ring Counter

The ring of five contains five stages. It divides by 5 and is a weighted counter similar to the decade ring. It is frequently called the 43210 counter.

8-4.3 Biquinary Ring Counter

Figure 8-5 shows a five-stage ring counter followed by a *T* flip-flop. As can be seen from the waveforms, the 1 progresses around the ring and toggles the *T* FF as it leaves stage 4. This is also a decade counter, dividing in the ring by a factor of 5

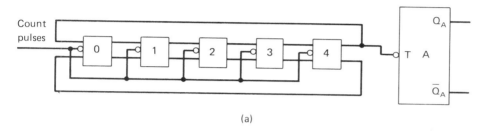

(a)

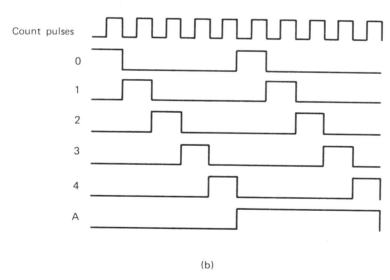

(b)

Figure 8-5 Biquinary ring counter: (a) block diagram; (b) waveforms

and then in the *T* FF by a factor of 2. Table 8-5 gives the count states for this counter. There are two ways of writing the count states. As can be seen, this counter is a weighted counter. Based upon the count-state representation, this counter is sometimes called the 5043210 or 543210 counter. Since the count states are not as simply defined as in the decade ring counter, it requires decoding gates to separate and define the counts.

Table 8-5 Biquinary ring counter count states

Count	7-Bit Weight	6-Bit Weight
	5043210	543210
0	0100001	000001
1	0100010	000010
2	0100100	000100
3	0101000	001000
4	0110000	010000
5	1000001	100001
6	1000010	100010
7	1000100	100100
8	1001000	101000
9	1010000	110000

8-4.4 Quibinary Ring Counter

The quibinary ring counter is like the biquinary except that it starts with a T FF and is then followed by a ring of five. Its diagram is given in Figure 8-6 and its count states in Table 8-6.

8-4.5 Unused Ring-Counter States

Although ring-counter operation requires that only one stage be in a state different from all other stages, it is quite possible for more than one stage to get into the desired state either because of noise or during startup. Provisions have to be made to sense such an unwanted state and force the counter into the wanted state if the counter is to operate successfully.

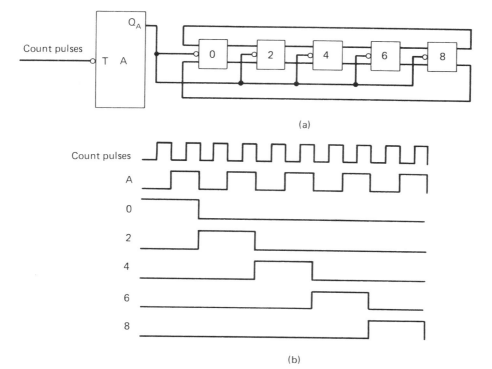

(a)

(b)

Figure 8-6 Quibinary ring counter: (a) block diagram; (b) waveforms

Table 8-6 Quibinary ring counter count states

Count	7-Bit Weight	6-Bit Weight
	1086420	186420
0	0100001	000001
1	1000001	100001
2	0100010	000010
3	1000010	100010
4	0100100	000100
5	1000100	100100
6	0101000	001000
7	1001000	101000
8	0110000	010000
9	1010000	110000

8-5 TWISTED-RING OR JOHNSON COUNTER

Figure 8-7 shows the diagram of a four-stage counter that differs from those of the previous section in one important feature. As can be seen, the connections from the output are interchanged or twisted. This causes stage A to go to a 1 whenever stage D has a 0, and stage A to go to a 0 whenever stage D has a 1. This type of counter, which is very popular with CMOS systems, is known by many names, such as the TWISTED-RING, JOHNSON, or RING-TAIL counter.

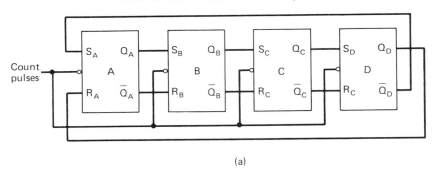

(a)

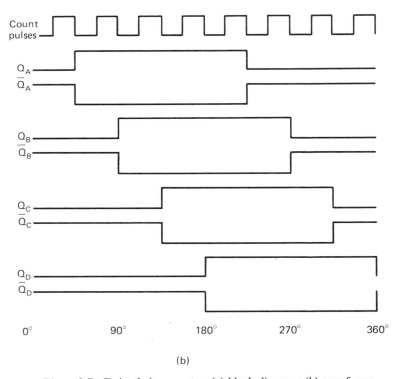

(b)

Figure 8-7 Twisted-ring counter: (a) block diagram; (b) waveforms

Table 8-7 Twisted ring counter

Count	Count State ABCD
0	0000
1	1000
2	1100
3	1110
4	1111
5	0111
6	0011
7	0001

Table 8-7 shows the count-state progression. The counter starts with a count state of 0000, fills up to all 1s, and then empties out to all 0s. This four-stage counter has eight count states and therefore divides an incoming frequency by 8. In general, a twisted-ring counter of N stages has $2N$ count states and divides an incoming frequency by $2N$. Although the count-state progression appears to be somewhat complex, separation of the individual counts (decoding) turns out to be quite simple. All TWISTED-RING counters can be decoded by two-input AND gates (Prob. 10-21).

The TWISTED-RING counter is being used to a great extent in CMOS ICs. The following are typical:

1 Type 4017: five stages. Decimal counter with 10 decoded decimal outputs
2 Type 4018: five stages. Jam (preset) inputs are brought out so that it can be made to divide by 10, 9, 8, 7, 6, 5, 4, 3, or 2.
3 Type 4022: like type 4017, but has four stages.

Let us take another look at the waveforms of Fig. 8-7b. At the bottom of the figure, angles are marked at every 90° for one cycle of the waveforms. The waveform at Q_D starts at 0°, is high for 180°, and low from 180 to 360°; Q_B is like Q_D but displaced by 90°. Q_D is displaced from Q_B again by 90°, and $\bar{Q}_B$ is displaced by another 90°. We can see that this twisted-ring counter can be used to generate a set of four-phase waveforms. This is true in general for the waveforms of the twisted-ring counter. As another example, a three-stage twisted-ring counter will divide by $6 = 2 \times 3$, and can be used to generate a set of three-phase waveforms.

8-6 TECHNOLOGY OF SHIFT REGISTERS

Shift registers having a small number of stages are available using bipolar technology in the standard logic families. Their use is limited by the number of pins available. Figure 8-8 shows the data sheet for type 10141, a 4-bit universal shift register in the ECL family. Serial entry inputs are available for left entry or right entry in order to shift data left or right and to perform serial-to-parallel conversion. Parallel data inputs (set inputs) permit parallel-to-serial data conversion.

Suppose that we are faced with the following problem. We want to store and recirculate 1000 bits for a CRT refresh memory at a rate of several megahertz and the register is to be supplied in a TO-99 case (0.32-in. diameter). The very fast rate requires that this be done electronically. The TO-99 case requires that each bit be very small and in addition dissipate very little power. A solution to this is to use a two-phase monolithic dynamic shift register with P-channel enhancement-mode field-effect transistors.

A primary advantage of MOS transistors is that they can be made on an area of the order of 1 mil², as compared to bipolar transistors, which require approximately 50 mil², a reduction of 50 to 1 in area required. MOS transistors require

Signetics

**ADVANCE INFORMATION
TO BE ANNOUNCED**

10141F: −30 to +85°C, CERDIP

DIGITAL 10,000 SERIES ECL

DESCRIPTION

The 10141 is a four bit universal shift register. The register performs shift left or right, serial/parallel in and serial/parallel out with no external gating. This device is useful for counting, temporary storage, and shifting in high speed digital communication systems, instrumentation, peripheral controllers and computers.

Inputs S1 and S2 control the four possible operations of the register without interfering with the clock. The flip-flops shift information on the positive edge of the clock. The four operations are: stop shift, shift left, shift right, and parallel entry of data. The other six inputs are all data type inputs: four for parallel data entry, one for shifting in from the left (DL), and one for shifting in from the right (DR). When the register is used for serial output only, the unused emitter-follower outputs can be left open.

The 10141 is capable of 200 MHz shift rate operation (typical).

BLOCK DIAGRAM

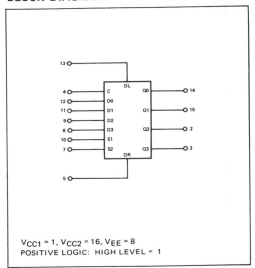

$V_{CC1} = 1$, $V_{CC2} = 16$, $V_{EE} = 8$
POSITIVE LOGIC: HIGH LEVEL = 1

TRUTH TABLE

PULSE	S1	S2	D3	D2	D1	D0	DR	DL	Q3	Q2	Q1	Q0
0	L	L	L	H	H	L	X	X	−	−	−	−
1	L	L	L	H	H	L	X	X	L	H	H	L
2	L	L	H	L	L	H	X	X	H	L	L	H
3	L	L	H	H	L	L	X	X	H	H	L	L
4	L	H	X	X	X	X	L	X	L	H	H	L
5	L	H	X	X	X	X	H	X	H	L	H	H
6	L	H	X	X	X	X	L	X	L	H	L	H
7	L	H	X	X	X	X	L	X	L	L	H	L
8	H	L	X	X	X	X	X	L	L	H	L	L
9	H	L	X	X	X	X	X	H	H	L	L	H
10	H	L	X	X	X	X	X	H	L	L	H	H
11	H	L	X	X	X	X	X	L	L	H	H	L
12	H	L	X	X	X	X	X	L	H	H	L	L
13	H	H	X	X	X	X	X	X	H	H	L	L
14	H	H	X	X	X	X	X	X	H	H	L	L

FUNCTION TABLE

FUNCTION TABLE		
SELECT		
S1	S2	OPERATING MODE
L	L	Parallel Entry
L	H	Shift Right
H	L	Shift Left
H	H	Stop Shift

*Outputs as exist after pulse appears at "C" input with input conditions as shown.

(Pulse = Positive transition of clock input)

X = Don't Care

TEMPERATURE RANGE

● −30 to +85°C Operating Ambient

PACKAGE TYPE

● F: 16-Pin CERDIP

Figure 8-8 Type 10141 universal 4-bit shift register. (Courtesy Signetics Corp.)

one-third of the process steps needed for the standard bipolar ICs, so manufacturing is simpler. In addition, MOS dissipation is very low, much lower than for bipolar transistors. Hence there are many reasons for using MOS technology to make LSI circuits, especially if the circuits are repetitive.

There are two basic types of MOS shift registers, the dynamic and the static. Let us consider the dynamic first.

The basic cell for a two-phase *P*-channel MOS dynamic shift register is given in Fig. 8-9. Two-phase voltage signals ϕ_1 and ϕ_2 are needed. Q_1 and Q_2 form an inverter that is activated when ϕ_1 goes negative (*P*-channel MOS). This also activates Q_3, charging C_1 to a logic level that is the invert of the input. ϕ_2, at a time later, repeats this for Q_4, Q_5, and Q_6, and C_2 is now charged to the level of the input. A cell therefore consists of a six-transistor unit that forms two inverters and two transmission gates Q_3 and Q_6. C_1 and C_2 are the parasitic capacitors of Q_3 and Q_6 on the IC. These capacitances are in parallel with reverse-biased junctions, which have some leakage. Therefore, we cannot wait too long or the charge on the capacitors will leak off and the data be lost. Hence there is a minimum rate (approximately 10 kHz) at which the shift register will operate. Other types of dynamic shift registers have been developed. Some do not require $-V_{DD}$ and some require four-phase clocks. These either dissipate less power or require less area. Most dynamic shift registers follow the method of Fig. 8-9 and operate between 10 kHz and 5 MHz.

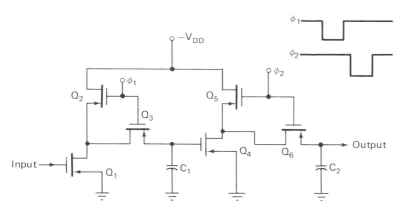

Figure 8-9   Two-phase dynamic shift register

The static shift register operates at low frequencies like the basic shift register shown in Fig. 8-1, and at high frequencies like the dynamic shift register of Fig. 8-9. The basic circuit is shown in Fig. 8-10. The inverters in Fig. 8-10 are the Q_2–Q_1 inverters of Fig. 8-9. Three phase signals are needed. At high frequencies, phases ϕ_1 and ϕ_2 complete paths through transistors *D* and *E* to charge capacitors C_1 and C_2, and ϕ_3 is not generated. The operation is similar to that of the dynamic register of Fig. 8-9. At low frequencies below 10 kHz, ϕ_3 is generated, and a simple very

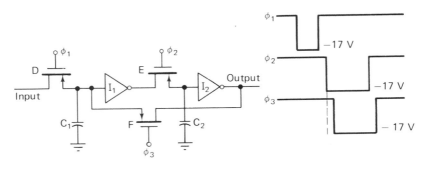

Figure 8-10 MOS static shift register

basic two inverter I_1I_2 bistable is completed through transistors E and F. ϕ_3 is generated on the chip and is developed when ϕ_1 is at a 0 and ϕ_2 is negative for more than 10 μs, thus completing the feedback path through transistor F. This type of shift register can operate from dc to 2.5 MHz. A typical static MOS shift register is the Signetics 2521. It contains two 128-bit shift registers with recirculate capability.

MOS shift registers are available in a large variety of bit lengths. Table 8-8 lists some of these. Some applications of MOS shift registers are† data handling, refresh memories, buffer memories, scratch-pad memories, delay line, desk-top calculators, display systems, computer peripherals, and radar systems.

Table 8-8 MOS shift registers

Bit Length	Style	Frequency Range (MHz)
1024	Dynamic	0.01–6
2–512	Dynamic	0.01–6
4–256	Dynamic	0.01–6
4–80	Dynamic	0.05–5
3–64	Dynamic	0.05–5
2–25	Static	0–1
2–50	Static	0–1
2–100	Static	0–2.5
2–128	Static	0–1
6–32	Static	0–1

8-7 *P*-CHANNEL MOS COMPATIBILITY WITH TTL AND *N*-CHANNEL MOS

TTL operates between $V_{CC} = +5$ V and 0. *P*-channel MOS operates with $V_{SS} = 0$ and V_{DD} and V_{GG} negative with voltages of the order of -17 V. This creates a compatibility problem. If we translate the V_{SS} voltage up to $+5$ V, this now allows the MOS transistors to function between $+5$ and -12 V. When the gate voltage

†MOS/LSI Standard Products Catalog, Texas Instruments, Inc.

is made $-12\ V$, this will turn on the MOS transistors, and the drain voltage now can swing between V_{CC} and ground to provide the logic swing needed for TTL. This is a generalized discussion of the interface between P-channel MOS and TTL. The actual details depend upon the TTL loading requirements and the TTL family being interfaced.

When MOS was first developed as a solution to large-scale integration, P-channel was the available technology. More recently, the more difficult technical problems of N-channel MOS have been solved, and N-channel shift registers have become available that are completely compatible with TTL. In addition to the prime advantage of compatibility, they take even less area than P-channel, require lower power, and have a lower drive input capacity for the clock input.

8-8 SHIFT COUNTERS

A shift counter is a modified ring counter or Johnson counter in which the feedback from the last stage back to the initial stage is modified by the logic levels of some of the intermediate stages. In series counters the count states in general sequence upward. To modify this progression makes the counter quite complex. When a nonincreasing sequence is required, it frequently can be accomplished in a simpler manner using a shift counter.

Consider the shift counter of Fig. 8-11. This might be a Johnson counter using D flip-flops with $\bar{Q}_C$ fed back to D_A. However, the feedback path is modified so that D_A is controlled by $\bar{Q}_C$ and $\bar{Q}_B$. Table 8-9 shows the count progression, from which we can see that the counter of Fig. 8-11 is a synchronous Mod-5 shift counter.

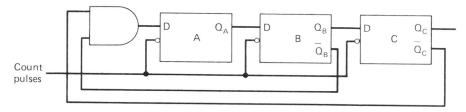

Count pulses

Figure 8-11   Shift counter

Table 8-9   Mod-5 shift counter

Count	Q_A	Q_B	Q_C	$\bar{Q}_B$	$\bar{Q}_C$	D_A
0	0	0	0	1	1	1
1	1	0	0	1	1	1
2	1	1	0	1	0	0
3	0	1	1	0	0	0
4	0	0	1	0	1	0

Problems

8-1. Define a shift register.

8-2. Draw a diagram of a five-stage shift register using D flip-flops.

8-3. Draw a diagram of a four-stage shift register with provision for parallel data entry and parallel output capability.

8-4. A four-stage shift register has the following initial conditions: $Q_D = 1$, $Q_C = 1$, $Q_B = 0$, $Q_A = 1$. Draw a table showing the logic levels as the data are shifted from Q_D toward Q_A.

8-5. Repeat Prob. 8-4 with data shifted from Q_A toward Q_D.

8-6. Define (a) parallel-to-serial data conversion, and (b) serial-to-parallel data conversion.

8-7. (a) It is desired to multiply a binary number by 16_{10}. How many shift pulses are needed to do this?
(b) Is this a shift left or shift right register?
(c) If the clock rate is 50 kHz, how long will it take to do this?

8-8. Repeat Prob. 8-7, but for division by 32_{10}.

8-9. How long will it take to fully load serially an 80-bit shift register if the clock rate is 2 MHz?

8-10. A 512-bit shift register is being used as a delay line. If the clock rate is 4 MHz, by how long are the data delayed?

8-11. Repeat Prob. 8-9, but with a clock rate of 10 kHz.

8-12. How does a circulating shift register differ from a ring counter?

8-13. Draw the diagram of a three-stage shift-right shift-left shift register.

8-14. (a) Draw a diagram of a three-stage JOHNSON counter.
(b) Give the count states for each count. The initial state is 000.
(c) If the clock frequency is 720 Hz, what frequency will appear at the Q terminals of each stage?
(d) Draw the Q and $\bar{Q}$ waveforms for each stage. Show an angular scale for a complete cycle of output voltage.
(e) Select the terminals needed to give three-phase outputs.

8-15. A 6000-Hz signal is applied to a five-stage ring counter. What is the frequency division of this counter?
(b) Repeat for a six-stage ring counter.

8-16. Why is there a minimum clock rate for a dynamic shift register?

8-17. Determine the count states of the shift counter and its modulus. Begin with $Q_A Q_B = 00$.

Figure P8-17 Shift counter

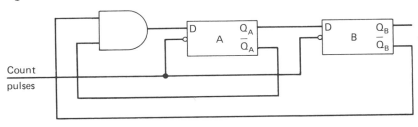

8-18. Determine the count states of the shift counter and its modulus. Begin with $Q_A Q_B Q_C = 000$.

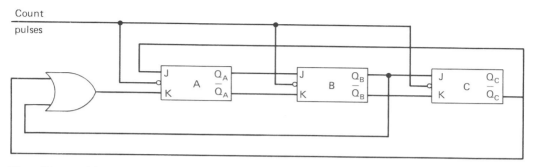

Figure P8-18 Shift counter

8-19. (a) Determine the count state of the shift counter and its modulus. Initial state is $Q_A Q_B Q_C = 100$.

(b) What happens if the counter should get into the 000 state?

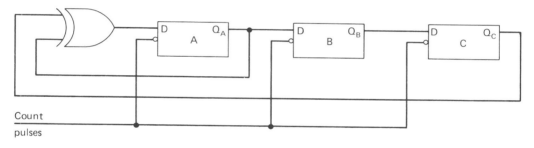

Figure P8-19 Shift counter

NINE

Pulse Generation, Pulse Shaping, and Schmitt Trigger

9-1 INTRODUCTION

In previous chapters we have made use of clocks. We have seen that a clock is a waveform of required shape and duration, which we, at this point, might rename a "pulse." In the gated flip-flop, a single pulse permits the transfer of data from the input of the FF to the output of the FF, where it is stored or placed in memory. In counters, reoccurring pulses represent events that are being counted. In the shift register, the clock or shift pulses determine the time when data are shifted one bit position in the register. Pulses of required shape, duration, timing, and repetition rate have to be generated. This chapter concerns itself with methods of generating and shaping clock pulses.

9-2 RC CHARGING NETWORKS

Since many pulse generators act to alternately charge and discharge capacitors through resistors, it is worthwhile discussing this charging and discharging of capacitors within the framework of the circuits we shall be studying.

Consider Fig. 9-1a with switch SW closed. The capacitor is across the battery

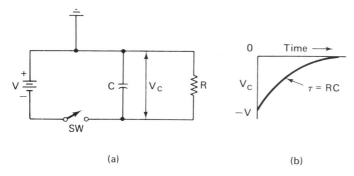

(a)   (b)

Figure 9-1 *RC* discharge: (a) *RC* circuit; (b) V_C versus time

and is charged to a voltage $= -V$. If the switch is opened at a time $t = 0$, the capacitor discharges through R, aiming toward zero exponentially with a voltage V_C given by

$$V_C = -V\epsilon^{-t/RC} = -V\epsilon^{-t/\tau} \tag{9-1}$$

where ϵ (2.72) is the base of natural logarithms. The discharge curve is shown in Fig. 9-1b. The term RC is the time constant τ of the circuit and is given in seconds if R is in ohms and C is in farads. As can be seen in Fig. 9-1b, the time it takes to get to 0 V is infinite. (For practical purposes, however, it takes about 5 time constants.) The time it takes to reach -0.37 V is given by one time constant and is equal to RC in seconds.

Figure 9-2 Universal *RC* charge—discharge curves

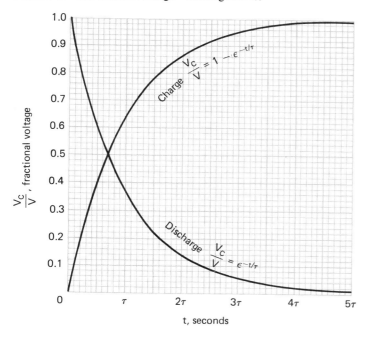

Figure 9-2 gives universal curves for the charging and discharging of capacitors through resistors. These curves plot the fractional voltage reached against the number of time constants it takes to reach this voltage. For example, in the charging curve, it takes 0.7τ to reach $0.5\,(V_C/V)$ and τ to reach $0.63\,(V_C/V)$. In the discharging curve it takes 0.7τ to reach $0.5\,(V_C/V)$ and τ to reach $0.37\,(V_C/V)$.

EXAMPLE 9-1   In the circuit of Fig. 9-1a, $V = -20$ V, $R = 500{,}000\ \Omega$, and $C = 0.1\ \mu$F.
   a. After the switch is opened, how long will it take the capacitor voltage to go from -20 to -10 V?
   b. to -5 V?

Solution

$$\tau = 500{,}000\ \Omega \times 0.1 \times 10^{-6}\ \text{F} = 0.05\ \text{s}$$

   a. This is a discharge circuit: $-10/-20 = 0.5$. From the discharge curve it takes 0.7τ to reach $0.5\,(V_C/V)$.

$$0.7\tau = 0.7(0.05) = 0.035\ \text{s}$$

   b. $-5/-20 = 0.25$. From the discharge curve it takes 1.4τ to reach $0.25\,(V_C/V)$.

$$1.4\tau = 1.4(0.05) = 0.07\ \text{s}$$

In Fig. 9-3a, the circuit of Fig. 9-1a has been modified so that the resistor is now connected to a positive voltage $+V$ equal to the negative voltage $-V$ of Fig. 9-1a, rather than to ground. What happens when the switch SW is now opened? It is apparent that if we wait long enough the capacitor will be charged through R, so that its final voltage at point A must be equal to $+V$. When the switch is opened, point A initially must be at $-V$. There is only R and C in the circuit, and these must determine the time constant τ. Hence the voltage V_C across C must start at $-V$ and go positive exponentially, aiming toward $+V$ as shown in Fig. 9-3b. While it still takes an infinitely long time for V_C to get up to $+V$, the time it takes the curve of Fig. 9-3b to reach 0 V is of great importance.

Figure 9-3   *RC discharge and charge: (a) circuit; (b) discharge—charge curve*

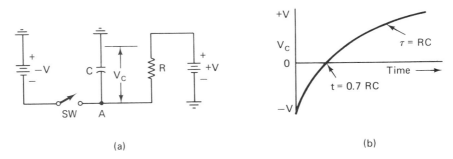

(a)   (b)

In the circuit of Fig. 9-3, the $+V$ voltage assists in the discharge from the $-V$ voltage condition, and then charges the capacitor toward its final $+V$ condition.

We can determine how many time constants are required to reach 0 V from the curves of Fig. 9-2. If we study the circuit of Fig. 9-3a, we realize that this is really the same type of discharging circuit as in Fig. 9-1a, but the *total* voltage is now equal to 2 V, and 0 V is *halfway* between $-V$ and $+V$. Figure 9-3b is the same as Fig. 9-1b, except that the total voltage equals 2 V and the curve has been displaced upward by $+V$. Hence the time to reach 0 V, the 50 percent [$0.5 (V_C/V)$] value, as obtained from Fig. 9-2, is equal to 0.7τ.

What if $+V$ is not equal to $-V$? How will this affect the time it takes to reach zero voltage? Figure 9-4 shows the effect. In this case, the positive voltage will not be able to charge the capacitor as fast, and it will take longer to reach the zero voltage level.

Figure 9-4   Effect of magnitude of $+V$ on discharge curves

EXAMPLE 9-2   Let $-V = -40$ V and $+V = +10$ V. How long will it take to reach 0 V? $R = 500$ kΩ, $C = 0.1$ μF.

Solution   The total voltage is 50 V. Zero volts is $\frac{10}{50} = 0.2$ on the discharge curve. From Fig. 9-2 we see that it takes 1.6 time constants.

$$T = 1.6(0.05) = 0.08 \text{ s}$$

Note: Alternatively, one can view this as a charging curve in which we reach 40 V/50 V $= 0.8$ (V_C/V). Again this gives 1.6 time constants from the charge curve.

9-3   ASTABLE MULTIVIBRATOR†

9-3.1   Introduction

Figure 9-5 is the diagram of a two-transistor pulse-generating circuit called an astable multivibrator. As we shall shortly see, it acts to alternately charge and discharge capacitors C_1 and C_2 in the manner of the circuit of Fig. 9-3. If we compare this to the circuit of Fig. 6-3, the transistor bistable, we see that, whereas

†M. E. Levine, *Digital Theory and Experimentation Using Integrated Circuits*. Englewood Cliffs, N.J.: Prentice-Hall, Inc., 1974, Expt. 11, Pulse Forming.

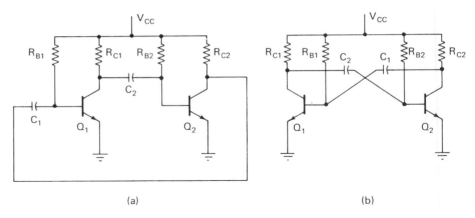

(a)   (b)

Figure 9-5   Transistor astable multivibrator: (a) series representation; (b) conventional representation

the bistable has dc coupling from collector to base, the astable is ac coupled by means of the capacitors C_1 and C_2.

In the bistable circuit, components are chosen so that one transistor is in saturation and the other cut off and the condition is *stable*. There are two *stable* states, with the saturated transistor and cutoff transistor interchanging operating conditions under an external forcing condition.

In the astable, the circuit components also are chosen so that the transistors are operated in saturation. The circuit again acts to keep one transistor in saturation and the other cut off, but the transistors interchange their operating conditions automatically and regularly at a repetition rate determined by resistors and capacitors *without need* for an external signal.

9-3.2   Circuit Operation

How does the circuit operate? To start, initially assume that Q_1 is in saturation, Q_2 is cut off, and the capacitor C_2 is charged so that there is a voltage $-V_{CC}$ across it, making the base of Q_2 negative. This is shown in Fig. 9-6. If Q_1 is in saturation, its V_{CE} is ≈ 0. The voltage across capacitor C_2 is of a polarity to make the base of Q_2 negative, so that Q_2 must be cut off. This makes I_{B2} and $I_{C2} = 0$. If we now compare Fig. 9-6 to Fig. 9-3a, we can see that they are the same at the time when SW in Fig. 9-3a was opened. One end of the capacitor goes to ground ($V_{CE\,sat}$); the other end is at a negative voltage $-V_{CC}$ and is returned to $+V_{CC}$ through resistor R_{B_2}. Therefore, the voltage at the base of Q_2 must aim toward $+V_{CC}$ as in Fig. 9-3b. This is shown in Fig. 9-6b.

At the same time, the cutoff bias at the base of Q_2 cuts Q_2 off, and $V_{CE}2 = V_{CC}$. Transistor Q_1 is in saturation, and its $V_{BE\,sat} \approx 0.7$ V ≈ 0 V. This is shown in Fig. 9-7. We can see that capacitor C_1 has a voltage across it equal to $V_{CC} - 0.7$V $\approx V_{CC}$, with the base side negative and collector side positive.

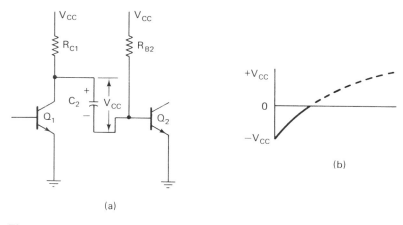

Figure 9-6   Initial conditions for astable multivibrator: (a) circuit operating conditions; (b) discharge—charge curve for C_2

Now, returning to Fig. 9-6, capacitor C_2 charges toward $+V_{CC}$, with a time constant $\tau = R_{B2}C_2$. The time constant is determined by the base resistor and capacitor connected to the *same* base. The charging voltage continues upward until the voltage at the base of Q_2 begins to conduct. A silicon transistor conducts when $V_{BE} = 0.7 \text{ V} \approx 0$. But we have seen from Fig. 9-3 that this time $= 0.7\tau$. Hence the time for the base voltage of Q_2 to go from $-V_{CC}$ to ≈ 0 is

$$T_2 = 0.7 R_{B2} C_2 \tag{9-2}$$

As soon as base current begins to flow in Q_2, collector current I_{c2} flows, and V_{CE2} drops from V_{CC}. This negative drop in voltage is coupled through capacitor C_1 to the base of Q_1 (Fig. 9-5) and is in a direction to cut off Q_1. This resultant decrease in collector current makes its collector voltage rise. This positive-going voltage is now coupled through C_2 and is in a direction to turn Q_2 on harder. It can be seen that the condition is regenerative. Once Q_2 starts to conduct, events

Figure 9-7   Conditions for capacitor C_1

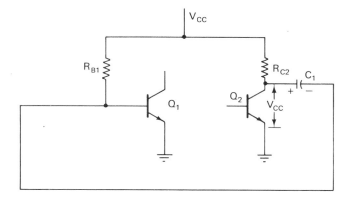

occur very rapidly, and transistor Q_2 is driven from cutoff into saturation very rapidly ($\approx$ 20 ns for small-signal transistors).

This sharp negative-going drop in voltage at the collector of Q_2 is coupled to the base of Q_1 through capacitor C_1 driving Q_1 quickly into cutoff. Now, what about the voltage across capacitor C_1? In Fig. 9-7 we saw that C_1 had a voltage across it equal to V_{CC}. To discharge any capacitor takes time and depends upon a time constant. However, the drop in voltage at the collector of Q_2 is so rapid that the charge on capacitor C_1 has no time to change. Hence its voltage must stay momentarily fixed at V_{CC}. Thus the voltage at the base side of Q_1 must be driven rapidly to $-V_{CC}$. But this is where we started, except that the two transistors have interchanged their operating conditions. Now capacitor C_1 discharges through R_{B1} in the same manner until V_{BE1} becomes approximate to 0. The time for the voltage at the base of Q_1 to go from $-V_{cc}$ to 0 V is now given by

$$T_1 = 0.7 R_{B1} C_1 \tag{9-3}$$

Note that this time is given by the base resistor and capacitor connected to the base of Q_1.

9-3.3 Frequency

The total time or period of a complete cycle for the base voltage of a transistor to return to its initial value is

$$T = T_1 + T_2 = 0.7(R_{B1}C_1 + R_{B2}C_2) \tag{9-4}$$

In many designs, $R_{B1} = R_{B2} = R_B$ and $C_1 = C_2 = C$. Hence T now becomes

$$T = 0.7(R_B C + R_B C)$$
$$= 1.4 R_B C \tag{9-5}$$

The frequency (f) or pulse repetition rate (PRR) is

$$f \text{ or PRR} = \frac{1}{T}$$

$$= \frac{1}{0.7(R_{B1}C_1 + R_{B2}C_2)} \tag{9-6a}$$

for unequal resistors and capacitors and

$$f = \frac{1}{1.4(R_B C)} = \frac{0.7}{R_B C} \tag{9-6b}$$

for equal resistors and capacitors.

EXAMPLE 9-3   In Fig. 9-5, $R_{B1} = 10$ kΩ, $R_{B2} = 10$ kΩ, $C_1 = 0.01$ μF, and $C_2 = 0.02$ μF. Find the pulse repetition rate (PRR).

Solution

$$T_1 = 0.7R_{B1}C_1 = 0.7(10,000)(0.01)(10^{-6}) = 0.7 \times 10^{-4} \text{ s}$$

This is the time Q_1 is cut off and Q_2 is in saturation.

$$T_2 = 0.7R_{B2}C_2 = 0.7(10,000)(0.02)(10^{-6}) = 1.4 \times 10^{-4} \text{ s}$$

This is the time Q_2 is cut off and Q_1 is in saturation.

$$T = 0.7 \times 10^{-4} + 1.4 \times 10^{-4} = 2.1 \times 10^{-4} \text{ s}$$

$$\text{PRR} = \frac{1}{T} = \frac{10^4}{2.1} = 4760 \text{ pulses per second (PPS)}$$

or

$$f = 4760 \text{ Hz}$$

EXAMPLE 9-4   In Example 9-3, find the pulse repetition rate if C_2 is changed to 0.01 μF.

Solution

$$R_{B1} = R_{B2} = R_B = 10,000 \ \Omega$$
$$C_1 = C_2 = C = 0.01 \ \mu\text{F}$$
$$T = 1.4R_BC = 1.4(10,000)(0.01)(10^{-6}) = 1.4 \times 10^{-4} \text{ s}$$
$$f = \frac{1}{T} = 7070 \text{ PPS}$$

9-3.4 Waveforms

Now let us consider the collector waveshape and return to Fig. 9-5. We began with Q_1 in saturation and Q_2 cut off. The voltage at the collector of Q_1 approximately equals 0. Capacitor C_2 has been discharging in a positive direction through R_{B2} until $V_{BE2} \approx 0$. At this time the following conditions exist:

1   Both sides of capacitor C_2 have a voltage approximate to 0, and the voltage across C_2 equals 0.
2   Q_2 conducts and V_{BE2} is clamped at approximately 0.
3   Q_1 cuts off, and its collector current becomes zero very quickly.

Under these conditions, Q_1 is an open circuit, and capacitor C_2 is connected to R_{C1}, which goes to V_{CC}. Hence capacitor C_2 charges toward V_{CC} with a collector time constant $\tau = R_{C1}C_2$, and the voltage at the collector rises toward V_{CC} with this time constant. Since $R_C \ll R_B$ ($R_B < h_{FE}R_C$ for good design), the collector time constant is much lower than the base time constant, and the collector voltage rises rapidly. Nevertheless, it does take time for the collector voltage to rise to V_{CC}. (Note that the collector time constant is given by the collector resistor and capacitor connected to its collector.)

When the collector turns on and goes into saturation, the time is short and is determined by the transistor's f_T. The transistor saturation resistance is very low. The collector rise time is much longer, and the resistance looking back toward the transistor is R_C since the transistor is cut off. This is high compared to $R_{CE\,sat}$ (transistor saturation resistance). Hence the useful edge is the turn-on edge, which is used for timing purposes. It is sharp and has low resistance.

Figure 9-8 shows the collector and base waveforms for both transistors.

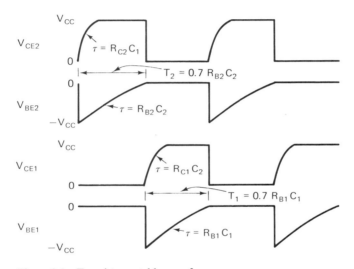

Figure 9-8   Transistor astable waveforms

9-4 FREQUENCY CHANGING OF THE TRANSISTOR ASTABLE

If we consider Eq. (9-4), we can see that to change the frequency we can change the base resistors R_B or capacitors C. There are limitations, however, to this. If we try to lower the frequency by increasing R_B, we have to be careful to maintain saturation. If we try to increase frequency by decreasing R_{B1}, we will put the transistor deeper into saturation and may get into storage-time problems, or the circuit may lock itself into a condition in which both transistors get into saturation simultaneously and not operate. In general, it is better to vary C once the base and collector resistors have been determined.

If we change just one capacitor, one collector now has a longer rise time constant, and this collector voltage may not be able to get up to V_{CC} before it is driven to saturation again. This upsets the accuracy of Eq. (9-4). A 10-to-1 unsymmetry in the capacitors is about as high as can satisfactorily be used.

Another approach is to modify Fig. 9-5 by removing the base resistors from their connection to $+V_{CC}$ and connecting them to a different positive voltage $+V_{BB}$. This then gives us the condition described in Fig. 9-5 for which we changed the time to get to 0 V. If $+V_{BB}$ is made variable, it is possible to change the frequency without changing the values of R_B and C. Such a variable-frequency oscillator is called a voltage-controlled oscillator (VCO).

9-5   FREQUENCY STABILITY OF THE TRANSISTOR ASTABLE

In deriving Eq. (9-4), we assumed that the transistor conducts at $V_{BE} = 0$. This made the PRR independent of V_{CC}. But we know that $V_{BE} \approx 0.7$ V before silicon transistors conduct. Hence it takes a little longer time to reach the $+0.7$-V condition, and the PRR is a little lower than that given by Eq. (9-4). The extent of the error is determined by the ratio of 0.7 V to V_{CC}. At about V_{CC} less than $+5$ V, corrections have to be made on the RC charging curve taking into account V_{BE}. If temperature changes, the change in V_{BE} will cause a change in PRR. In Fig. 9-8, note that the circuit operates by driving V_{BE} to $-V_{CC}$. The transistor must have a V_{BE} breakdown rating greater than this or Eq. (9-4) becomes quite inaccurate, and the frequency will change markedly as V_{CC} is changed.

One way to increase the frequency stability of pulse generators is to have them controlled by a stable element such as a quartz crystal. This will be discussed in a later section.

9-6   INTEGRATED-CIRCUIT ASTABLE

Where it is possible to make a connection to a transistor base, such as in RTL and DTL, IC gates can be substituted for the transistors in the circuit of Fig. 9-5. This is shown in Fig. 9-9 for a DTL gate. Note that the capacitors are connected to expander inputs, which connect to transistor bases. Similarly with RTL, the capacitors can be connected to gate inputs since these go directly to bases. The collector time constant is long because the output collector resistor is high. In Fig. 9-9, resistors R are pull-up resistors, which reduce the collector time constant and "square-up" the collector waveshape.

Equation (9-4) is quite inaccurate for determining the PRR of IC astables. Consider Fig. 9-10 for the DTL gate with expander mode. Before Q_1 conducts, point N has to get to $+2.1$ V. This is large compared to $+5$ V. In the negative direction, the voltage cannot get to $-V_{CC}$ because it is limited to -0.7 V by a parasitic diode (shown in Fig. 9-10) caused by the method of making the IC. It is not possible to derive a simple expression for the period.

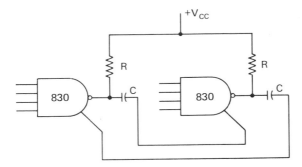

Figure 9-9 DTL astable multivibrator

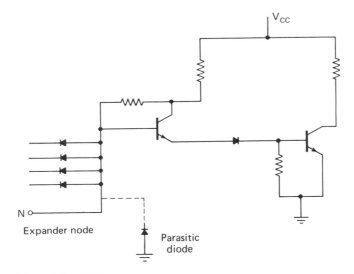

Figure 9-10 DTL gate

It is common practice therefore to express the PRR by an empirical equation,

$$PRR = KC \qquad (9\text{-}7)$$

where K is determined experimentally and C is the capacitance.

The repetition rate of this pulse generator changes considerably as V_{cc} and temperature change.

In TTL, a similar circuit can be built, as shown in Fig. 9-11. In TTL no internal bases are available. Collector feedback through resistors R will put the gates into the active region and the circuit will multivibrate. No simple formula can be derived for the period of oscillation.

A different approach to the IC astable multivibrator using CMOS inverters is

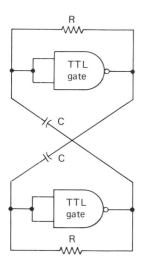

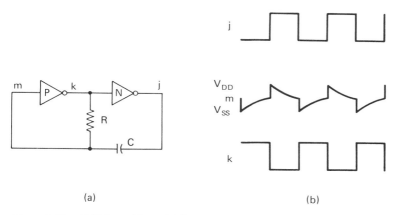

Figure 9-11 TTL astable multivibrator

(a)

(b)

Figure 9-12 CMOS astable multivibrator

shown in Fig. 9-12a. In this circuit only one R and one C are needed, and the circuit alternately charges and discharges C through R.

Initially, suppose that point j is at ground or V_{SS}, and capacitor C has no charge on it or voltage across it. This makes the voltage at point m equal to V_{SS}. By inverter action point k is at $+V_{DD}$. The positive voltage at k charges C through R, and the voltage at point m goes positive exponentially at a rate determined by the time constant RC. At $0.7RC$, the voltage at m reaches half V_{DD}, the inverter transfer voltage, and gate p switches, making k go to logic 0 or V_{SS} (ground). Point j goes to logic 1 or $+V_{DD}$. Point m is limited from going more positive than $+V_{DD}$ by an internal protective damping diode. Now we have points j and m at $+V_{DD}$ and point k at ground equal to V_{SS}. This is how we started (charge on C equal to 0), but the voltages are interchanged. Capacitor C (and point m) is now charged downward in a negative direction toward k through R. This exponential charge lasts until the transfer voltage of $V_{DD}/2$ is again reached. Again the time is 0.7RC. When the transfer voltage is reached, gate N again switches as before, but in the opposite polarity, and points j and k (again limited by another protective internal damping diode) are at V_{SS}; the cycle is ready to begin again. The waveforms are shown in Fig. 9-12b. The total time for a cycle is

$$T = 0.7RC + 0.7RC = 1.4RC \qquad (9\text{-}8)$$

If gates are used in place of inverters N and P of Fig. 9-12a, a separate control to a gate input can be used to start and stop the multivibrator.

9-7 UNIJUNCTION TRANSISTOR (UJT)

The UJT oscillator is a very simple RC charging–discharging circuit that is used quite widely as a clock generator in digital work. The basic construction of the UJT is shown in Fig. 9-13a, and its symbol is shown in Fig. 9-13b.

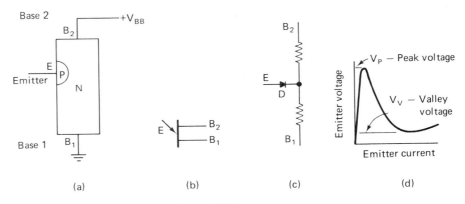

Figure 9-13 The unijunction transistor—UJT

Into a bar of N-type silicon material (approximately 10,000 Ω) is diffused a P region to form a diode. Contacts B_2 and B_1 are made to the ends of the bar and are called the bases. The equivalent circuit is shown in Fig. 9-13c. If, for example, $R_2 = 3$ kΩ, $R_1 = 7$ kΩ, and interbase voltage V_{BB} is made equal to $+10$ V, the voltage at point

$$X = \frac{R_1}{R_1 + R_2} V_{BB} = +7 \text{ V}$$

We now define the *intrinsic standoff ratio* η:

$$\eta = \frac{R_1}{R_1 + R_2} \qquad (9\text{-}9)$$

If the voltage at point E, the emitter input, is at 0, diode D is reverse biased. Diode D will be reverse biased until the voltage at E exceeds $\eta V_{BB} + 0.7$ V. If we apply a voltage in excess of this at point E, holes from the P region are injected into the bar of silicon, which decreases resistance R_1. As a result we get into a condition where, as current into the emitter terminal increases, the voltage from emitter to B decreases, a *negative resistance*. The emitter voltage at which this begins is called V_P, the *peak voltage*.

$$V_P = \eta V_{BB} + V_D \approx \eta V_{BB} + 0.7 \tag{9-10}$$

The voltage decreases with increasing current until a minimum voltage, the *valley voltage* (V_V), is reached. Beyond the valley voltage, the voltage increases.

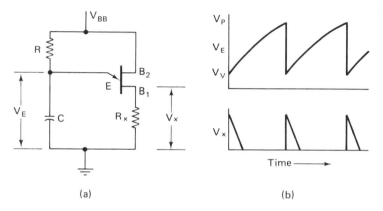

(a)   (b)

Figure 9-14 The UJT oscillator: (a) circuit; (b) waveform

Figure 9-14 shows a UJT oscillator. V_E starts from a low voltage, and C is charged exponentially toward $+V_{BB}$ through R. When V_P is reached, the capacitor is discharged through the emitter B_1 path to V_V. This occurs quite rapidly. This discharging current develops a sharp short-duration pulse of several microseconds across the B_1 resistor R_X. This voltage, V_X, is used for timing purposes in most applications. The process repeats as shown in Fig. 9-14b. The period of the wave is given by

$$T = 2.3RC \log_{10} \frac{V_{BB} - V_V}{V_{BB} - \eta V_{BB} - V_D} \tag{9-11a}$$

If $V_V \ll V_{BB}$, and $V_D \ll \eta V_{BB}$, as occurs quite frequently,

$$T = 2.3RC \log_{10} \frac{V_{BB}}{V_{BB} - \eta V_{BB}}$$

$$= 2.3RC \log_{10} \frac{1}{1 - \eta} \tag{9-11b}$$

If $\eta = 0.63$, a value that occurs frequently,

$$T = 2.3RC \log_{10} \frac{1}{0.37}$$

$$= RC \qquad\qquad\qquad (9\text{-}11c)$$

EXAMPLE 9-5   In a UJT oscillator, $V_{BB} = 15$ V, $\eta = 0.72$, $V_V = 1.2$ V, $R = 10 \text{ k}\Omega$, and $C = 0.2 \ \mu\text{F}$. Find the frequency of oscillation.

Solution

a. From Eq. (9-11a),

$$T = 2.3(10,000)(0.2 \times 10^{-6}) \log_{10} \frac{15 - 1.2}{15 - 0.72(15) - 0.7}$$

$$= 4.6 \times 10^{-3} \log_{10} 3.94$$

$$= 2.73 \times 10^{-3} \text{ s}$$

$$f = \frac{1}{T} = 363 \text{ PPS}$$

b. From Eq. (9-11b),

$$T = 2.3RC \log_{10} \frac{1}{1 - 0.72}$$

$$= 2.3(10,000)(0.2 \times 10^{-6}) \log_{10} 3.57$$

$$= 2.54 \times 10^{-3} \text{ s}$$

$$f = \frac{1}{T} = 394 \text{ PPS}$$

This does not agree with part a because of the assumption that V_V and V_D can be neglected.

c. From Eq. (9-11c),

$$T = RC = (10,000)(0.2 \times 10^{-6})$$

$$= 2 \times 10^{-3} \text{ s}$$

$$f = \frac{1}{T} = 500 \text{ PPS}$$

This is in even greater disagreement with part a because $\eta \neq 0.63$.

9-8   RING OSCILLATOR

Figure 9-15 is the circuit diagram of an IC ring oscillator. It has three inverters connected in a ring. Assume that $K = 1$. Then $L = 0$, $M = 1$, and $K = 0$. But this is in opposition to our original assumption and at first glance appears impossible. Actually, it is possible. The logic levels do occur as given but at different times. The

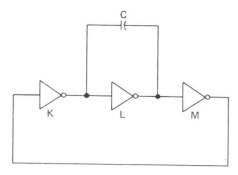

Figure 9-15   IC ring oscillator

ones and zeros circulate around the ring, changing at a rate determined by the switching times of the gates, plus the time required to charge and discharge capacitor *C*.

9-9   CRYSTAL-STABILIZED OSCILLATORS

There are many applications for pulse generators, for example, in accurate timekeeping, which require a constancy of the PRR well beyond the capability of the oscillators we have been discussing. The PRR of previous oscillators changes because of temperature changes, power supply voltage variation, and component drift. Even under excellent conditions of temperature and power supply stability, the best PRR constancy that can be attained is of the order of 0.1 percent.

To obtain a PRR constancy of the order of 0.001 percent or better, oscillators whose frequency is controlled by a bar of quartz are used. These are called quartz crystal oscillators.

Quartz is a material with different properties along its three axes, 90° apart from each other. One is the optical axis, with which we are not concerned. The other two have piezoelectric properties. If mechanical pressure is applied to the crystal along one axis (the mechanical), electric charges appear on the crystal surfaces at an axis (the electrical) 90° from the mechanical. If the crystal is mounted between two plates on its electrical faces, a capacitor is formed. Where voltage is applied to the plates, charges appear because of the capacitance, and mechanical stresses are developed along the mechanical axis. When alternating current is applied to the crystal, the crystal vibrates. The crystal is mechanically resonant at a frequency determined by its dimensions. This resonance is quite sharp, much sharper than can be attained by practical electrical components. As a result, the crystal can be used to fix the frequency of an electrical circuit because of resultant coupling between an electrical circuit and the mechanical resonant equivalent circuit of the quartz crystal.

There are many different circuits of quartz crystal oscillators. Figure 9-16 shows two that can be used with ICs. Figure 9-16a is a modification of the IC ring oscillator. Figure 9-16b is a circuit for use with CMOS ICs. For crystal-controlled

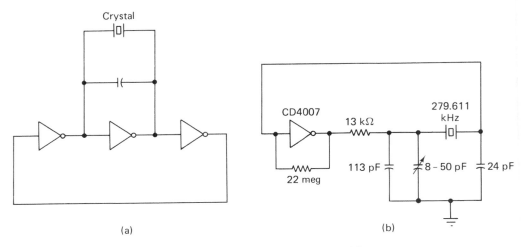

Figure 9-16   IC crystal oscillators: (a) crystal ring oscillator; (b) CMOS crystal oscillator (Courtesy RCA Manufacturing Co.)

oscillators with even greater stability, compensating techniques are used or the crystals are placed in temperature-controlled ovens.

9-10   THE MONOSTABLE

Many applications require the generation of a single pulse of adjustable width in response to a pulse command, which in many cases is short or different in shape from the desired pulse. For example, in the transmission of data over long distances on cables, the pulse shape deteriorates because of cable characteristics or spurious noise. Before this becomes too severe, a "single-pulse" generator is used to reform or regenerate the pulse so that it can again be transmitted down the cable. This is continued again and again over long distances, thus preserving the original waveshape. Other examples would be for a timer for a photographic enlarger for which it is desired to control exposure time, or for use in an application when a delay of a second operation is required some time after the initial pulse occurs.

The single-pulse generator has many names, the most popular being monostable, single shot, and one-shot. The basic principle behind the monostable is a combination of bistable and astable action. The circuit is in a stable condition because of bistable conditions. The triggering pulse takes the monostable out of the stable condition into a quasi-stable (astable) condition. As in the astable, a charged capacitor is allowed to discharge until the stable condition is again reached. The time is a function of the product of an R and C. In monostable circuits, either R or C or both are adjustable to vary the time of the pulse. Many complete ICs are available with an external R or C connected to the IC for adjusting the time. In these ICs the internal circuitry has been developed to make the timing independent of all ambient conditions such as voltage supply and temperature. In such ICs, the time of the pulse depends only upon the product of *R* and *C*.

As we have indicated, a monostable is a combination of half of an astable and half of a bistable. The astable consisted of two ac or capacity-coupled inverters, and the bistable consisted of two dc-coupled inverters. The circuit design was such that transistors in both were either in saturation or cut off. This is again the desired operating condition. In the monostable two inverters are again used and coupled back to back. In one coupling, the inverters are dc coupled; in the second coupling, they are ac coupled through a capacitor. To illustrate the principle, Fig. 9-17 shows two gates, such as DTL gates, having extender nodes used in a monostable.

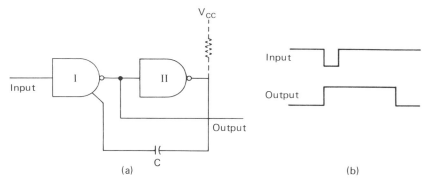

Figure 9-17   Two-gate monostable: (a) circuit; (b) waveforms

In Fig. 9-17a, gate I is dc coupled to gate II, whereas gate II is ac coupled to gate I. The operation of this circuit is as follows:

1   The input to gate I is high before the triggering pulse. This makes the output of gate I low, which in turn makes the output of gate II high.
2   Initially the capacitor is charged to the voltage difference between V_{OH} ($\approx V_{cc}$) and the internal logic node voltage of DTL gates (≈ 2.1 V).
3   The input triggering pulse goes quickly low. This makes the output of gate I go high and the output of gate II go low.
4   The rapid change at the output of gate II goes low too rapidly for capacitor *C* to discharge, driving the node voltage negative. This cuts off the transistors of gate I, making the output of gate I go high.
5   Capacitor *C* now charges toward V_{cc} through the internal DTL base resistors until the expander turn-on node voltage is reached (≈ 2.1 V). The time for this is determined by the product of the internal base resistance and *C*.
6   When *C* has charged up to the node threshold voltage, the output of gate I goes low again.

7 Note that it takes only a short trigger pulse to generate a long output pulse.

8 The time is difficult to determine mathematically. Parasitic internal diodes limit the amount the capacitor can drive the node negatively. The node voltage is approximately 2.1 V and is quite temperature sensitive. In general, all that can be said is that the pulse width of this type of IC monostable is determined by the switching time of the IC transistors plus some capacitor charging time.

Integrated-circuit monostables are available in many of the IC families. It is only necessary to add external R and C to them to determine the timing. They come in two basic modes: (1) nonretriggerable, and (2) retriggerable. In the nonretriggerable type, typified by the TTL 54/74121, pulse triggering occurs at a particular input voltage level and is not related to the rise time of the input pulse. Once fired, the outputs are independent of input changes. The length of the output pulse is determined only by the external R and C. This IC can generate a pulse width from 30 ns to 40 s by the choice of R and C.

The retriggerable type is typified by the TTL type 9601/8601. Pulse duration is also determined by an external R and C. This unit can be retriggered or restarted by means of a new input pulse after it has once been triggered. By doing this it is possible to obtain an output pulse width from 50 ns to ∞.

Figure 9-18 shows the logic symbol and external components required to operate an IC monostable such as the type 9601/8601.

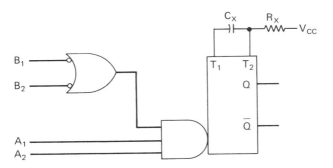

Figure 9-18 Retriggerable IC monostable: R_x and C_x are external components used to determine the pulse width

Whenever the input conditions for initiating the timing are met, C_x is discharged and recharges through R_x. In this IC, it is possible to discharge again a partially charged C_x in the retriggerable mode or the retriggerable feature can be inhibited.

Figure 9-19 is a chart giving the pulse width for values of C_x less than 1000 pF.

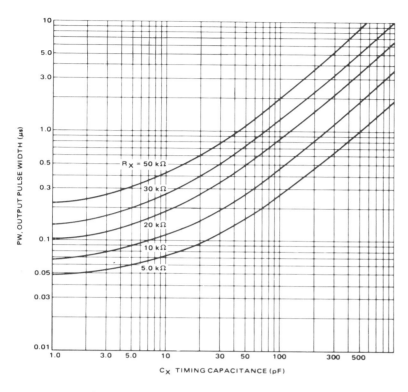

Figure 9-19   Pulse width: $C_x < 1000$ pF, type 9601/8601. Pulse initiation: output pulses are initiated by any combination of inputs satisfying the logic expression A1·A2·(B1 + B2). A1 and A2 must be high and either B1 or B2 must be low. Output pulses are generated by transitions from low to high on A1 or A2, and by transitions from high to low on B1 or B2. (Courtesy Motorola Semiconductor Products Inc.)

When $C_x > 1000$ pF, the pulse width is given by

$$\text{PW} = 0.32 R_x C_x \left(1 + \frac{0.7}{R_x}\right) \text{ns} \qquad (9\text{-}12)$$

where R_x is in kilohms and C_x is in picofarads.

In timing circuits it is important to know *exactly* when the timing pulse begins. Inputs A_1, A_2, B_1, and B_2 are used to determine whether the output pulse is introduced at the leading or trailing edge of the initiating pulse. This is shown in Fig. 9-20.

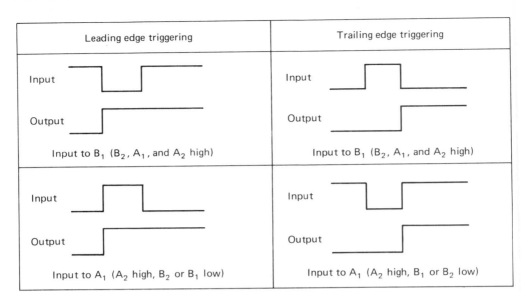

Leading edge triggering	Trailing edge triggering
Input Output Input to B_1 (B_2, A_1, and A_2 high)	Input Output Input to B_1 (B_2, A_1, and A_2 high)
Input Output Input to A_1 (A_2 high, B_2 or B_1 low)	Input Output Input to A_1 (A_2 high, B_1 or B_2 low)

Figure 9-20 Output pulse timing: type 9601/8601

9-11 THE 555 TIMER

The development of the IC type 555† timer has been of great importance. Diagrams and waveforms describing its operation are shown in Fig. 9-21. This unit has the following features:

1 Wide operating voltage range (up to 18 V).
2 Timing from microseconds to hours.
3 Astable or monostable modes.
4 High output current capability.
5 Can drive TTL.
6 Excellent temperature stability (0.005 percent per °C).

9-11.1 Monostable Operation

Figure 9-21b shows operation as a monostable. Initially, the external capacitor is held discharged by a transistor. A negative trigger pulse is applied. This triggers the flip-flop to release the short circuit across C. The voltage across the capacitor charges exponentially, with time constant equal to $R_A C$. When the voltage across the capacitor reaches $\frac{2}{3}V_{cc}$, the comparator resets the flip-flop discharging the capacitor. The time delay is given by Eq. (9-13) and by the chart of Fig. 9-21b:

$$T = 1.1 R_A C \tag{9-13}$$

†Data on the type 555 timer provided by the Signetics Corp.

236

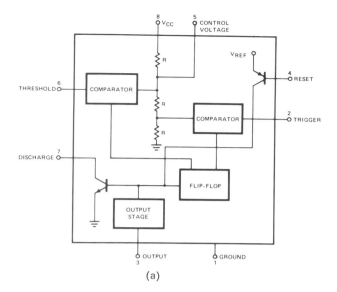

(a)

Figure 9-21 555 Timer operation: (a) block diagram showing internal operation;

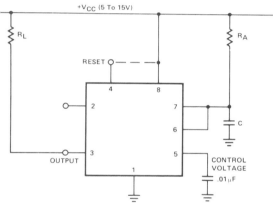

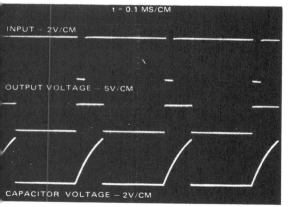

$R_A = 9.1$ KΩ, C = .01 μF, $R_L = 1$ KΩ

(b)

TIME DELAY
vs R_A, R_B AND C

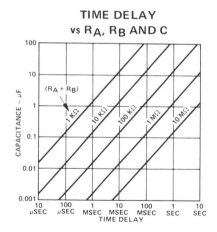

Figure 9-21 (continued)
(b) monostable operations;

9-11.2 Astable Operation

Figure 9-21c shows operation as an astable multivibrator. In this connection the circuit retriggers itself. The external capacitor charges through R_A and R_B, and

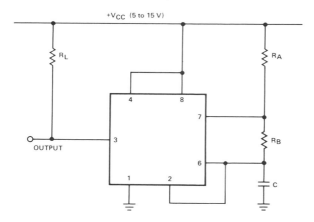

(c)

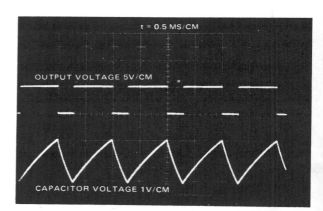

$R_A = 4\ K\Omega,\ R_B = 3\ K\Omega,\ R_L = 1\ K\Omega$

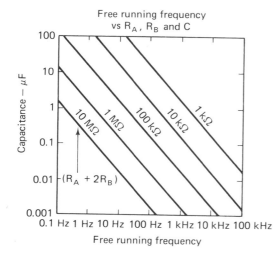

Figure 9-21 (continued) (c) astable operation (Courtesy Signetics Corp.)

discharges through R_B only. This enables the circuit designer to set the duty cycle by the ratio of these two resistors. In this mode, the capacitor charges and discharges between $\frac{1}{3}V_{CC}$ and $\frac{2}{3}V_{CC}$. The frequency of oscillation f and duty cycle D are given by Eq. (9-14). The frequency is also given on the chart of Fig. 9-21c.

$$f = \frac{1}{T} = \frac{1}{(R_A + 2R_B)C} \tag{9-14a}$$

$$D = \frac{R_B}{R_A + 2R_B} \tag{9-14b}$$

9-12   PULSE STRETCHER†

The pulse stretcher is used to extend a pulse width. The circuit is quite simple and is shown in Fig. 9-22. A short-duration pulse to inverter 1 discharges capacitor C. The input to inverter 2 is low and the output of inverter 2 high until C has charged up to the threshold voltage. Pull-up resistor R may be external, as is required for CMOS, or it may be the internal input biasing resistor, as in DTL and TTL. The increased pulse width can be used to provide additional time for logic operations or to provide some delay. This circuit has become quite useful to "capture" short-duration pulses and display their presence with light-emitting diodes in the digital tester called a *logic probe*.

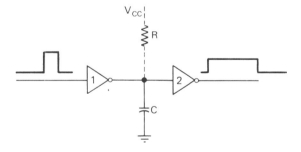

Figure 9-22   Integrated circuit pulse stretcher

9-13   SCHMITT TRIGGER

Some waveforms have a long rise time and will cause difficulties in digital circuitry. For example, TTL gates and flip-flops require that the input waveforms keep the IC in the active region for very short times of the order of tens of nanoseconds. If

†M. E. Levine, *Digital Theory and Experimentation Using Integrated Circuits.* Englewood Cliffs, N.J.: Prentice-Hall, Inc., 1974, Expt. 12, Pulse Shaping/Schmitt Trigger.

kept in the active region too long, noise or undesired parasitic oscillation may give a false output from a flip-flop. To couple a long rise time wave through a capacitor requires a large capacitor.

In many cases, all that is required is that a voltage level be sensed, and that this level be used to operate a gate or flip-flop. The logic block that senses a level and converts a slow-changing waveform to a fast-changing waveform when a voltage level is sensed is called the Schmitt trigger.

Like the astable and monostable multivibrator, the Schmitt trigger† is a two-inverter regenerative circuit with feedback between its output and input, as shown in Fig. 9-23. The circuit operates in the following manner. When the signal voltage is low, the voltage at point A, the input to inverter 1, is below the threshold voltage. This is helped by the low zero output level of inverter 2 through resistor R_4. As the signal voltage increases, it forces current through R_1 to point A, raising its voltage level until the threshold voltage of inverter 1 is reached. The output of inverter 1 now goes to a zero level, and inverter 2 goes to a high output level. This forces current through R_4 to make the voltage at point A even higher. The action is regenerative and occurs rapidly, resulting in a very fast change in voltage from a low level to a high level at the output of inverter 2.

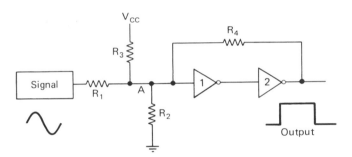

Figure 9-23   Integrated circuit Schmitt trigger

If we start to decrease the signal voltage, it must now have to get to a lower voltage (because of the R_4 current) to reach the threshold voltage. At this signal voltage level, the circuit regeneratively switches again rapidly to its zero voltage level.

The signal voltage at which the Schmitt trigger switches from a low to a high level is called the

$$\text{Upper trip level} = \text{UTL in volts}$$

---

†The two-gate IC Schmitt trigger differs from the classic two-transistor version. With two transistors, feedback occurs through a common emitter resistor. In the transistor version, the zero level does not get to ground unless a positive and negative power supply is used, a serious shortcoming. This is not the case for the IC version.

The signal voltage at which the Schmitt trigger switches from the high level to the low level is lower than the UTL, and is called the

Lower trip level = LTL in volts

The difference between the two is called the hysteresis and is defined as

Hysteresis = UTL − LTL in volts   (9-15)

EXAMPLE 9-6 In the Schmitt trigger of Fig. 9-24, let the inverters be TTL (open collector) and the resistors have the following values: $R_1 = 1\,\text{k}\Omega$, $R_2 = 1\,\text{k}\Omega$, $R_3 = 4.7\,\text{k}\Omega$, $R_4 = 12\,\text{k}\Omega$, and $R_C = 3.3\,\text{k}\Omega$.

Solution

a. For the UTL: the threshold voltage at point $A = 1.4\,\text{V}$. We have to find E_{UTL} if the output of inverter 2 is 0 V.

$$I_3 = \frac{(5 - 1.4)\,\text{V}}{4.7\,\text{k}\Omega} = 0.765\,\text{mA}$$

$$I_4 = \frac{1.4\,\text{V}}{12\,\text{k}\Omega} = 0.117\,\text{mA}$$

$$I_2 = \frac{1.4\,\text{V}}{1\,\text{k}\Omega} = 1.4\,\text{mA}$$

$$I_5 = \frac{5 - 1.4 - 0.7\,\text{V}}{4\,\text{k}\Omega} = 0.725\,\text{mA}$$

Summing the currents, we find

$$I_1 = I_2 - I_3 + I_4 - I_5$$
$$= (1.4 - 0.765 + 0.117 - 0.725)\,\text{mA}$$
$$= 0.027\,\text{mA}$$

Figure 9-24 TTL Schmitt trigger

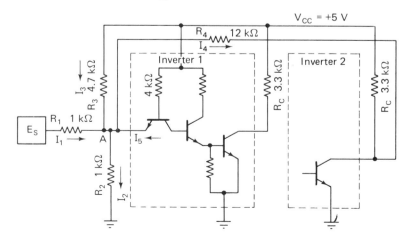

Hence

$$E_{\text{UTL}} = 1.4 \text{ V} + 0.027 \text{ mA} \times 1 \text{ k}\Omega = 1.427 \text{ V}$$

b. For the LTL: Inverter 2 is at a high output level and its output transistor does not conduct. We again require a voltage at point $A = 1.4$ V. I_4 now sends current in the opposite direction through the 3.3-kΩ inverter load resistance R_C and the 12-kΩ feedback resistor:

$$I_4 = -\frac{(5 - 1.4) \text{ V}}{(3.3 + 12) \text{ k}\Omega} = 0.236 \text{ mA}$$

I_2, I_3, and I_5 are the same as before. Summing the currents,

$$\begin{aligned} I_1 &= I_2 - I_3 - I_4 - I_5 \\ &= (1.4 - 0.765 - 0.236 - 0.725) \text{ mA} \\ &= -0.326 \text{ mA} \end{aligned}$$

Hence

$$\begin{aligned} \text{LTL} &= 1.4 \text{ V} - 0.326 \text{ mA} \times 1 \text{ k}\Omega \\ &= 1.074 \text{ V} \end{aligned}$$

c. The hysteresis is $1.427 \text{ V} - 1.074 \text{ V} = 0.353$ V.
d. If E_S were a sine wave with a 4-V p–p amplitude, we would obtain the input and output waveshapes shown in Fig. 9-25.
e. The high output voltage level of inverter 2 is not equal to V_{CC} but is lower due to the voltage drop in R_C caused by I_4. This is equal to

$$5 \text{ V} - 0.236 \text{ mA} \times 3.3 \text{ k}\Omega = 4.22 \text{ V}$$

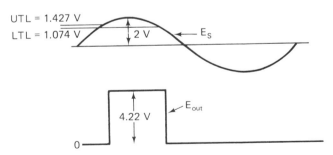

Figure 9-25 Input sine wave—output square wave-Example 9-6

It must be realized that the Schmitt trigger is a level-detecting circuit. The circuit triggers at the UTL and LTL and is independent of the frequency of the incoming signal.

Integrated-circuit Schmitt triggers are available in almost all the logic families. For example, in TTL the type 7414 is a hex inverter Schmitt trigger and the 7413 is a quad two-input NAND Schmitt trigger. A hysteresis loop is shown inside an IC logic gate symbol to indicate the Schmitt trigger function (Fig. 9-26).

Figure 9-26 Logic symbol for 2-input NAND Schmitt trigger

Problems

9-1. (a) Determine time Q_1 is in saturation or Q_2 is cut off.
   (b) Time Q_2 is in saturation or Q_1 is cut off.
   (c) Period of the oscillation.
   (d) Frequency or pulse repetition rate (PRR).

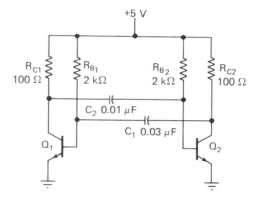

9-2. Draw the collector and base waveforms for Prob. 9-1.

9-3. Repeat Prob. 9-1 if C_1 is made 0.02 μF and $C_2 = 0.04$ μF.

9-4. Repeat Prob. 9-1 if C_2 is made equal to $C_1 = 0.01$ μF.

9-5. What is the minimum value of h_{FE} that will make transistors Q_1 and Q_2 in Prob. 9-1 saturate. (Neglect the effect of V_{BE}.)

9-6. The following components are used in the circuit of Prob. 9-1. Determine the frequency of oscillation: $R_{B1} = 50$ kΩ, $R_{B2} = 100$ kΩ, $C_1 = C_2 = 0.1$ μF, $R_{C1} = R_{C2} = 1$ kΩ.

9-7. If the transistors used in Prob. 9-6 are specified to have a minimum h_{FE} of 75, is this a good design? Explain.

9-8.

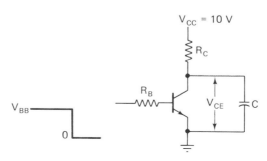

V_{BB} is high enough to put the transistor in saturation. Using the curves of Fig. 9-2, how many time constants after the voltage V_{BB} is made equal to 0 will the voltage V_{CE} get to (a) $V_{CE} = 1$ V (10 percent V_{CC}); (b) $V_{CE} = 9$ V (90 percent V_{CC}); (c) from 10 to 90 percent of V_{CC}. *Note:* This is the standard way of defining "rise time" (10 to 90 percent of final voltage) in switching circuits.

9-9. In Prob. 9-1,
   (a) What is the Q_1 collector time constant?
   (b) How long does it take the V_{CE} of Q_1 to go from 0.5 to 4.5 V (rise time) after Q_1 is cut off?
   (c) Repeat (a) for Q_2.
   (d) Repeat (b) for Q_2.

9-10. In the circuit of Fig. 9-3, $-V = +V = 5$ V, $R = 2$ kΩ, and $C = 0.01$ μF. Using the curves of Fig. 9-2,
   (a) How long (time constants and time) after the switch is opened will it take V_C to reach 0 V?
   (b) How long will it take to reach $+0.7$ V?

9-11. Repeat Prob. 9-8, but with $-V$ and $+V$ equal to 3 V.

9-12. The solution of Prob. 9-4 assumed that $V_{BE} = 0$. Using the results of Prob. 9-10, determine the multivibrator oscillation frequency, taking into account the effect of V_{BE} ($= 0.7$ V).

9-13. Repeat Prob. 9-12, but let $V_{CC} = 3$ V.

9-14. In a transistor, V_{BE} decreases approximately 2 mV per °C increase in junction temperature. What effect will an increase in junction temperature from 25 to 200°C have on the PRR of the oscillator of Prob. 9-13?

9-15. In a UJT (unijunction transistor) oscillator, $V_{BB} = 10$ V, $\eta = 0.6$, $V_V = 0.8$ V, $R = 50$ kΩ, and $C = 1$ μF. Determine the frequency of oscillation (a) using the exact formula, (b) assuming that V_V and V_D can be neglected, and (c) assuming that $\eta = 0.63$.

9-16. Determine the pulse width of a type 9601/8601 monostable with $R_X = 20$ kΩ and $C_X = 100$ pF. (*Hint:* use Fig. 9-19.)

9-17. Determine the pulse width of a type 9601/8601 monostable with $R_X = 20$ kΩ and $C_X = 0.02$ μF.

9-18. In the Schmitt trigger of Fig. 9-24, R_2 is changed to 500 Ω. Assume that the threshold voltage is 1.4 V. Determine (a) the UTL, (b) the LTL, and (c) the hysteresis of the circuit.

9-19. A Schmitt trigger has the following characteristics: UTL $= 2.4$ V, LTL $= 1.1$ V, and the output levels are 0.3 V when the input signal is below the LTL and 3.6 V when the input signal is above the UTL. A 10-V p–p sine wave signal at a frequency of 400 Hz is applied to the Schmitt trigger:
   (a) Determine the angles at which the output level changes.
   (b) During one cycle, how long is the output voltage at its high level?
   (c) Sketch both the input and output waveforms on the same scale.

TEN

Binary Codes, Encoding, Decoding, Multiplexing†

10-1 INTRODUCTION

The computer and many other electronic systems are binary operated and use 1s and 0s. However, we communicate with each other with decimal numbers and alphabetic characters. We need an interface between the two systems. To accomplish this, standard binary codes have been developed. The process of generating these binary codes is called *encoding*. The process of recognizing and converting the binary codes to decimal and alphabetic characters is called *decoding*.

Actually, in somewhat disguised form, we have already had considerable experience with codes, particularly in Chapters 4, 7, and 8. In Chapter 4 we discussed the octal and hexadecimal systems. They are really codes. Octal is a 3-bit binary code for decimals 0 to 7, and hexadecimal is a 4-bit binary code for decimal 0 to 16. In Chapters 7 and 8 we discussed counters and developed the count-state concept as a binary way, actually a binary code, for expressing the number of counts.

†M. E. Levine, *Digital Theory and Experimentation Using Integrated Circuits.* Englewood Cliffs, N.J.: Prentice-Hall, Inc., 1974, Expt. 13, Decoding–Encoding.

10-2 FOUR-BIT DECIMAL CODES

There is an enormous number of ways to encode decimals 0 to 9 using 4-bit codes. Only a few have ever been used.

10-2.1 The 8421, NBCD, or BCD Code

The most frequently used code today is the 8421 code. This is a weighted code with each bit carrying the code weight. It progresses in natural binary from 0000 to 1001 (0 to 9 decimal) and therefore is also called natural binary coded decimal (NBCD). This code is also called binary coded decimal (BCD). Although all decimal binary codes should be called BCD, and some have in the past, it is now becoming common practice to call only the 8421 code BCD. This code is given in Table 10-1. This

Table 10-1   The 8421, NBCD, or BCD code

Decimal	Code
	Weight 8421
0	0000
1	0001
2	0010
3	0011
4	0100
5	0101
6	0110
7	0111
8	1000
9	1001

code has the property that each bit of the binary code has a weight, and the conversion to its decimal equivalent is straightforward. Example 10-1 shows how this code is used.

EXAMPLE 10-1   Convert the number 963 to a binary equivalent using 8421 (NBCD) code.

Solution

$$\underbrace{9}_{1001}\ \underbrace{6}_{0110}\ \underbrace{3}_{0011}$$

Ans.   100101100011

We see that each digit of the original number is expressed in its 4-bit 8421 equivalent and the result expressed as a binary word.

To recover the decimal equivalent all that has to be done is to separate the binary word into groups of 4 bits, and then determine the decimal equivalent of each 4-bit grouping.

EXAMPLE 10-2   Convert the 8421 binary word 1000011100010101 to its decimal equivalent.

Solution   Separate the word into 4-bit groups and express as a decimal equivalent.

$$\underbrace{1000}_{8}\ \underbrace{0111}_{7}\ \underbrace{0001}_{1}\ \underbrace{0101}_{5} \equiv 8715$$

10-2.2   The 4221 and 2'421 Codes

These codes are also weighted decimal codes (See Table 10-2). At one time they were in common use, but in recent times they have been superseded by the 8421 code. There are two 2'421 codes. They are generated by two different decade counters. 2'421b has the property that the 9s complement in decimal is represented by complements in the code. For example, 2 and 7, which are 9s complements, are given by the respective complements 0010 and 1101.

Table 10-2   The 4221 and 2'421 codes

Decimal	0	1	2	3	4	5	6	7	8	9
4221	0000	0001	0010	0011	0110	0111	1100	1101	1110	1111
2'421a	0000	0001	0010	0011	0100	0101	1100	1101	1110	1111
2'421b	0000	0001	0010	0011	0100	1011	1100	1101	1110	1111

EXAMPLE 10-3   Express the number 6478 in (a) 4221 and (b) 2'421a codes.

Solution

a. 4221: $\underbrace{6}_{1100}\ \underbrace{4}_{0110}\ \underbrace{7}_{1101}\ \underbrace{8}_{1110} \equiv 1100011011011110$

b. 2'421a: $\underbrace{6}_{1100}\ \underbrace{4}_{0100}\ \underbrace{7}_{1101}\ \underbrace{8}_{1110} \equiv 1100010011011110$

EXAMPLE 10-4 Convert the 2'421a coded number 00010011110101001111 to its decimal equivalent.

Solution

$$\underbrace{0001}_{1} \quad \underbrace{0011}_{3} \quad \underbrace{1101}_{7} \quad \underbrace{0100}_{4} \quad \underbrace{1111}_{9} \equiv 13749$$

10-2.3 Excess-3 Code

The excess-3 code is a nonweighted binary code that is 3 more than the 8421 code. The code is given in Table 10-3, along with the 8421 code for comparison. The

Table 10-3 Excess-3 code

Decimal	Excess-3	8421
0	0011	0000
1	0100	0001
2	0101	0010
3	0110	0011
4	0111	0100
5	1000	0101
6	1001	0110
7	1010	0111
8	1011	1000
9	1100	1001

excess-3 code has two properties:

1 No number is expressed as 0000. This provides means for checking for errors.
2 The 9s complement numbers in decimal, when expressed in excess-3, are also expressed in complements. For example, 2 and 7, which are 9s complements, in excess-3 code are given by 0101 and 1010, which are also complements.

The excess-3 code is used for 9s complement subtraction problems.

10-2.4 Binary Arithmetic with Numbers Expressed in 4-Bit Codes

It is possible to perform arithmetic with numbers expressed in the 4-bit codes. Rules (algorithms) have been generated for performing addition. For example, in excess-3 the addition of two numbers results in a number that is excess-6, and also may result in a nonallowed code condition. However, techniques have been developed for taking these into account.

10-3 BIQUINARY AND QUIBINARY CODES

The biquinary and quibinary are decimal codes with more than 4 bits. The biquinary is generated by a five-stage (quinary) ring counter followed by a T flip-flop, and the quibinary is generated by a T flip-flop followed by a five-stage ring counter. These codes are given in Tables 10-4 and 10-5, respectively. They are weighted codes and are used in 7- and 6-bit arrangements. The 7-bit codes have the feature of being a 2 out of 7 code. Only two 1s are used, and this yields a simple error-checking code. The 6-bit codes are sometimes referenced by their weighting as 543210 and 864201 codes.

Table 10-4 Biquinary code

(a) Seven-bit code

Decimal	Code
	Weight 50 43210
0	01 00001
1	01 00010
2	01 00100
3	01 01000
4	01 10000
5	10 00001
6	10 00010
7	10 00100
8	10 01000
9	10 10000

(b) Six-bit code

Decimal	Code
	Weight 543210
0	000001
1	000010
2	000100
3	001000
4	010000
5	100001
6	100010
7	100100
8	101000
9	110000

TABLE 10-5 Quibinary code

(a) Seven-bit code

Decimal	Code
	Weight 86420 10
0	00001 01
1	00001 10
2	00010 01
3	00010 10
4	00100 01
5	00100 10
6	01000 01
7	01000 10
8	10000 01
9	10000 10

(b) Six-bit code

Decimal	Code
	Weight 864201
0	000010
1	000011
2	000100
3	000101
4	001000
5	001001
6	010000
7	010001
8	100000
9	100001

Table 10-6 9876543210 code

Decimal	Code
	Weight 9876543210
0	0000000001
1	0000000010
2	0000000100
3	0000001000
4	0000010000
5	0000100000
6	0001000000
7	0010000000
8	0100000000
9	1000000000

EXAMPLE 10-5   Express the number 643 in 7-bit biquinary code.

Solution

$$\underbrace{10\ 00010}_{6}\ \ \underbrace{01\ 10000}_{4}\ \ \underbrace{01\ 01000}_{3} \equiv 100001001100000101000$$

10-4 THE 9876543210 OR RING-COUNTER CODE

The 9876543210 or ring-counter code is generated by a 10-stage ring counter (see Table 10-6). This code is represented by a single 1 and therefore is a self-error-checking code. It is a weighted code, and the location of the 1 corresponds to the decimal equivalent. Since the position of the 1 corresponds to the equivalent decimal number, no additional circuitry is required to separate the numbers. Later in this chapter we would call this code a self-decoding code.

10-5 PARITY CODE†

The parity code is an error-checking code in which binary numbers are transmitted with an additional bit that is used only for error checking. In this code all data are transmitted either with an odd number of 1s (odd parity) or with an even number of 1s (even parity). Odd parity is preferred since it is never possible to transmit a number or word with all 0s.

The following are examples of data with parity bits added:

Odd Parity			Even Parity		
Data	Parity Bit		Data	Parity Bit	
001001	1		001001	0	
111011	0		111011	1	
011111	0		011111	1	
101010	0		(01010	1)	error
(101010	1)	error	101010	1	
100111	1		100111	0	

The parity bit addition only serves as a warning that that binary word is in error. It does not detect the error. A technique that both detects the location of the error and from this the actual error is the use of parity bits for both the rows and

†See Sec. 5-8.

columns, as shown in the following set of data. This uses odd parity, but even parity could also have been used.

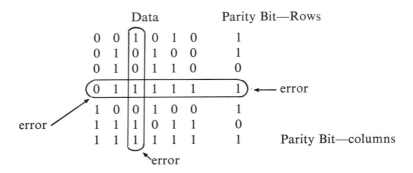

10-6   TWISTED-RING CODE

The twisted-ring code is generated by the twisted-ring counter. A five-stage twisted-ring counter generates a decimal code. Decoding this counter is relatively simple and requires 10 two-input AND gates (Prob. 10-20). This code is given in Table 10-7.

Table 10-7   Twisted-ring code

Decimal	Code
0	00000
1	10000
2	11000
3	11100
4	11110
5	11111
6	01111
7	00111
8	00011
9	00001

A code that changes by only 1 bit is needed in the important digital application of shaft encoders or positioners. This is the Gray code, given in Table 10-8 for

Table 10-8 Gray code

Decimal	Gray	Binary
0	0000	0000
1	0001	0001
2	0011	0010
3	0010	0011
4	0110	0100
5	0111	0101
6	0101	0110
7	0100	0111
8	1100	1000
9	1101	1001
10	1111	1010
11	1110	1011
12	1010	1100
13	1011	1101
14	1001	1110
15	1000	1111

decimals 0 to 15. For comparison, the comparable binary code is also given. The Gray code is generated from the binary code by means of an EXCLUSIVE–OR comparison of successive bits of the equivalent binary number. For the most significant bit, an additional 0 is added for this comparison.

EXAMPLE 10-6 Determine the Gray code for decimal 13.

Solution

$$13_{10} = 1101_2$$

The Gray code finds its primary application in the location of angles on a rotating shaft. This is called a shaft encoder. To locate the angular position, a wheel is mounted on the shaft with sections of rings on it coded light and dark in Gray code. By means of photosensors, one can determine the shaft position code and from it the angular position. As we shall see, the Gray code minimizes possible error in angular position.

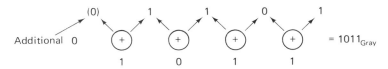

Figure 10-1 shows two possible encodings for shaft encoder wheels. A light area represents a 1 and a dark area a 0. In Fig. 10-1a, the wheel rings and segments are encoded in binary code, in Fig. 10-1b, the wheel is encoded in Gray code. If the wheel position is such that the center of a segment is opposite the photosensors, the sensors will read the position code correctly. Suppose, however, as is shown in Fig. 10-1a, that the wheel stops at the dividing line between the 111 and 000 segments. The sensors will become confused and may read each bit as a 1 or a 0 and give any possible position location. It is quite conceivable that the sensors may give a 100 reading, which would be 180° in error. Now consider the same problem in the Gray coded wheel of Fig. 10-1b. Note that the two outer rings have the same color in adjacent segments and the only possible source for confusion is in the inner ring. The maximum possible error therefore is only in *one* segment position, because the Gray code changes by 1 bit for each decimal number change.

Practical shaft encoders may have as many as 12 Gray encoded rings. Since $2^{12} = 4096$, such an encoding can resolve angular positions to within approximately 0.1°.

A Gray code application we have used in the past is the Karnaugh map. In it, the order of the variables is Gray coding.

Figure 10-1 Eight-segment shaft encoder: (a) binary coded wheel; (b) Gray coded wheel

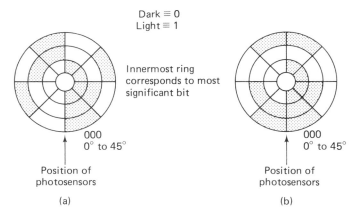

10-8 ALPHANUMERIC CODES

There are two multibit codes in common use for handling both digits and alphabetic characters. With these codes it is possible to code numbers, upper- and lowercase letters, and many characters. These two codes are the following:

1 ASCII: American standard code for information interchange, a 7-bit code that can code 2^7 or 128 characters.
2 EBCDIC: Extended binary-coded decimal interchange code, an 8-bit code that can code 2^8 or 256 characters.

Portions of these codes are given in Table 10-9.

Table 10-9 Portions of ASCII and EBCDIC codes

Character	ASCII	EBCDIC
0	011 0000	1111 0000
1	011 0001	1111 0001
2	011 0010	1111 0010
3	011 0011	1111 0011
4	011 0100	1111 0100
5	011 0101	1111 0101
6	011 0110	1111 0110
7	011 0111	1111 0111
8	011 1000	1111 1000
9	011 1001	1111 1001
A	100 0001	1100 0001
B	100 0010	1100 0010
C	100 0011	1100 0011
D	100 0100	1100 0100
E	100 0101	1100 0101
F	100 0110	1100 0110
G	100 0111	1100 0111
H	100 1000	1100 1000
I	100 1001	1100 1001
J	100 1010	1101 0001
K	100 1011	1101 0010
L	100 1100	1101 0011
M	100 1101	1101 0100
N	100 1110	1101 0101
O	100 1111	1101 0110
P	101 0000	1101 0111
Q	101 0001	1101 1000
R	101 0010	1101 1001
S	101 0011	1110 0010
T	101 0100	1110 0011
U	101 0101	1110 0100
V	101 0110	1110 0101
W	101 0111	1110 0110
X	101 1000	1110 0111
Y	101 1001	1110 1000
Z	101 1010	1110 1001

EXAMPLE 10-7   Express 3N27 in ASCII code.

Solution

$$\underbrace{011\ 0011}_{3}\quad\underbrace{100\ 1110}_{N}\quad\underbrace{011\ 0010}_{2}\quad\underbrace{011\ 0111}_{7}$$

10-9   OTHER CODES

There are many other codes. The teletypewriter code is a 5-bit code. Touch-tone telephones use two tones out of a possible five to transmit telephone numbers. There are more complex codes that are self-checking.

10-10   DECODING

Consider data being transmitted in parallel in 8421 code. We wish to determine when a 5 is being transmitted. What is different about a 5 in 8421 code from the codes for the other decimals? A 5 in 8421 code is 0101; a 0 *and* a 1 *and* a 0 *and* a 1. The *ands* in the previous statement indicate that to detect the presence of a 5 requires an AND gate. If we assign the letters D, C, B, and A to the bit locations, with D the most significant bit and A the least significant bit, we can see that to obtain a 1 from the AND gate requires the circuit of Fig. 10-2. We call this detection process *decoding*. The decoding technique therefore consists of using an AND gate and connecting to its inputs an input from each bit. A 1 is a direct connection, and a 0 requires an inverted input. In Chapter 7, considerable emphasis was placed on count-state tables. We see now that to *decode*, or separate the counts of a counter, requires AND gates with *inputs* corresponding to the *count states*. A 1 in the count state is an input from the Q output of the counter, and 0 requires a $\bar{Q}$ input to the AND gate.

Figure 10-2   Decoding gate for 8421 coded 5

EXAMPLE 10-8   A computer requires eight separate repetitive clock generators. Show how to do this.

Solution   Use a three-stage binary counter with decoding gates for each of the eight count states as shown in Fig. 10-3. The Q outputs for the three stages and the output waveforms of each decoding gate are shown.

Now let us try a more complex problem. Let us attempt to decode 0101_{8421}. From a count-state standpoint, the decoding AND gate inputs are obtained by expressing the count state in its equivalent letter form $\bar{D}C\bar{B}A$ and then connecting

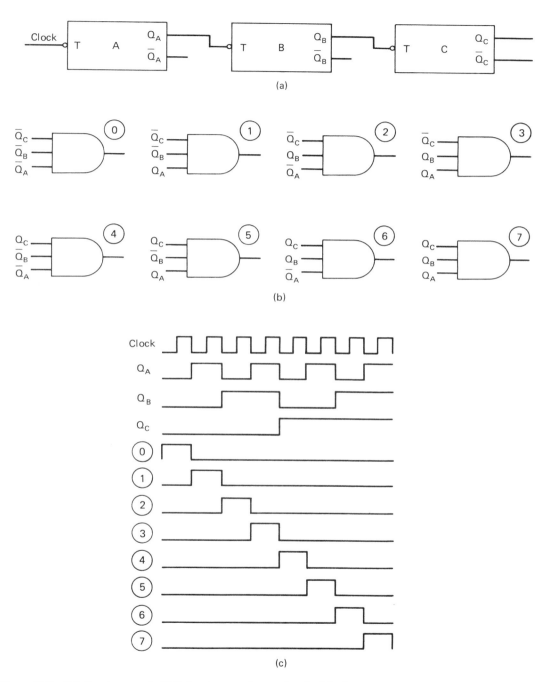

Figure 10-3 Solution to example 10-8 showing waveforms of each of the 8 decoding gates: (a) circuit diagram; (b) decoding gates; (c) waveforms

these inputs to a four-input AND gate shown in Fig. 10-4d. But, is this the simplest form of decoding? In the 8421 code, six states are unused. Perhaps they can be used to simplify the decoding by treating them as "don't care" states. To do this, the following steps are taken:

1   Express the code for each decimal in equivalent letter form, as was just done for 0101_{8421} $(= \bar{D}C\bar{B}A)$. These are tabulated in Fig. 10-4.

Figure 10-4   Karnaugh map "don't care" simplification for 0101_{8421}: (a) used states; (b) unused states plotted as "don't cares"; (c) unused states assigned for simplification and $\bar{D}C\bar{B}A$; (d) count state decoding; (e) simplified decoding

Used States		
Decimal	Binary	Letter
0	0000	$\bar{D}\bar{C}\bar{B}\bar{A}$
1	0001	$\bar{D}\bar{C}\bar{B}A$
2	0010	$\bar{D}\bar{C}B\bar{A}$
3	0011	$\bar{D}\bar{C}BA$
4	0100	$\bar{D}C\bar{B}\bar{A}$
5	0101	$\bar{D}C\bar{B}A$
6	0110	$\bar{D}CB\bar{A}$
7	0111	$\bar{D}CBA$
8	1000	$D\bar{C}\bar{B}\bar{A}$
9	1001	$D\bar{C}\bar{B}A$

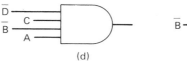
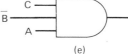

2 Karnaugh map all these used forms, as shown in Fig. 10-4a.

3 For the unused states, plot a second Karnaugh map with ×s in the boxes not used in step 2. This is shown in Fig. 10-4b. When it comes to simplification, these ×s as "don't cares" can be assigned a 1 or a 0, whatever level will simplify.

4 Plot $\bar{D}C\bar{B}A$ (0101_{8421}) as a 1 on the map and then look at the ×s. If any

Figure 10-5 NBCD decoding

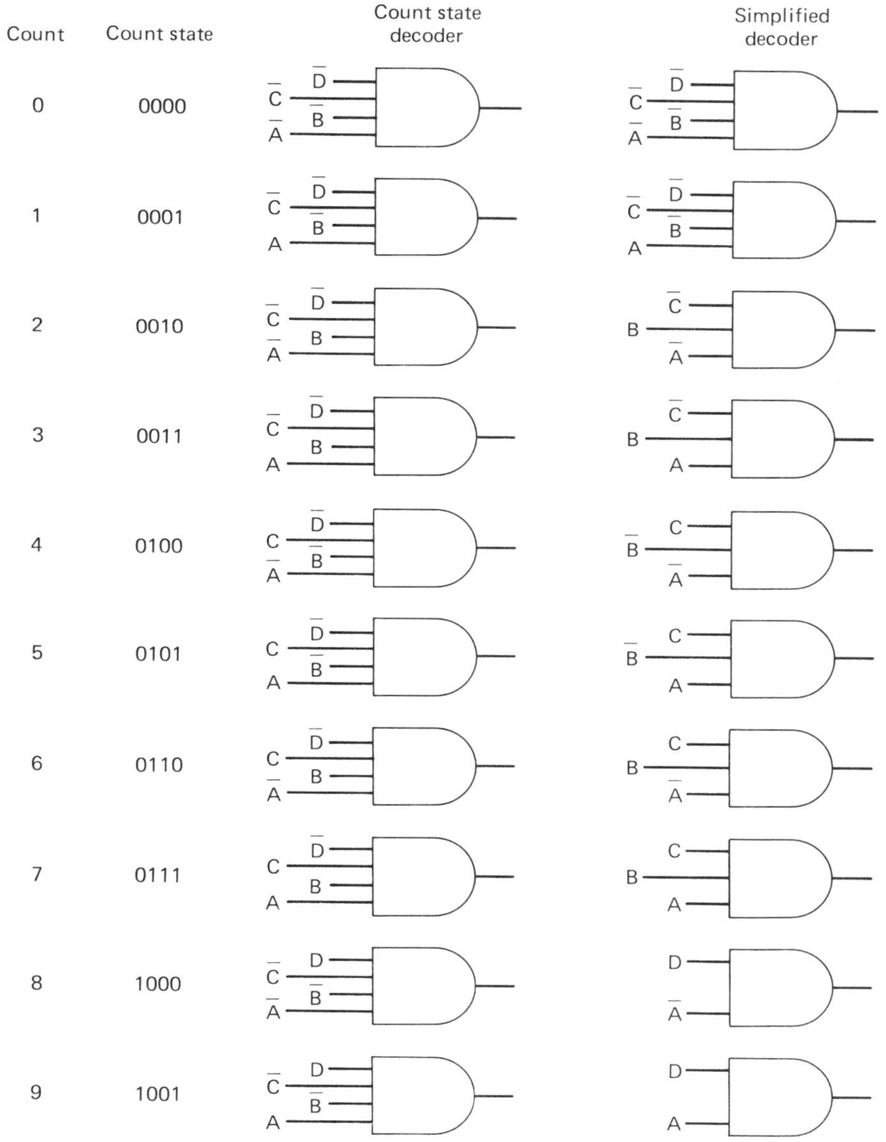

Count	Count state	Count state decoder	Simplified decoder
0	0000		
1	0001		
2	0010		
3	0011		
4	0100		
5	0101		
6	0110		
7	0111		
8	1000		
9	1001		

×s are adjacent to the 1 just plotted for the used state, assign it a 1; then assign all other ×s the level 0. This is shown in Fig. 10-4c. In this case the box $DC\bar{B}A$ is next to $\bar{D}C\bar{B}A$ and is assigned a 1. All other ×s are assigned 0.

5 Simplify as shown in Fig. 10-4c. In this case, the decoding can be accomplished by a three-input AND gate $C\bar{B}A$ rather than a four-input AND. This is shown in Fig. 10-4e.

Since 8421 is very important, the basic count state and the simplified decoding gates are given in Fig. 10-5. 8421 decoding is used quite frequently. TTL types 7441 and 74141 are one out of ten 8421 decoders that provide 0-level outputs at the output terminals. They are designed to drive cold-cathode gas decimal indicator tubes (NIXIE);† therefore, the output transistors in these decoders have the relatively high output voltage breakdown of 55 V to meet this requirement. In practice they are driven from type 7490 8421 decade counters through an intermediate memory quad latch type 7475, as shown in Fig. 10-6.

Figure 10-7 is a data sheet describing the type 74141.

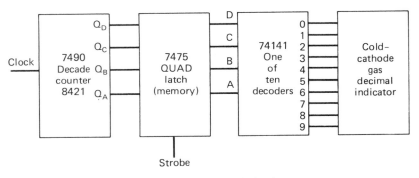

Figure 10-6   Counter-decoder-display: 8421 to decimal

10-11 ENCODING

Let us now consider the problem of encoding. Encoding covers a general technique for code generation. We have already discussed in previous chapters the generation of codes as count states of counters. The 8421, 4221, 2′421, and shift counter codes are generated at the output terminals of counters as the response to input pulses and essentially are representations in binary codes of the count number. Here, however, we want to treat a more basic problem. Using logic devices, how do we generate one code from another when the binary coding for both are given? We have already seen earlier in this chapter a method of going from binary to Gray by

†Burroughs Corporation.

- Drives gas-filled cold-cathode indicator tubes directly
- Fully decoded inputs ensure all outputs are off for invalid codes
- Input clamping diodes minimize transmission-line effects

FUNCTION TABLE

INPUT				OUTPUT
D	C	B	A	ON[†]
L	L	L	L	0
L	L	L	H	1
L	L	H	L	2
L	L	H	H	3
L	H	L	L	4
L	H	L	H	5
L	H	H	L	6
L	H	H	H	7
H	L	L	L	8
H	L	L	H	9
H	L	H	L	NONE
H	L	H	H	NONE
H	H	L	L	NONE
H	H	L	H	NONE
H	H	H	L	NONE
H	H	H	H	NONE

H = high level, L = low level
[†]All other outputs are off

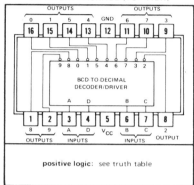

J OR N DUAL-IN-LINE
OR W FLAT PACKAGE (TOP VIEW)

positive logic: see truth table

description

The SN74141 is a second-generation BCD-to-decimal decoder designed specifically to drive cold-cathode indicator tubes. This decoder demonstrates an improved capability to minimize switching transients in order to maintain a stable display.

Full decoding is provided for all possible input states. For binary inputs 10 through 15, all the outputs are off. Therefore the SN74141, combined with a minimum of external circuitry, can use these invalid codes in blanking leading- and/or trailing-edge zeros in a display. The ten high-performance, n-p-n output transistors have a maximum reverse current of 50 microamperes at 55 volts.

Low-forward-impedance diodes are also provided for each input to clamp negative-voltage transitions in order to minimize transmission-line effects. Power dissipation is typically 80 milliwatts. The SN74141 is characterized for operation over the temperature range of $0°C$ to $70°C$.

functional block diagram

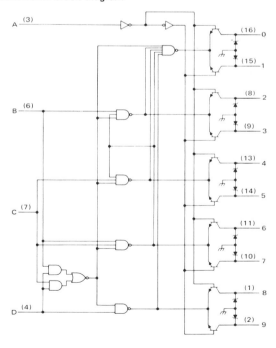

Figure 10-7 Type 74141 BCD-to-decimal decoder-driver. (Courtesy Texas Instruments, Inc.)

absolute maximum ratings over operating free-air temperature range (unless otherwise noted)

Supply voltage, V_{CC} (see Note 1)	7 V
Input voltage	5.5 V
Current into any output (off-state)	2 mA
Operating free-air temperature range	$0°C$ to $70°C$
Storage temperature range	$-65°C$ to $150°C$

NOTE 1: Voltage values are with respect to network ground terminal.

recommended operating conditions

	MIN	NOM	MAX	UNIT
Supply voltage, V_{CC}	4.75	5	5.25	V
Off-state output voltage			60	V
Operating free-air temperature, T_A	0		70	$°C$

electrical characteristics over recommended operating free-air temperature range (unless otherwise noted)

	PARAMETER		TEST CONDITIONS†		MIN	TYP‡	MAX	UNIT
V_{IH}	High-level input voltage				2			V
V_{IL}	Low-level input voltage						0.8	V
V_I	Input clamp voltage		V_{CC} = MIN,	$I_I = -5$ mA			-1.5	V
$V_{O(on)}$	On-state output voltage		V_{CC} = MIN,	$I_O = 7$ mA			2.5	V
$V_{O(off)}$	Off-state output voltage for input counts 0 thru 9		V_{CC} = MAX,	$I_O = 0.5$ mA	60			V
$I_{O(off)}$	Off-state reverse current		V_{CC} = MAX,	$V_O = 55$ V			50	μA
$I_{O(off)}$	Off-state reverse current for input counts 10 thru 15		V_{CC} = MAX, $V_O = 30$ V	$T_A = 55°C$			5	μA
				$T_A = 70°C$			15	
I_I	Input current at maximum input voltage		V_{CC} = MAX,	$V_I = 5.5$ V			1	mA
I_{IH}	High-level input current	A input	V_{CC} = MAX,	$V_I = 2.4$ V			40	μA
		B, C, or D input					80	
I_{IL}	Low-level input current	A input	V_{CC} = MAX,	$V_I = 0.4$ V			-1.6	mA
		B, C, or D input					-3.2	
I_{CC}	Supply current		V_{CC} = MAX,	See Note 2		16	25	mA

†For conditions shown as MIN or MAX, use the appropriate value specified under recommended operating conditions.
‡This typical value is at $V_{CC} = 5$ V, $T_A = 25°C$.
NOTE 2: I_{CC} is measured with all inputs grounded and outputs open.

schematics of inputs and outputs

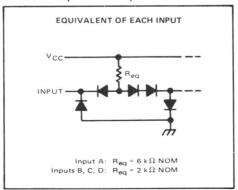

EQUIVALENT OF EACH INPUT

Input A: R_{eq} = 6 kΩ NOM
Inputs B, C, D: R_{eq} = 2 kΩ NOM

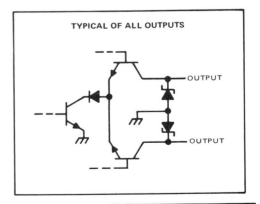

TYPICAL OF ALL OUTPUTS

Figure 10-7 (continued)

means of exclusive-OR comparison of successive bits of the binary code. This is a special case. What we want to do now is develop a technique that is quite general and holds for all transformations.

Consider the case when we have a 9876543210 code. This means that we have 10 lines, each line representing a digit. All lines are at 0 except for the high line at the desired digit. This coding can occur as the result of a 10-stage ring counter or as the decoded outputs of any of the decade codes we have discussed earlier in this chapter, or for that matter decoded outputs of any decade code. Now suppose that we want to generate or encode the excess-3 code for these ten digits. Table 10-10 gives the excess-3 code for each of the ten digits.

Table 10-10 Decimal to excess-3 code

Decimal	Excess-3
	NMLK
0	0011
1	0100
2	0101
3	0110
4	0111
5	1000
6	1001
7	1110
8	1011
9	1100

When do we get a 1 for bit N? Bit N will be a 1 at the digits 5 *or* 6 *or* 7 *or* 8 *or* 9. When do we get a 1 for bit M? Bit M will be a 1 at the digits 1 *or* 2 *or* 3 *or* 4 *or* 9. We can see that fundamentally encoding is an OR process. Following this procedure, Fig. 10-8 shows how to encode excess-3 from decimal.

Let us go one step further. Suppose that our 9876543210 code had been generated from a NBCD 8421 code. Let us now consider the problem of going from 8421 to excess-3. Using the results of Fig. 10-5, we can write

$$\text{⑤} + \text{⑥} + \text{⑦} + \text{⑧} + \text{⑨} = N \qquad (10\text{-}1)$$

$$C\bar{B}A + CB\bar{A} + CBA + D\bar{A} + DA = N \qquad (10\text{-}2)$$

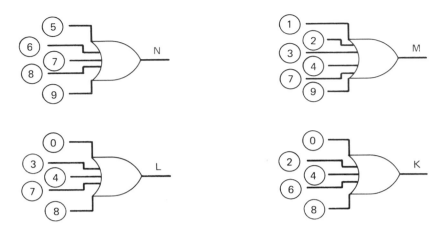

Figure 10-8 Encoding: decimal to excess-3

Can this be simplified? Let us Karnaugh map and try to add the don't-care conditions for the 8421 code (see Fig. 10-4). This is shown in Fig. 10-9a. It turns out in this case that the don't-care conditions were already used in the NBCD simplification. Similar procedures can be followed for bits K, L, and M. The simplification from the Karnaugh map and logic diagram are given in Fig. 10-9b. If one carefully compares the results of the simplification and the truth tables for 8421 and excess-3, as given in Table 10-3, one can see that $N = 1$ when

1 $D = 1$. This occurs only at counts 8 and 9.
2 C and B are both 1. This occurs only on counts 6 and 7.
3 C and A are both 1. This occurs only at counts 5 and 7.

Hence, the simplification procedure is correct. Although it might have been possible to arrive at these simplification results by the type of careful comparison just given between the two codes, the Karnaugh map technique systematizes the procedure.

Figure 10-9 Simplification NBCD-8421 to bit N excess-3: (a) map; (b) logic diagram

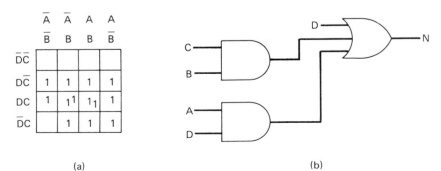

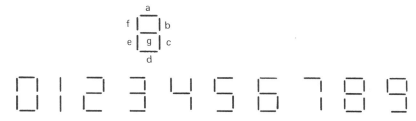

Figure 10-10   Seven-segment display

Figure 10-10 shows a seven-segment arrangement used for digital display. By lighting specific segments, any of the 10 digits can be displayed as shown. Table 10-11 shows the display coding between the decimal code and the segments.

Table 10-11   Decimal to seven-segment code

Decimal	a	b	c	d	e	f	g
0	1	1	1	1	1	1	
1		1	1				
2	1	1		1	1		1
3	1	1	1	1			1
4		1	1			1	1
5	1		1	1		1	1
6	1		1	1	1	1	1
7	1	1	1				
8	1	1	1	1	1	1	1
9	1	1	1	1		1	1

The encoding from a logic standpoint consists of seven OR gates with inputs following Table 10-11. For example,

$$a = \; 0 \; + \; ② + ③ + ⑤ + ⑥ + ⑦ + ⑧ + ⑨$$

Figure 10-11 shows the pin assignments and logic interconnections for the type 7449. This is an 8421 to seven-segment IC that generates output levels of 1, following Table 10-11. It is intended for use as a driver for transistors, which in turn drive lamps in a seven-segment display. In this IC there is also a blanking input that blanks all the lamps when a low input is applied to it. This is used for intensity modulation of lamps and for multiplexing. Study of the logic diagram will show that the 8421 outputs are first decoded and then encoded following Fig. 10-5 and Table 10-11.

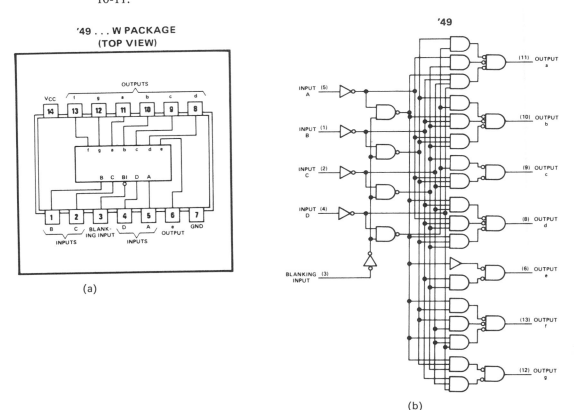

(a)

(b)

Figure 10-11   Type 7449 BCD to 7-segment decoder-driver: (a) pin assignments; (b) logic interconnections

10-12   MULTIPLEXING–DEMULTIPLEXING

Consider the problem of transmitting data from four signal sources to four receivers. One method, the obvious one, is to send the data over four separate wires. However, another technique is to use a single wire for the transmission but to time share, multiplex (MUX) this wire between the four signals. At the transmitting end the data are combined in a time-shared sequence or multiplexed, and at the receiving end the data are separated into four separate channels and demultiplexed (DMUX). Figure 10-12 shows a multiplexer for four inputs in which the inputs are selected one at a time by binary decoding and ORed.

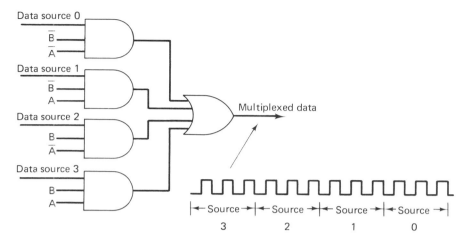

Figure 10-12   Four-line multiplexer

To recover the data, they must be selected, again by decoding gates. In this case the total serial data are applied to every one of four AND gates. These AND gates are selected by a binary code and the gates are enabled when commanded by the binary gating code. This is shown in Fig. 10-13.

Figure 10-14 is the data sheet and logic diagram of the TTL type 251 eight-line multiplexer. The logic diagram is essentially the same as that of Fig. 10-12. This IC has tristate outputs. When the strobe input is high, the outputs are at a high impedance level, and the outputs of additional multiplexers can be connected to the common bus output line for further multiplexing.

The type 54/74154 TTL IC is a four- to 16-line decoder–demultiplexer. Part of its data sheet is shown in Fig. 10-15. It uses binary decoding to select one of 16 out-

Figure 10-13   Demultiplexer

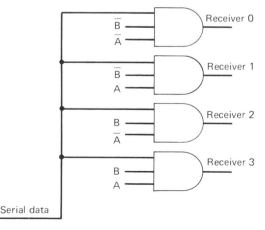

TTL
MSI

TYPES SN54251, SN54LS251, SN54S251,
SN74251, SN74LS251, SN74S251
DATA SELECTORS/MULTIPLEXERS WITH 3-STATE OUTPUTS
BULLETIN NO. DL-S 7211834, DECEMBER 1972

- Three-State Versions of '151, 'LS151, 'S151
- Three-State Outputs Interface Directly with System Bus
- Perform Parallel-to-Serial Conversion
- Permit Multiplexing from N-lines to One Line
- Complementary Outputs Provide True and Inverted Data
- Fully Compatible with Most TTL and DTL Circuits

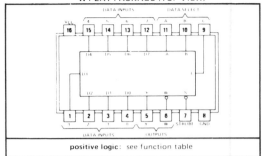

J OR N DUAL-IN-LINE OR
W FLAT PACKAGE (TOP VIEW)

positive logic: see function table

TYPE	MAX NO. OF COMMON OUTPUTS	TYPICAL AVG PROP DELAY TIME (D TO Y)	TYPICAL POWER DISSIPATION
SN54251	49	17 ns	250 mW
SN74251	129	17 ns	250 mW
SN54LS251	19	17 ns	35 mW
SN74LS251	19	17 ns	35 mW
SN54S251	39	8 ns	275 mW
SN74S251	129	8 ns	275 mW

functional block diagram

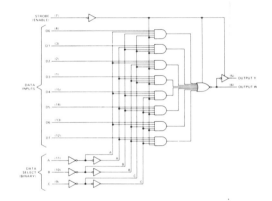

description

These monolithic data selectors/multiplexers contain full on-chip binary decoding to select one-of-eight data sources and feature a strobe-controlled three-state output. The strobe must be at a low logic level to enable these devices. The three-state outputs permit a number of outputs to be connected to a common bus. When the strobe input is high, both outputs are in a high-impedance state in which both the upper and lower transistors of each totem-pole output are off, and the output neither drives nor loads the bus significantly. When the strobe is low, the outputs are activated and operate as standard TTL totem-pole outputs.

To minimize the possibility that two outputs will attempt to take a common bus to opposite logic levels, the output control circuitry is designed so that the average output disable time is shorter than the average output enable time. The SN54251 and SN74251 have output clamp diodes to attenuate reflections on the bus line.

FUNCTION TABLE

INPUTS				OUTPUTS	
SELECT			STROBE	Y	W
C	B	A	S		
X	X	X	H	Z	Z
L	L	L	L	D0	$\overline{D0}$
L	L	H	L	D1	$\overline{D1}$
L	H	L	L	D2	$\overline{D2}$
L	H	H	L	D3	$\overline{D3}$
H	L	L	L	D4	$\overline{D4}$
H	L	H	L	D5	$\overline{D5}$
H	H	L	L	D6	$\overline{D6}$
H	H	H	L	D7	$\overline{D7}$

H = high logic level, L = low logic level
X = irrelevant, Z = high impedance (off)
D0, D1 . . . D7 = the level of the respective D input

Figure 10-14 Type 251 multiplexer (Courtesy Texas Instruments, Inc.)

DESCRIPTION

The 54/74154 decodes 4 binary-coded inputs to one of 16 mutually exclusive outputs when each of the two strobe inputs are low. The demultiplexing function is achieved by using the 4 input lines for output addressing and data from one strobe input while the other strobe input is held low.

PIN CONFIGURATIONS

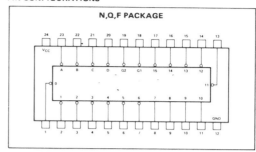

LOGIC DIAGRAM

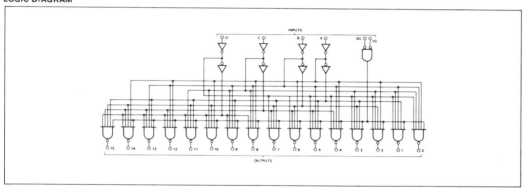

TRUTH TABLE

INPUTS						OUTPUTS															
G1	G2	D	C	B	A	0	1	2	3	4	5	6	7	8	9	10	11	12	13	14	15
L	L	L	L	L	L	L	H	H	H	H	H	H	H	H	H	H	H	H	H	H	H
L	L	L	L	L	H	H	L	H	H	H	H	H	H	H	H	H	H	H	H	H	H
L	L	L	L	H	L	H	H	L	H	H	H	H	H	H	H	H	H	H	H	H	H
L	L	L	L	H	H	H	H	H	L	H	H	H	H	H	H	H	H	H	H	H	H
L	L	L	H	L	L	H	H	H	H	L	H	H	H	H	H	H	H	H	H	H	H
L	L	L	H	L	H	H	H	H	H	H	L	H	H	H	H	H	H	H	H	H	H
L	L	L	H	H	L	H	H	H	H	H	H	L	H	H	H	H	H	H	H	H	H
L	L	L	H	H	H	H	H	H	H	H	H	H	L	H	H	H	H	H	H	H	H
L	L	H	L	L	L	H	H	H	H	H	H	H	H	L	H	H	H	H	H	H	H
L	L	H	L	L	H	H	H	H	H	H	H	H	H	H	L	H	H	H	H	H	H
L	L	H	L	H	L	H	H	H	H	H	H	H	H	H	H	L	H	H	H	H	H
L	L	H	L	H	H	H	H	H	H	H	H	H	H	H	H	H	L	H	H	H	H
L	L	H	H	L	L	H	H	H	H	H	H	H	H	H	H	H	H	L	H	H	H
L	L	H	H	L	H	H	H	H	H	H	H	H	H	H	H	H	H	H	L	H	H
L	L	H	H	H	L	H	H	H	H	H	H	H	H	H	H	H	H	H	H	L	H
L	L	H	H	H	H	H	H	H	H	H	H	H	H	H	H	H	H	H	H	H	L
L	H	X	X	X	X	H	H	H	H	H	H	H	H	H	H	H	H	H	H	H	H
H	L	X	X	X	X	H	H	H	H	H	H	H	H	H	H	H	H	H	H	H	H
H	H	X	X	X	X	H	H	H	H	H	H	H	H	H	H	H	H	H	H	H	H

H = High, L = Low, X = Irrelevant

Figure 10-15 Type 54/74154 decoder-demultiplexer (Courtesy Signetics, Corporation)

put lines to enable the data input. In this IC there are two data inputs, 1G and 2G. Either one may be used as a data input with the other as an enable. The logic output of these two inputs goes to the input of all 16 NAND gates very much like that shown in Fig. 10-13. Each output NAND is enabled by decoding four-line binary inputs.

An interesting application of multiplexing and demultiplexing is on modern jet airplanes. The passenger can select any one of eight music channels, yet the music for all the lines is being transmitted over a single line on a multiplexed basis. Although these are analog data, the basic principles are the same as we have just discussed. Each music line is sampled at a rate much higher than the response of the human ear. When demultiplexed, the ear ignores the rapid sampling rates since these rates are too high for it to follow.

The type 4016A is a CMOS IC switch intended for service in multiplexing digital or analog (variable voltage) signals. It constains four bilateral (transmit in both directions) switches. When the control voltage is at level 1 (V_{DD}), the switch-on resistance is approximately 300 Ω, and when the control voltage is at level 0 (V_{SS}), the switch resistance is extremely high. Hence logic levels can be used to control the time when any switch can transmit. This IC therefore can be used to multiplex four signals.

Problems

10-1. Express the following decimal numbers in 8421 code: (a) 27; (b) 628; (c) 4372.

10-2. Decode the following numbers expressed in 8421 code: (a) 0111 1000 0101; (b) 1001 1001 0001 0101.

10-3. Express the following decimal numbers in 4221 code: (a) 648; (b) 2643.

10-4. Decode the following numbers expressed in 4221 code: (a) 0011 1100 1110; (b) 0010 0110 1110 1111.

10-5. Express the following decimal numbers in excess-3 code: (a) 821; (b) 6243.

10-6. (a) Express the decimal number 37 in excess-3 code.
(b) Take the complement of part (a).

10-7. Express the decimal number 62 in (a) 7-bit biquinary code; (b) 7-bit quibinary code.

10-8. Express the decimal number 47 in (a) 543210 code; (b) 864201 code.

10-9. Express the decimal number 18 in (a) 7-bit biquinary code; (b) 7-bit quibinary code.

10-10. Express the decimal number 74 in (a) 543210 code; (b) 864201 code.

10-11. Express the following decimal numbers in 8421 code and add a parity bit to make them odd parity: (a) 64; (b) 87; (c) 135.

10-12. Express the following decimal numbers in excess-3 code and add a parity bit to make them even parity: (a) 24; (b) 79; (c) 456.

10-13. The following data are being sent in even parity and are encoded in rows and columns. Determine the location of the error.

Data						Row Parity Bit
0	1	0	1	0	1	1
0	0	0	0	1	0	1
1	0	1	0	1	1	0
1	0	0	0	0	0	0
1	0	0	0	1	0	0
1	1	0	1	1	0	Column parity bit

10-14. Determine the Gray code for the decimal number 47 using the method of Example 10-6.

10-15. Repeat Prob. 10-14 for decimal 28.

10-16. A shaft is encoded in Gray code with eight rings. How many degrees does each segment represent?
(b) Repeat for 13 rings.

10-17. Express the following in ASCII code: (a) PAGE 28; (b) CMOS.

10-18. Repeat Prob. 10-17 but use EBCDIC coding.

10-19. (a) Draw the waveforms at the clock Q_A, $\bar{Q}_A$, Q_B, and $\bar{Q}_B$.
(b) Draw the decoding gates for the four counts.
(c) Draw the decoded waveforms under these for part (a), showing the decoded outputs in their proper time.

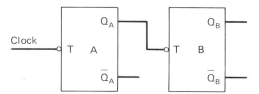

Figure P10-19   Two-stage ripple counter

10-20. Show that the count-state decoding for decimal 9 ($= 1001$ or $D\bar{C}\bar{B}A$) in the 8421 code simplifies to DA.

10-21. In the twisted-ring counter, the count 0 has the code $0000 = \bar{D}\bar{C}\bar{B}\bar{A}$:
(a) Tabulate all eight count states in the logic level form and in letter form.
(b) Show that to decode it requires only two-input AND gates. (This is true for all twisted-ring counters.) Determine the AND gate inputs and draw the decoding AND gates.

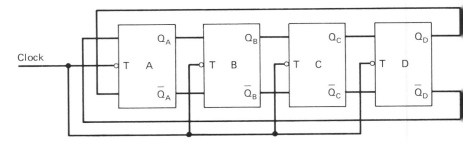

Figure P10-21   Four-stage twisted-ring counter

10-22. (a) Show that decimal 0 in 4221 is decoded by $\bar{C}\bar{B}\bar{A}$.
   (b) Show that decimal 6 in the 4221 code can be decoded by $C\bar{B}\bar{A}$ or $D\bar{B}\bar{A}$.
   (c) Show that all 10 decimals in the 4221 can be decoded by three-input AND gates, and give the inputs to these gates.

10-23. Show that to decode 2′421a requires 10 three-input AND gates. Draw the gates giving all the inputs.

10-24. Show that to decode count 5 in code 2′421b requires $D\bar{C}$.

10-25. Code X is an arbitrary 2-bit code. It is to be converted to arbitrary code Y. Show how to do this and give the logic diagrams.

	X	Y
	AB	CDE
1	11	100
2	01	001
3	00	011
4	10	101

ELEVEN

Memory

11-1 INTRODUCTION

Most of the fundamental concepts of memory have already been developed in previous chapters. A memory is basically a bistable. In this chapter we shall discuss the style and organization of both semiconductor and magnetic memories.

There are two types of memory

1 Random access: any element of the memory can be reached in the same time.
2 Limited access: memory systems in which the memory elements move (mechanically or electronically), and the access time depends upon the memory element's distance from the read–write location.

11-2 BASIC SEMICONDUCTOR MEMORY ELEMENT

Figure 11-1 is the diagram of a basic semiconductor memory element, an *RS* bistable. It is like the two-input NAND bistable of Chapter 6 except for the inverter in series with the *R*–reset input and the two-input NAND gate in series

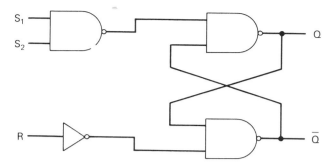

Figure 11-1 *RS* bistable: the basic semicoductor memory element

with the *S*–set input. To reset this bistable, the *R* input has to be made a 1. This stores (puts in memory) a 0 in this memory element ($Q = 0$, $\bar{Q} = 1$). A two-input NAND gate is in series with the set input. One input to this gate, for example S_1, can be used as an ENABLE or SELECT control. When it is made a 1, the bistable can be set with a 1 to the S_2 input.

In the bistable of Fig. 11-1, the level at the Q and $\bar{Q}$ terminals can be measured or sensed as many times as desired without affecting or destroying the logic levels at these terminals. Semiconductor memories are *NDRO* (nondestructive read out).

However, if power is turned off or lost, the voltage levels at Q and $\bar{Q}$ go to 0. When power is recovered, the levels that Q and $\bar{Q}$ now assume are determined by the transistor characteristics, components, or noise, not by the levels previously stored in the memory. This bistable may or may not come back to the desired state, and we have lost the stored information. This *RS* bistable semiconductor memory element is *volatile* (memory is lost when power is turned off).

To overcome the voltatility problem, temporary power sources are sometimes connected to the memory to provide standby power. These power sources may take the form of a large storage capacitor or a battery.

11-3 ORGANIZATION OF A SEMICONDUCTOR RANDOM-ACCESS MEMORY (RAM)†

Figure 11-1 is the diagram of a single memory element. In Chapter 8 we discussed the shift register, which is a multibit memory. However, in many cases the shift register is a memory with limited access to the memory elements and the memory stored in it. When only input and output terminals are available, we have to shift the data to the putput read terminal to determine what is stored in a specific location, and this takes time. What we have to do is show how a memory can be

---

†M. E. Levine, *Digital Theory and Experimentation Using Integrated Circuits.* Englewood Cliffs, N.J.: Prentice-Hall, Inc., 1974, Expt. 14, Random Access (Ram)–Scratch Pad Memories.

organized in which any element can be reached in the same amount of time. Such a memory is called a random-access memeory (RAM), a read–write memory, or a scratch-pad memory.

Figure 11-2 shows the basic principles of a 2-bit RAM with the added features of a READ/$\overline{\text{WRITE}}$ line and an ENABLE line.

Figure 11-2   Two-bit RAM with READ/$\overline{\text{WRITE}}$ and ENABLE

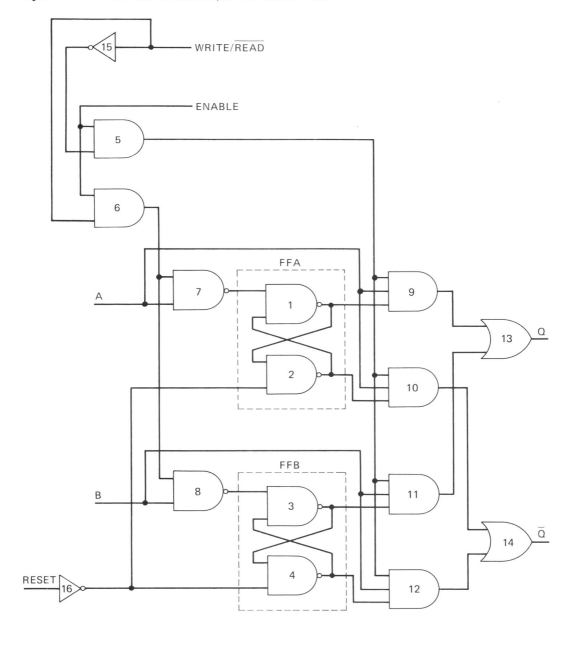

1   Flip-flop A is composed of NAND gates 1 and 2 and control gate 7.

2   Flip-flop B is composed of NAND gates 3 and 4 and control gate 8.

3   There is a common reset line that resets both bits to the 0 state when it is made a 1.

4   There is an ENABLE line. When made a 1, it will permit either a WRITE or $\overline{\text{READ}}$ operation by means of AND gates 5 and 6.

5   With the ENABLE line at a 1, if the WRITE/$\overline{\text{READ}}$ line is made a 1, data levels of 1 can be stored in *A* or *B* by selecting lines *A* or *B* and making them equal to 1. (Although it is possible to select both *A* and *B* simultaneously in the illustration of Fig. 11-2, in general only *A* or *B* can be selected in ICs.)

6   With the WRITE/$\overline{\text{READ}}$ line at a 1, inverter 15 disables output gates 9–14.

7   To read the state of the FFs, the WRITE/$\overline{\text{READ}}$ line is made a 0. This disables the write function and enables the read output gates 9–12, which again can be selected by lines *A* or *B*. The level of the stored data can be measured at outputs Q and $\bar{Q}$.

8   In Fig. 11-2, a 1 stored in an element leads to a 1 at Q and a 0 at $\bar{Q}$. However, in some ICs the stored data levels are inverted, and a stored 1 results in $Q = 0$ and $\bar{Q} = 1$. This is to make the outputs compatible with the *PR* and *CLR* inputs of FFs such as the TTL type 7476.

11-4 TERMINAL ORGANIZATION OF A RAM

In Fig. 11-2 the memory is organized so that each bit requires a separate input terminal to write a 1. Since the number of output terminals to an IC is limited, several methods have been developed to reduce the number of required terminals. Figure 11-3 shows an *XY* method of bit selection for a 16-bit (4 × 4) RAM. Each bit is selected by an *X* and a *Y* line, which becomes the S_1 input of Fig. 11-1. We see that it takes eight lines to select one of 16 bits, rather than the expected 16 individual lines.

Another common method of reducing the number of input terminals needed is to address the input terminals with binary numbers and internally decode. For example, with internal decoding only two input lines are needed to develop a four-input selection. The type 7489 (Fig. 11-4) is a TTL 64-bit RAM (16 words of 4 bits each). In this IC, four data select lines, *A*, *B*, *C*, and *D*, are decoded to give 16 selected words of 4 bits per word. Four input lines now provide data input to be stored, and four sense lines are used for reading the outputs of the stored data. Figure 11-4 shows the pin assignment and internal organization of this IC.

In another type of organization, the type 2102A is a static 1024- × 1-bit NMOS RAM. Ten input lines, as binary input with internal decoding, select any one of 1024 bits. Figure 11-5 shows the technical data for this IC.

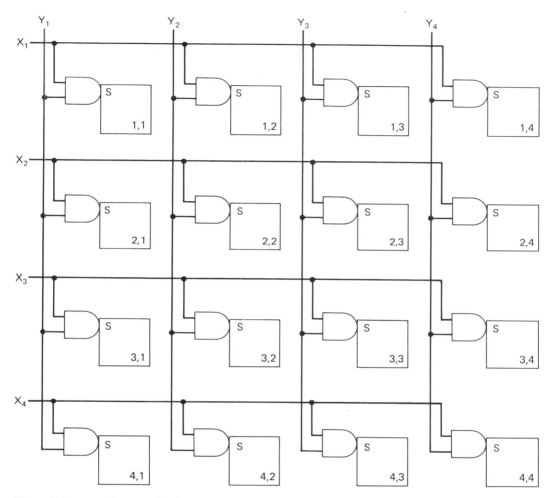

Figure 11-3 *XY* RAM organization

The type 7489 is a 16- × 4-bit memory and the type 2102A is a 1024- × 1-bit memory. To solve a problem with any complexity requires memories with much greater capacity. How is this done? The basic method is the WIRED-COLLECTOR logic that was discussed in Chapter 3. The type 7489 has an open collector. By using the WIRED-COLLECTOR technique with the outputs of other similar ICs (a pull-up resistor is needed), the system storage capacity can be increased. In the same way, the 2102A has a tristate output, and the output terminal of this IC can be directly connected to the output terminals of other ICs. For example, eight type 2102A's can be connected together to obtain an 8192 × 1 (8K × 1) memory.

- For Application as a "Scratch Pad" Memory with Nondestructive Read-Out
- Fully Decoded Memory Organized as 16 Words of Four Bits Each
- Fast Access Time . . . 33 ns Typical
- Diode-Clamped, Buffered Inputs
- Open-Collector Outputs Provide Wire-AND Capability
- Typical Power Dissipation . . . 375 mW
- Compatible with Most TTL and DTL Circuits

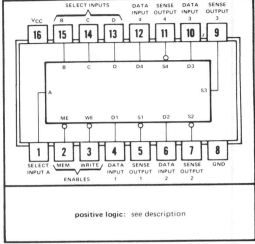

J OR N DUAL-IN-LINE OR W FLAT PACKAGE (TOP VIEW)†

positive logic: see description

†Pin assignments for these circuits are the same for all packages.

description

This 64-bit active-element memory is a monolithic, high-speed, transistor-transistor logic (TTL) array of 64 flip-flop memory cells organized in a matrix to provide 16 words of four bits each. Each of the 16 words is addressed in straight binary with full on-chip decoding.

The buffered memory inputs consist of four address lines, four data inputs, a write enable, and a memory enable for controlling the entry and access of data. The memory has open-collector outputs which may be wire-AND connected to permit expansion up to 4704 words of N-bit length without additional output buffering. The open-collector outputs may be utilized to drive external loads directly; however, dynamic reponse of an output can, in most cases, be improved by using an external pull-up resistor in conjunction with a partially loaded output. Access time is typically 33 nanoseconds; power dissipation is typically 375 milliwatts.

FUNCTION TABLE

ME	WE	OPERATION	CONDITION OF OUTPUTS
L	L	Write	Complement of Data Inputs
L	H	Read	Complement of Selected Word
H	L	Inhibit Storage	Complement of Data Inputs
H	H	Do Nothing	High

write operation

Information present at the data inputs is written into the memory by addressing the desired word and holding both the memory enable and write enable low. Since the internal output of the data input gate is common to the input of the sense amplifier, the sense output will assume the opposite state of the information at the data inputs when the write enable is low.

read operation

The complement of the information which has been written into the memory is nondestructively read out at the four sense outputs. This is accomplished by holding the memory enable low, the write enable high, and selecting the desired address.

Figure 11-4 Type SN7489 64-bit READ/WRITE memory. (Courtesy Texas Instruments, Inc.)

functional block diagram

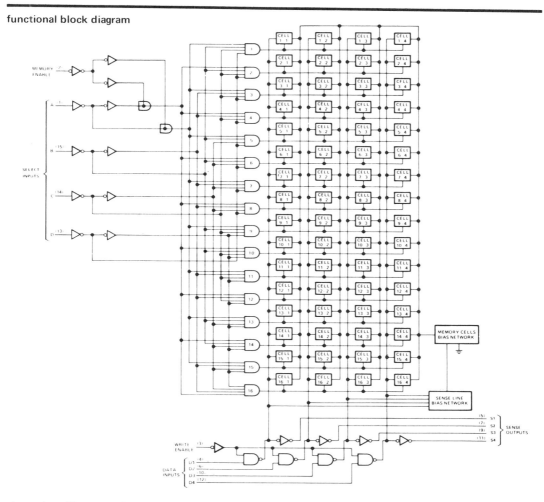

schematics of inputs and outputs

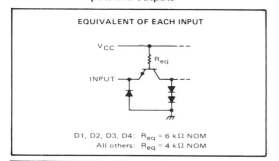

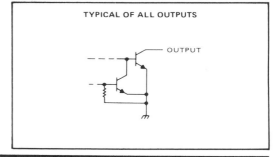

Figure 11-4 (continued) Type SN7489 64-bit READ/WRITE memory.
(Courtesy Texas Instruments, Inc.)

Silicon Gate MOS **2102A**

1024 BIT FULLY DECODED STATIC MOS RANDOM ACCESS MEMORY

*Fast Access Time -- 350 ns max.

- Single +5 Volts Supply Voltage
- Directly TTL Compatible — All Inputs and Output
- Static MOS — No Clocks or Refreshing Required
- Low Power — Typically 150 mW
- Three-State Output — OR-Tie Capability

- Simple Memory Expansion — Chip Enable Input
- Fully Decoded — On Chip Address Decode
- Inputs Protected — All Inputs Have Protection Against Static Charge
- Low Cost Packaging — 16 Pin Plastic Dual-In-Line Configuration

The Intel 2102A is a high speed 1024 word by one bit static random access memory element using N-channel MOS devices integrated on a monolithic array. It uses fully DC stable (static) circuitry and therefore requires no clocks or refreshing to operate. The data is read out nondestructively and has the same polarity as the input data.

The 2102A is designed for memory applications where high performance, low cost, large bit storage, and simple interfacing are important design objectives. *A low standby power version (order as a 2102A/S1172) is also available. It has all the same operating characteristics of the 2102A with the added feature of 42 mW maximum power dissipation in standby.*

It is directly TTL compatible in all respects: inputs, output, and a single +5 volt supply. A separate chip enable ($\overline{CE}$) lead allows easy selection of an individual package when outputs are OR-tied.

The Intel 2102A is fabricated with N-channel silicon gate technology. This technology allows the design and production of high performance easy to use MOS circuits and provides a higher functional density on a monolithic chip than either conventional MOS technology or P-channel silicon gate technology.

Intel's silicon gate technology also provides excellent protection against contamination. This permits the use of low cost silicone packaging.

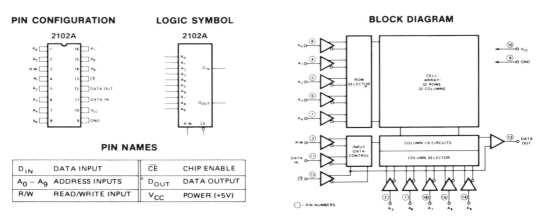

PIN NAMES

D_{IN}	DATA INPUT	$\overline{CE}$	CHIP ENABLE
$A_0 - A_9$	ADDRESS INPUTS	D_{OUT}	DATA OUTPUT
R/W	READ/WRITE INPUT	V_{CC}	POWER (+5V)

Figure 11-5 Type 2102A static 1024 × 1-bit NMOS RAM

SILICON GATE MOS 2102A

Absolute Maximum Ratings*

Ambient Temperature Under Bias	0°C to 70°C
Storage Temperature	−65°C to +150°C
Voltage On Any Pin With Respect To Ground	−0.5V to +7V
Power Dissipation	1 Watt

*COMMENT:
Stresses above those listed under "Absolute Maximum Rating" may cause permanent damage to the device. This is a stress rating only and functional operation of the device at these or at any other condition above those indicated in the operational sections of this specification is not implied. Exposure to absolute maximum rating conditions for extended periods may affect device reliability.

D. C. and Operating Characteristics

$T_A = 0°C$ to $+70°C$, $V_{CC} = 5V \pm 5\%$ unless otherwise specified.

Symbol	Parameter	Limits			Unit	Test Conditions
		Min.	Typ.[1]	Max.		
I_{LI}	Input Load Current (All Input Pins)			10	μA	$V_{IN} = 0$ to 5.25V
I_{LOH}	Output Leakage Current			5	μA	$\overline{CE} = 2.0V$, $V_{OUT} = 2.4$ to V_{CC}
I_{LOL}	Output Leakage Current			−10	μA	$\overline{CE} = 2.0V$, $V_{OUT} = 0.4V$
I_{CC1}	Power Supply Current		30	60	mA	All Inputs = 5.25V Data Out Open $T_A = 25°C$
I_{CC2}	Power Supply Current			70	mA	All Inputs = 5.25V Data Out Open $T_A = 0°C$
V_{IL}	Input "Low" Voltage	−0.5		0.8	V	
V_{IH}	Input "High" Voltage	2.0		V_{CC}	V	
V_{OL}	Output "Low" Voltage			0.4	V	$I_{OL} = 2.1mA$
V_{OH}	Output "High" Voltage	2.4			V	$I_{OH} = -100\mu A$

Standby Characteristics (For 2102A Device Types—S1172, S1173, S1174)

$T_A = 0°C$ to $70°C$

Symbol	Parameter	Limits			Unit	Test Conditions
		Min.	Typ.[1]	Max.		
V_{PD}	V_{CC} in Standby	1.5			V	
V_{CES}	$\overline{CE}$ Bias in Standby	V_{PD}			V	
I_{PD1}	Standby Current Drain		15	28	mA	All Inputs = V_{PD1} = 1.5V
I_{PD2}	Standby Current Drain		20	38	mA	All Inputs = V_{PD2} = 2.0V
t_{CP}	Chip Deselect to Standby Time	0			ns	
t_R	Standby Recovery Time	T_{RC}			ns	

STANDBY WAVEFORMS

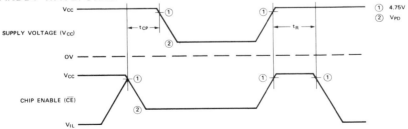

NOTE: 1. Typical values are for $T_A = 25°C$ and nominal supply voltage.

Figure 11-5 (continued)   Type 2102A static 1024×1-bit NMOS RAM

SILICON GATE MOS 2102A

A. C. Characteristics T_A = 0°C to 70°C, V_{CC} = 5V ±5% unless otherwise specified

Symbol	Parameter	Min.	Typ.[1]	Max.	Unit
READ CYCLE					
t_{RC}	Read Cycle	350			ns
t_A	Access Time			350	ns
t_{CO}	Chip Enable to Output Time			180	ns
t_{OH1}	Previous Read Data Valid with Respect to Address	40			ns
t_{OH2}	Previous Read Data Valid with Respect to Chip Enable	0			ns
WRITE CYCLE					
t_{WC}	Write Cycle	350			ns
t_{AW}	Address to Write Setup Time	20			ns
t_{WP}	Write Pulse Width	250			ns
t_{WR}	Write Recovery Time	0			ns
t_{DW}	Data Setup Time	250			ns
t_{DH}	Data Hold Time	0			ns
t_{CW}	Chip Enable to Write Setup Time	250			ns

NOTE: 1 Typical values are for T_A = 25°C and nominal supply voltage.

A. C. CONDITIONS OF TEST

Input Pulse Levels:	0.8 Volt to 2.0 Volt
Input Rise and Fall Times:	10nsec
Timing Measurement Inputs:	1.5 Volts
Reference Levels Output:	0.8 and 2.0 Volts
Output Load:	1 TTL Gate and C_L = 100 pF

Capacitance[2] T_A = 25°C, f = 1 MHz

SYMBOL	TEST	LIMITS (pF)	
		TYP.[1]	MAX.
C_{IN}	INPUT CAPACITANCE (ALL INPUT PINS) V_{IN} = 0V	3	5
C_{OUT}	OUTPUT CAPACITANCE V_{OUT} = 0V	7	10

NOTE: 2. This parameter is periodically sampled and is not 100% tested.

Waveforms

READ CYCLE

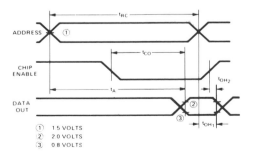

①	1.5 VOLTS
②	2.0 VOLTS
③	0.8 VOLTS

WRITE CYCLE

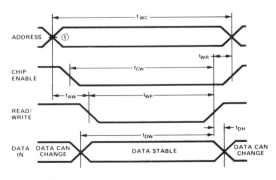

Figure 11-5 (continued)   Type 2102A static 1024 × 1-bit NMOS RAM

11-5 READ ONLY MEMORY (ROM)

Problems that require a RAM occur very frequently. Examples of this are look-up tables, such as trigonometric and logarithmic tables. These could be computed every time they are needed, or transferred from another memory into a RAM, but these actions take time and equipment. In many cases, it is better to have this information stored in a RAM in which the necessary data have been permanently written. Such a *fixed* memory is called a read-only memory (ROM). Another example of the need for such a memory table is in code conversion. A typical example of a code conversion requirement occurs when we go from the selectric typewriter code to ASCII, and vice-versa.

The generation of ROMs to specific needs occurs so frequently that IC manufacturers have developed a set of charts and programs to tell the IC manufacturer exactly how the customer wants the ROM to be written. Such charts can be found in IC manufacturers' data books.

To provide even greater versatility in the availability of such permanent random-access memories, ROMs are available that can be programmed by the user. They come to the user either as all 1s or all 0s. Suppose that the ROM is all 1s. The user selects the bit location where his memory requires a 0 in the usual manner and then applies a high pulse of current to the IC. This current flows through a metallic link (nickel–chromium). The heat generated vaporizes the link, opens it, and a 0 has been programmed by the user in the desired location. Such a memory is called programmable read-only memory (PROM). This is a permanent modification of the memory, and the 1 cannot be restored.

These ROMs are NDRO and nonvolatile.

A most remarkable development by INTEL has been the erasable read-only memory (EROM). This an MOS circuit with a floating gate. To program a 1 in a specific location, the location is selected and a programming voltage is applied. This programming voltage produces a charge on the gate, and there is no discharge path once the voltage has been removed. To reprogram it, the IC is exposed to ultraviolet light through a quartz window, which is part of the case of the IC. This removes the charge and the IC can now be reprogrammed.

11-6 SEMICONDUCTOR MEMORY TECHNOLOGY

The technology of making semiconductor memories follows the same principles used in making integrated circuits. In bipolar logic the memory elements are modifications of the basic gates, and with MOS technology the memories developed have considerable similarity to the static and dynamic shift registers discussed in Chapter 8.

11-6.1 The TTL Memory Element

A basic bipolar memory element in TTL construction is shown in Fig. 11-6. As can be seen, it is basically a bistable. To select a particular cell, both the X and Y lines

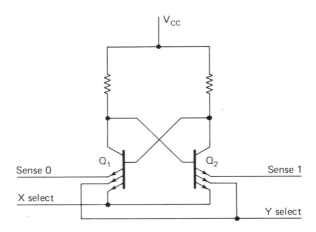

Figure 11-6 TTL basic memory cell

are made high. If a 1 is stored in the cell, current flows in the Sense-1 line of Q_2 and is sensed in a sense amplifier. If a 0 is stored in the cell, current flows in the Sense-0 line of Q_1. To write a 1 into the cell, the X and Y lines are selected and made high. The 0 line is forced high. This writes a 1 into the cell. Similarly, to write a 0 into the cell, the 1 line is forced high. The TTL type 7481 has 16 such cells.

11-6.2 MOS Memory Elements

MOS memory elements have many advantages over the bipolar memories just discussed. They dissipate less power and are less expensive per memory element. They are more economical in chip area, and require less processing steps to manufacture when compared to bipolar. The first memory elements used PMOS technology. This caused interfacing problems with TTL. Recent technological developments have enabled NMOS ICs to be made (Fig. 11-5) that are directly interfaceable with TTL. MOS memory elements are either static or dynamic. The static is more complex and has reduced speed and higher dissipation. The dynamic is simpler, faster, occupies less chip area, and has lower dissipation. However, the storage element in the dynamic is a charged capacitor, and its charge must be restored or refreshed. The rate of refreshing is temperature dependent and is about once every 2 ms at 70°C.

MOS Static Element

The basic MOS static memory element is the MOS P-channel eight-transistor bistable shown in Fig. 11-7.

1 Q_1Q_3 is a two-transistor inverter, as is Q_2Q_4.
2 The inverters are cross-coupled to form a basic bistable.
3 Bistable output D can be coupled through transistors Q_5 *and* Q_7 to the 0 output line.

4   Bistable output E can be coupled through transistors Q_6 *and* Q_8 to the 1 output line.

5   If the X select line is at ground potential, transistors Q_5 and Q_6 are non-conducting. To turn on transistors Q_5 and Q_6, pulse the X select line negative.

6   If the Y select line is at ground potential, transistors Q_7 and Q_8 are nonconducting. To turn on transistors Q_7 and Q_8, pulse the Y select line negative.

7   If the state of the FF is to be read, sense the voltage levels at the 0 and 1 output lines.

8   If data are to be stored in the FF, apply ground voltages as needed at the 0 or 1 output lines. This forces the FF into either the 0 or 1 state.

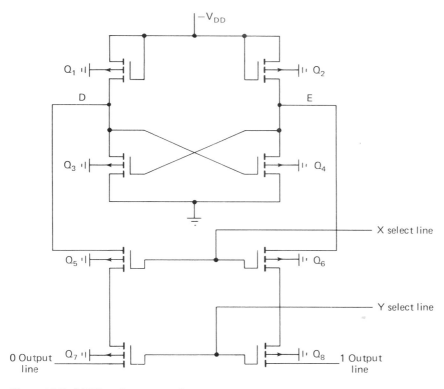

Figure 11-7   MOS static memory element

MOS Dynamic Element

The basic MOS dynamic memory element is the MOS P-channel three-transistor cell shown in Fig. 11-8. Data are stored in dynamic storage memories by charging a capacitor. In Fig. 11-8 the charge storage capacitor is the gate capacitance of transistor Q_2.

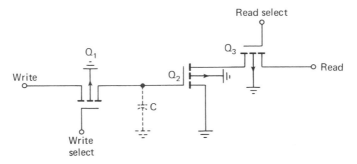

Figure 11-8 Three MOS transistor dynamic memory cell

1 To write, make the write select line negative. This turns on Q_1 (P-channel MOS).

    a To write a 0, make the write line negative. This charges up capacitor C to a negative voltage.

    b To write a 1, make the write line go to ground. This makes the charge on C equal to zero and the voltage at the gate of Q_2 equal to ground.

2 Transistor Q_2 is an inverter. This means that ultimately another inverter is needed to reinvert the level.

3 To read, make the read select line go negative. This turns on Q_3. The inverted voltage level at the gate of Q_2 appears at the output read line.

4 Since capacitor C is part of Q_2, there is some leakage current into the gate of Q_2 and the voltage across C decays. It is necessary to periodically refresh the charge in C.

5 If the write select and read select lines are at ground potential, the memory element is isolated.

A very popular LSI memory is the type 1103. This is a 1024-word by 1-bit dynamic memory using the memory cell of Fig. 11-8. Ten input lines are decoded to select any cell in the memory. Refreshing is required every 2 ms, and the time required for a write cycle or read/write cycle is 580 μs. For memory expansion, the output of this IC can be WIRE-ORed with other 1103's, and a chip enable control permits selection of any one of the 1103's.

11-7 RANDOM-ACCESS MAGNETIC MEMORIES

The most popular type of random-access memory has for many years been the magnetic core (a small doughnut) memory. It is only recently that semiconductor technology has progressed to the point where semiconductor memories are beginning to compete with core memory. Magnetic-core memories are nonvolatile, but, as we shall see, they are destructive read out (DRO).

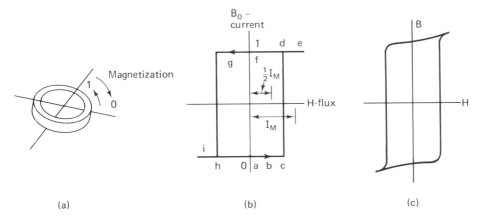

Figure 11-9   Principle of the magnetic-core memory: (a) cores; (b) ideal square loops; (c) actual magnetization curve

Figure 11-9 shows the basic principles of the magnetic-core memory. It can be magnetized in one direction, corresponding to a 1, or in the reverse direction, corresponding to a 0, as shown in Fig. 11-9a. The core is a square loop material whose ideal magnetization curve is shown in Fig. 11-9b. Figure 11-9c shows the actual magnetization curve for such a material. In a core memory, currents (called "half-select currents"; $\frac{1}{2}I_M$) in two wires are needed to reverse the magnetization of the core.

Suppose that the core is magnetized in the 0 state and no current flows in core wires X and Y, Fig. 11-7b, position a. If current $+\frac{1}{2}I_M$ is now applied only to X (or Y), we reach position b on the curve, not enough to change the direction of magnetic flux. It is only when currents are applied to both wires X and Y that we can move from $a \rightarrow b \rightarrow c \rightarrow d \rightarrow e$. If the currents are now removed from the wires, we now move on the curve from $e \rightarrow d \rightarrow f$. Point f represents a change in flux direction, and the core has been magnetized into the 1 state. In the same way, to reverse the direction of magnetization again requires "half-currents" in the reverse direction through wires X and Y. We now travel along the magnetization curve from $f \rightarrow g \rightarrow h \rightarrow i$, and from $i \rightarrow h \rightarrow a$ when the currents are removed from X and Y. The "half-currents" needed are about 0.25 ampere (A) and are produced in transistors called *core drivers.*

Actually, the cores are more complex. Four wires go through the cores; Fig. 11-10 shows such a core. We have already discussed the X and Y wires, which are used to write a 1 into a core. The SENSE wire is used to detect the presence or absence of a 1 in a core. Suppose that there is a 1 in a core. If we now put, for a short time, reverse currents into X $(-\frac{1}{2}I_M)$ and Y $(-\frac{1}{2}I_M)$, the core will now switch from point f to point i and to point a. This change in magnetic flux will be sensed by the SENSE wire as an induced voltage. Suppose that the core had been at a 0. The reverse currents would move us along the magnetization curve from a to i and back to a, no change in flux. We can see that the presence of a stored 1

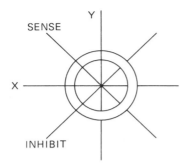

Figure 11-10 Core element with four wires

develops a voltage on the SENSE wire during the read reversal process, whereas a stored 0 results in no voltage. Integrated circuits called sense amplifiers are used to amplify the voltage developed in the sense wire since this voltage is quite small.

However, we see that the way in which we detect the presence of a 1 is to convert it to a 0. This detection method is a destructive read out (DRO). This is not good since to read a 1 we have to destroy it. It is the function of the fourth wire, the *INHIBIT* wire, to help recover the 1 that was lost. It does this in the following way. After the read cycle, which is destructive, write currents $(= +\frac{1}{2}I_M)$ are again put into X and Y to restore a 1. If the core had originally stored a 1, no current is put through the INHIBIT wire, and a 1 is rewritten. However, if a 0 had been stored in the core, a reverse current $(= -\frac{1}{2}I_M)$ is put into the INHIBIT wire and this cancels one of the currents of X and Y; the net result is that a 0 is left in the core.

To summarize, data are stored in a core as 1 or 0 depending upon the direction of magnetization. The process of reading the data from the core is a DRO procedure, and data must be rewritten into the core in a read/write cycle, which proceeds as follows:

1   If the data stored is a 1:
   a   The core switches its magnetic field.
   b   The change in magnetic field is sensed by the SENSE wire as a 1.
   c   Currents of magnitude $+\frac{1}{2}I_M$ are put into X, $+\frac{1}{2}I_M$ into Y, and 0 into INHIBIT.
   d   The net magnetizing current is I_M and a 1 is rewritten into the core.
2   If the data stored is a 0:
   a   The core does not switch its magnetic field.
   b   The SENSE wire does not sense a change in flux, and the SENSE wire senses a 0.
   c   Currents of magnitudes $+\frac{1}{2}I_M$ are put into X, $+\frac{1}{2}I_M$ into Y, and $-\frac{1}{2}I_M$ into INHIBIT.
   d   The net magnetizing current is $\frac{1}{2}I_M$, and the core does not switch. A 0 remains in the core.

Figure 11-11 shows the sequence of the read/write cycle. The time required for a read/write cycle is approximately 0.5 μs. The cores used for magnetic memory are

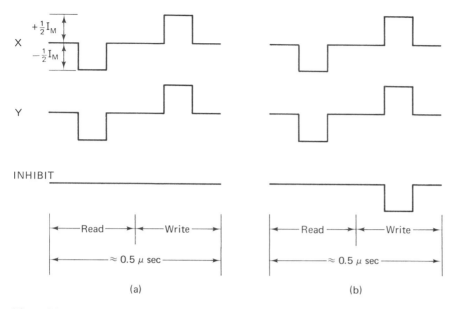

Figure 11-11   Magnetic-core read/write cycle: (a) stored 1; (b) stored 0

very small, and it is difficult to imagine that four wires can be threaded through such a small opening, especially in an array as discussed in the next section. For a long time 50-mil cores were used; more recently, 18-mil cores have been used, and even smaller 13-mil cores are being developed. The objectives in core-size reduction are reductions in cost, currents required, and time required for the read/write cycle. To give you some idea of how small these cores are, Fig. 11-12 shows a group of 30,000 18-mil cores.

Figure 11-12   Thirty thousand 18-mil memory cores. (Courtesy Ferroxcube Corporation)

Figure 11-13 shows a 4 × 4 array of 16 memory cores called a memory plane. The arrangement is similar to that of the semiconductor memory array of Fig. 11-4 and, like Fig. 11-4, it is also a RAM. In this array, common SENSE and INHIBIT wires are threaded through all the cores. As in Fig. 11-4, eight wires are needed to locate any one of the 16 memory cores in a random-access arrangement. This is called coincident-current selection. It is effectively a magnetic AND gate. For example, to write a 1 in cell $X = 2$, $Y = 3$ requires that $+\frac{1}{2}I_M$ be put into X_2 and $\frac{1}{2}I_M$ into Y_3. The read/write procedure follows that discussed in Sec. 11.7. Memory plane arrays of this type have been very popular. The most economical way from the standpoint of wires required is to make a square array. When this is done, the number of wires needed to select a core of K bits is given by

$$N = X_n + Y_n = 2\sqrt{K} \tag{11-1}$$

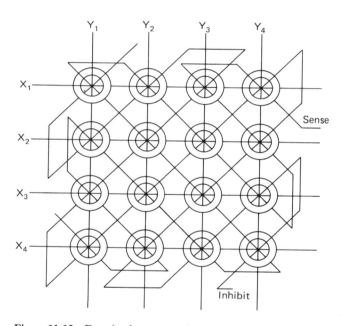

Figure 11-13 Four-by-four magnetic-core RAM

EXAMPLE 11-1 How many wires are needed to select any bit in a 4K memory (actually 4096 bits)?

Solution:

$$N = 2\sqrt{4096} = 2(64) = 128$$

There will be 64 wires for the X-select lines and 64 wires for the Y-select lines; the memory plane will be 64 × 64 = 4096 cores.

In addition a single SENSE and a single INHIBIT wire will be needed on the plane.

Memories using magnetic cores are organized so that each plane stores information for a single bit of a word, and the complete word is composed of a single bit from several planes. For example, a 16-bit word will have the same XY location on each of 16 planes. With 16 planes, the memory of Example 11-1 can store 4096 16-bit words. In this case a read/write cycle will select the same $X_i Y_j$ location for each plane and read the data stored on the 16 bits. Then the write cycle takes place with $-\frac{1}{2}I_M$ currents in the INHIBIT plane, where 0s had been stored. For example, if the word stored was 1000110011001111, INHIBIT currents of $-\frac{1}{2}I_M$ would be applied to planes 2, 3, 4, 7, 8, 11, and 12 and no INHIBIT currents to the other planes. Figure 11-14 shows the organization of a multiplane RAM core memory capable of storing 16 ($= 4 \times 4$) 5-bit words.

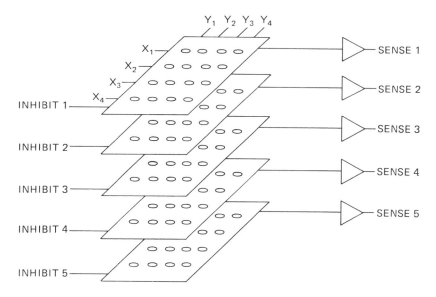

Figure 11-14   Multiplane core memory capable of storing 16 ($= 4 \times 4$) 5-bit words

11-9   BULK MAGNETIC DATA STORAGE

In Sec. 11-8 we made use of a changing magnetic field for a random access memory. In the magnetic RAM, the magnetic field was changed by means of electric currents; this change in magnetic flux was sensed in a wire as an induced voltage. Another way to induce a voltage into a wire is to move a magnet in the vicinity of the curve. If the strength of the magnetic field changes in the neighborhood of the

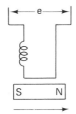

Figure 11-15 Moving magnet induces voltage in sense coil

Direction of motion

wire, this change in magnetic flux is sensed as a voltage in the wire, as shown in Fig. 11-15. In Fig. 11-15, if the magnetic direction is changed (S before N), the voltage induced in the coil is reversed in polarity.

Whereas in the core RAM, the only way to sense a change in flux was to destroy the magnetic information in Fig. 11-12, the magnet is not changed by its passage past the sense coil. In the core RAM, it was possible to change the magnetic flux rapidly. In Fig. 11-15, the magnet has to be moved mechanically, and we are involved with longer times. In the core RAM a change in flux during the read cycle represented a 1, and no change in flux represented a 0. In Fig. 11-15, a 1 or a 0 (represented by opposite magnetization directions) is sensed by opposite polarities in the sense voltage.

The principle of Fig. 11-15 serves as the basis for magnetic bulk data storage. For storing extremely large quantities of digital data, such as megabits, moving magnetic systems have no competitors. The disc, drum, and tape are all similar; all have surfaces coated with a magnetic oxide. Small areas of the surface can be magnetized in either one direction or the other (as a 1 or 0) by means of WRITE HEADs, which are coils with a magnetizing current in one direction or the other (similar to Fig. 11-15). Magnetic bulk memories differ from core RAMs in their long and finite access time, the time it takes for the data to reach the READ HEAD. In the case of the disc and drum, it may be the time it takes for one rotation of the memory; this is measured in milliseconds. In tape memories it may take minutes. In the core RAM, any element could be reached in the same time, whereas in bulk moving memories the access time depends upon the element's distance from the read head. However, in the disc, drum, and tape, the process of reading the memory is NDRO.

The disc (Fig. 11-16a) is very much like a phonograph record except that the surface is magnetically coated. Many discs are frequently mounted in pack form, one above another on the same spindle. There is a read/write head for the upper and lower surface of each disc. The disc surface stores the data in many tracks arranged in circles.

The magnetic drum (Fig. 11-16b) is a rotating cylinder whose surface is coated with a magnetic oxide. Data are stored on the surface of the drum in many tracks. The storage density on a drum is quite high; data up to the order of 10 megabits can be stored on a single drum. Drums rotate quite rapidly, and data can be read into and out of a drum quite rapidly. In addition, many read heads make possible the reading and writing of multibit words simultaneously.

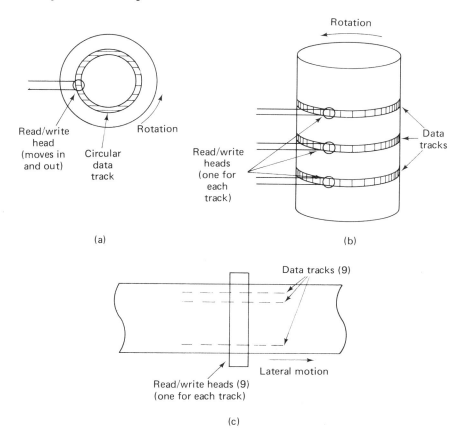

Figure 11-16 Moving bulk magnetic memories: (a) disc; (b) drum; (c) tape

The magnetic tape (Fig. 11-16c) is a plastic tape coated with a magnetic oxide. The common tape used in the computer is typically about 2400 ft long and $\frac{1}{2}$ in. wide. Data are recorded in the tape in nine parallel *tracks*. This permits data to be recorded with 7 bits in ASCII code. One of the two additional bits is used as a parity check; the other can be used for timing purposes. Data are recorded quite closely along the tape. Densities of 200, 556, 800, and 1600 bits per in. on each track are used. The tape speeds vary from 50 to 200 in. per s.

For storing smaller quantities of data, cassette tapes have come into use. They are like audio cassettes, but of higher quality.

EXAMPLE 11-2 An 8-in.-diameter magnetic drum memory rotates at 1200 rpm. Data are recorded on it with a density of 100 bits per in. per track.
  a. What is the average access time?
  b. What is the maximum access time?
  c. What is the minimum access time?
  d. What is the data rate in bits per second per track?

Solution

$$1200 \text{ rpm} = \frac{1200}{60} = 20 \text{ rps} \quad \text{(revolutions per second)}$$

a. The average access time is the time it takes to sense data at the opposite side of the drum. This is the time for a half revolution.

$$\text{Time per revolution} = \tfrac{1}{20} \text{ s} = 50 \text{ ms}$$
$$\text{Access time} = \tfrac{50}{2} \text{ ms} = 25 \text{ ms}$$

b. The maximum access time is the time for data to go completely around the drum. The data had just passed the sense head.

$$T = 50 \text{ ms}$$

c. Total data in the track is $\pi \times 8 \times 1000 = 2512$ bits. Minimum time is the time for 1 bit:

$$\frac{50 \times 10^{-3} \text{ s}}{2512} \approx 20 \times 10^{-6} \text{ s}$$

d. Data rate per track is

$$20 \text{ rps} \times 2512 \text{ bits per revolution} \approx 50{,}000 \text{ bits per s}$$

EXAMPLE 11-3   The drum of Example 11-2 has 100 tracks:
    a. How much data can be stored on the drum?
    b. What is the data rate in bits per second?

Solution
    a. $100 \times 2512 = 251{,}200$ bits.
    b. $100 \times 50{,}000 = 5{,}000{,}000$ bits per s.

EXAMPLE 11-4   A magnetic tape with nine tracks of data has a recording density of 200 bits per in. and is 2400 ft long. The tape speed is 75 in./s.
    a. How much data could be stored on the tape?
    b. What is the average access time?
    c. What is the maximum access time?
    d. What is the data rate in bits per second?

Solution
    a. Total length $= 2400 \times 12 = 28{,}800$ in. Per track: 200 bits per in. $\times$ 28,800 in. $= 5.76 \times 10^6$ bits. The tape can store 5.76×10^6 9-bit words if 100 percent utilization of the tape is assumed.
    b. Average access time is the time to reach half the tape length:

$$T = \frac{28{,}800}{2 \times 75} = 193 \text{ s} \approx 3.25 \text{ min}$$

c. Maximum access time is time to reach the total tape:

$$T = 385 \text{ s} \approx 6.5 \text{ min}$$

d.

$$\text{Data rate} = \frac{5.76 \times 10^6 \text{ words}}{385 \text{ s}} \approx 1500 \text{ words per s}$$

$$15,000 \times 9 = 135,000 \text{ bit per s}$$

Random-access memories such as those we have discussed in this chapter are expensive and are therefore used only to communicate directly with a computer since time is of great importance. Because of the expense, their storage capacity is limited. Disc, drum, and tape are much less expensive in cost per bit of storage and have the capacity to store large quantities of memory. Data from the disc, drum, and tape are transferred to RAMs for direct use with the computer; after the computer has performed its operation, the data are transferred back to bulk storage.

Table 11-1 is a price–performance comparison chart of three different types of mangetic tape systems. In addition, comparative specifications are given for paper tape and disc storage.

Table 11-1   Price–performance comparisons of computer tape systems

	Cartridge	Cassette	Half-Inch Reel-to-Reel	Paper Tape	Floppy Disc
Storage capacity (megabytes)	to 2.88	0.1–0.2	7–14	0.12–0.36	0.16–0.4
Packing density (bpi)	1600	556–1600	556–1600	10	1000–4000
Number of tracks	1, 2, or 4	1–4	7 or 9	5–8	32–77
Transfer rate (K bytes/s)	5–6	0.5–1.6	5–120	10–300 ch/s	200–400
Average access time (ms)	to 20	to 15	to 40	NA	0.06–1.6
Search speed (ips)	90	to 120	to 150	NA	NA
Encoding technique	NRZ/PE	NRZ/PE	NRZ/PE	NA	Mod. P/E
Media life (passes)	5000	300	10,000	300	to 2M
Approx. transport volume (in.³)	300–800	50–300	4000–6000	300–2500	700–3000
Storage cost (cents/byte)	0.3–0.5	2–3	0.1–0.15	1.6	0.15–0.25
Media price (dollars)	12–18	8–15	10–20	1.50	4–18
Transport price (dollars)	750–2000	600–2000	2000–30,000	200–2000	600–6000

D. Rasmussen, "Tape cartridges: large scale storage of small scale prices," *Electronic Products*, Vol. 18, No. 1, June 16, 1975, pp. 55–58. Reprinted by permission of *Electronic Products Magazine*, 645 Stewart Ave., Garden City, New York 11530, 1975, United Technical Publications Inc., a division of Cox Broadcasting Corporation.

Problems

11-1. Define the following terms: volatile memory, nonvolatile memory, DRO, NDRO.

11-2. Define the following terms: RAM, ROM, PROM, EROM.

11-3. A 64×1-bit semiconductor RAM can be organized in different ways. How many select terminals are needed for each of the following ways?
(a) Each bit has a separate select input.
(b) Organized on an XY basis: 2 rows $\times$ 32 columns.
(c) Organized on an XY basis: 4 rows $\times$ 16 columns.
(d) Organized on an XY basis: 8 rows $\times$ 8 columns.

11-4. Repeat Prob. 11-3. However, internal decoding is used to select the bits.

11-5. Semiconductor RAMs are organized on an XY basis with equal numbers of rows and columns. How many input terminals are needed to select any one bit? (a) 16 bits; (b) 256 bits; (c) 1024 bits; (d) 4096 bits.

11-6. Repeat Prob. 11-5, but with decoding of the input terminals.

11-7. What are two methods of changing magnetic fields?

11-8. What do we mean by "square-loop" material?

11-9. Why must you simultaneously apply currents to the X and Y wires in a magnetic core?

11-10. Why do we need an inhibit wire in a core memory?

11-11. In a core memory, why must you have a write cycle after the read cycle?

11-12. (a) How many wires are needed for a 400-bit square-matrix core memory?
(b) How many for a 576 bit?

11-13. It is desired to store 1024 words, 9 bits per word, in a core memory. Describe the memory and indicate how many wires are needed for the memory.

11-14. What are the advantages and disadvantages of moving magnetic memories versus core memories?

11-15. A 10-in.-diameter drum rotates at a speed of 2400 rpm. What is the average access time in milliseconds for the drum?

11-16. Data are stored on the drum of Prob. 11-15 on tracks with a density of 800 bits per in. How many bits of data are stored on a track?

11-17. If the drum of Probs. 11-15 and 11-16 has 180 tracks, how many bits of data can be stored on the drum?

11-18. What is the data rate of the drum of Prob. 11-17 in bits per second?

11-19. A nine-track tape is 1200 ft long. It is magnetized with a density of 1600 bits per in. per track. The tape speed is 120 in. per s.
(a) How many 9-bit words can be stored on the tape?
(b) How many bits of data can be stored in the tape?
(c) What is the average access time?
(d) What is the maximum access time?
(e) What is the data rate in words per second?
(f) What is the data rate in bits per second?

TWELVE

The Operational Amplifier†

12-1 INTRODUCTION

The operational amplifier, commonly called the OP-AMP, is possibly the most important single building block in digital systems. In this chapter we shall develop the basic principles of how to use the OP-AMP; in Chapter 13 we shall show how it is used to change analog signals, such as voltages that occur continuously in nature, to a digital representation, and digital signals to their equivalent analog voltages.

As its name suggests, initially the function of the OP-AMP was to perform mathematical operations upon voltages, such as sign changing (inversion), addition, subtraction, multiplication, division, differentiation, and integration. This was done using feedback. However, the amplifier is used by itself so frequently without feedback that it has become common to call the amplifier itself an operational amplifier.

†M. E. Levine, *Digital Theory and Experimentation Using Integrated Circuits*. Englewood Cliffs, N.J.: Prentice-Hall, Inc., 1974, Expt. 15, The Operational Amplifier.

The OP-AMP is a linear dc voltage amplifier with very high gain. Table 12-1 gives the parameters of the ideal OP-AMP and those of practical OP-AMPS.

Table 12-1 OP-AMP parameters

	Ideal	Practical
Voltage gain	∞	50,000–10,000,000
Input resistance	∞	500,000 Ω with bipolar transistor Multimegohms with FET transistor
Output resistance	0	$\approx 100\ \Omega$
Output voltage level	0 when input voltage equals 0	Zero adjust provision
Output voltage limits	$\pm \infty$	Limits at approximately 90% of $+V_{CC}$ and 90% of $-V_{CC}$ Note: two power supplies, $+V_{CC}$ and $-V_{CC}$, are usually required
Frequency response	DC to infinite hertz	DC to approx 100 Hz (3 dB)
Phase of output voltage	Inverting, but in many amplifiers a noninverting output is also available	

As can be seen from Table 12-1, with the exception of frequency response, the parameters of practical OP-AMPs approach those of the ideal amplifier. However, in many OP-AMP applications, a considerable amount of inverse voltage feedback is used, which reduces the voltage gain. Since in an amplifier with feedback the gain bandwidth product is conserved, practical amplifiers have bandwidths considerably beyond the 100-Hz frequency indicated in the table. However, there are applications in which their low value of frequency cutoff can cause difficulties, especially if the amplifier is used without feedback (open loop) or with small amounts of feedback.

The logic symbol of an OP-AMP is shown in Fig. 12-1, with inversion indicated by the minus (−) and plus (+) symbols. The input to an inverting OP-AMP is to one side of a differential amplifier, with the other side connected to ground.

Figure 12-1   Logic symbol: inverting
OP-AMP

Frequently, an OP-AMP has two inputs, one inverting and the other noninverting. The logic symbol for this type of amplifier is shown in Fig. 12-2. In this amplifier both sides of the input differential amplifier are brought out to input terminals. Signals can be applied to either or both inputs. The output voltage is given by the gain multiplied by the difference of the voltages applied to the two inputs.

Operational amplifiers are operated with two power supplies, $+V_{CC}$ and $-V_{CC}$. They operate linearly until limiting occurs at approximately 90 percent of the supply voltage. A common value of supply voltage for monolithic amplifiers is $+15$ and -15 V. Hence such an amplifier can deliver output voltages up to approximately ±14 V before limiting occurs.

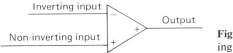

Figure 12-2   Logic symbol: Inverting and noninverting OP-AMP

12-3 FUNDAMENTALS OF OPERATIONAL-AMPLIFIER OPERATION: MULTIPLYING AND SUMMING

Consider an ideal amplifier with a resistor feedback as shown in Fig. 12-3. As will be shown shortly, point A, the input to the amplifier, is at a voltage almost equal to 0 V, and point A is at a *virtual ground*.

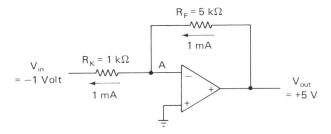

Figure 12-3   Amplifier with resistive feedback

Apply a voltage $V_{in} = -1$ V. If point A is at ground, 1 mA must flow into V_{in} through the 1-kΩ resistor. However, the input at A to the OP-AMP is a very high resistance, and no current can flow into the OP-AMP. The only place current can come from is the 5-kΩ resistor, and since point A is at ground potential, the output voltage must be $+5$ V (1 mA $\times$ 5 kΩ). We can therefore write

$$\frac{V_{\text{out}}}{V_{\text{in}}} = \frac{+5\text{ V}}{-1\text{ V}} = -5 = 5\underline{/180°} \tag{12-1}$$

But

$$\frac{R_F}{R_K} = \frac{5\text{ k}\Omega}{1\text{ k}\Omega} = 5$$

also. Hence we see that the gain is determined only by the ratio of the two resistors and not by the amplifier itself.

What about the assumption that point A was at a ground potential? Suppose that the amplifier had an open-loop gain of 50,000. If $V_{\text{out}} = 5$ V, then $V_A = (V_{\text{out}}/10,000) = (5/50,000) = 0.0001$ V, which is very nearly equal to zero. This is close enough to 0 compared to the 1-V input signal and 5-V output signal to justify the assumption that the voltage at A equals 0. If the amplifier has an input resistance of 500,000 Ω, this is large enough compared to the circuit resistance values of 1000 and 5000 Ω to justify neglecting its effect (to a first degree of approximation). Moreover, with point A at ground potential, no current flows into the 500,000-Ω amplifier input resistance.

Consider Fig. 12-4 with two inputs and equal input resistors. With point A at ground potential, the currents into points K and L are 1 mA and 0.5 mA, respectively. This current must come from the 5-kΩ resistor and the output voltage must become $+7.5$ V. At point A, the currents $I_K = 1$ mA and $I_L = 0.5$ mA are added together, and point A is known as the summing point. The circuit of Fig. 12-4 can be used to add two analog voltages and also multiply them by a scale factor determined by the resistor values used.

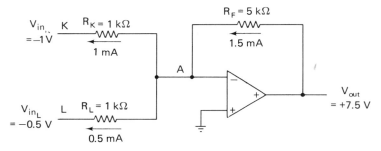

Figure 12-4 Feedback amplifier with two inputs and equal summing resistors

Consider Fig. 12-5 with two equal inputs but unequal summing resistors. Circuit considerations again show that the currents are the same as in Fig. 12-4, and the output voltage is again the same and equal to $+7.5$ V. It is obtained as follows:

$$V_{out} = -\left[\frac{5\ K\Omega}{1\ K\Omega} V_{in\ K} + \frac{5\ k\Omega}{2\ k\Omega} V_{in\ L}\right]$$
$$= -[5(-1\ V) + 2.5(-1\ V)]$$
$$= +7.5\ V \tag{12-2}$$

We see now that by choice of feedback resistors we can change the scale factors by which signal voltages can be multiplied and summed.

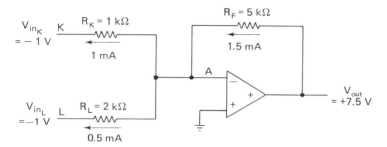

Figure 12-5 Feedback amplifier with two inputs and unequal summing resistors

EXAMPLE 12-1 For the amplifier of Fig. 12-5, determine the output voltages for the following input voltages:

Solution

$V_{in\ K}$	$V_{in\ L}$	V_{out} (V)
0	0	$0 + 0 = 0$
0	-1	$0 + 2.5 = 2.5$
-1	0	$5 + 0 = 5.0$
-1	-1	$5 + 2.5 = 7.5$

12-4 INTEGRATION AND DIFFERENTIATION

Figure 12-6 shows an OP-AMP with a capacitor as the feed-back element. Let the input voltage V_{in} change suddenly from zero to a value $-V_{in}$. As in the previous case, a current $I = -(V_{in}/R)$ flows through resistor R. As before, the current must come from the capacitor. But, current flowing for a time t corresponds to a charge $q = i \times t$, which must have come from capacitor C. But this charge could have come only if V_{out} had increased from 0 to a voltage V_{out} given by the expression $q = CV_{out}$.

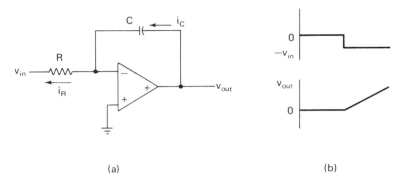

(a)   (b)

Figure 12-6   Feedback amplifier with capacitor feedback integrator: (a) amplifier; (b) waveforms of v_{in} and v_{out}

Combining the expressions

$$q = CV_{out} = i \times t = \frac{V_{in}}{R} \times t \tag{12-3}$$

Since the amplifier inverts, we can now write

$$V_{out} = -\left(\frac{t}{RC} V_{in}\right) \tag{12-4}$$

Equation (12-4) can be interpreted to be a positive-going voltage that increases linearly with time, or a ramp, as shown in Fig. 12-6. The inverting property of the OP-AMP converts the negative input step voltage to a positive-going ramp.

EXAMPLE 12-2   The voltage V_{in} in the integrator OP-AMP of Fig. 12-7a changes from 0 V to +3 V at a time $t = 0$.

a. What is V_{out} 2.5 ms later?
b. How rapidly does the output voltage change?
c. How long after the input voltage changes will the amplifier limit if $V_{CC} = \pm 20$ V and limiting occurs at 90 percent of V_{CC}?

Solution

a. $V_{out} = -\left(\frac{t}{RC} V_{in}\right)$

$$= -\frac{2.5 \times 10^{-3}}{2 \times 10^3 \times 1 \times 10^{-6}} \times 3 = -4.75 \text{ V}$$

b. In 1 ms

$$V_{out} = \frac{1 \times 10^{-3}}{2 \times 10^3 \times 1 \times 10^{-6}} \times 3 = -1.5 \text{ V}$$

The voltage is changing at a rate of -1.5 V per ms or

$$\frac{-1.5}{1 \times 10^{-3}} = -1500 \text{ V per s}$$

c. $0.9 \times -20 = -18$ V for limiting. Since the voltage is changing at a rate of -1.5 V per ms, $t = -18/-1.5 = 12$ ms. Figure 12-7b shows a plot of the input and output voltages.

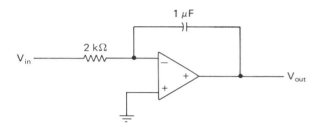

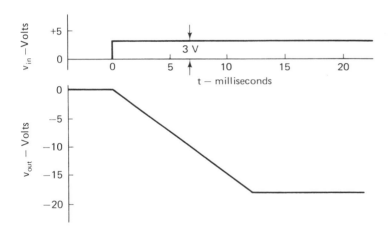

Figure 12-7   Integrator OP-AMP

Another way of looking at the circuit of Fig. 12-6 is to realize that in a capacitor

$$i_C = C \frac{dv}{dt}$$

Hence, in Fig. 12-6, since $i_R = i_C$,

$$\frac{V_{in}}{R} = C \frac{dV_{out}}{dt} \tag{12-5}$$

Solving for V_{out}, we obtain

$$V_{out} = \frac{1}{RC} \int V_{in}\, dt \qquad (12\text{-}6)$$

The OP-AMP with resistor input and capacitor feedback is therefore an integrating circuit and gives an output proportional to the integral of V_{in}.

Still another way of looking at the circuit of Fig. 12-6 is to consider its frequency response. Comparing this circuit with that of Fig. 12-3, we see that a capacitor with reactance $(1/j2\pi fC)$ takes the place of R_F. We can therefore write

$$\frac{V_{out}}{V_{in}} = \frac{1/j2\pi fC}{R}\ \underline{/180^\circ} = \frac{1}{2\pi fCR}\ \underline{/90^\circ} \qquad (12\text{-}7)$$

From this we see that the gain *increases* as the frequency goes *down*. Theoretically, it is infinite at $f = 0$ (direct current); practically, it is the open-loop gain. This means that if we build an integrating amplifier, as in Fig. 12-6, we have to deal with the problem of high gain at direct current. Even very small unbalances in the differential amplifier or small input currents flowing through the input resistor R can cause a small voltage unbalance. When multiplied by the open-loop gain, the output voltage can become large enough to drive the amplifier into saturation and make the circuit unusable. Hence, in practice the circuit of Fig. 12-6 is modified as shown in Fig. 12-8 with a feedback resistor R_J to reduce the dc voltage gain. R_J is frequently chosen to make the dc voltage gain approximately 100.

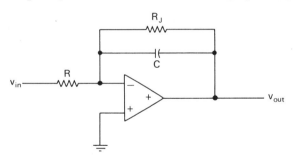

Figure 12-8 Integrator modification with R_J to reduce dc gain

If the square wave of Fig. 12-9 is applied to the input of Fig. 12-6, the feedback capacitor is alternately charged and discharged linearly to provide a triangular output.

Figure 12-9 Integrator with squarewave input and triangular output

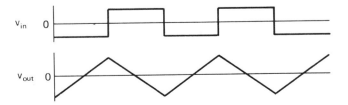

If the resistor and capacitor are interchanged as shown in Fig. 12-10,

$$i_{in} = C \frac{dv_{in}}{dt} \tag{12-8a}$$

$$i_R = \frac{V_{out}}{R} \tag{12-8b}$$

and

$$V_{out} = RC \frac{dv_{in}}{dt} \tag{12-8c}$$

The output voltage is proportional to the derivative of the input.

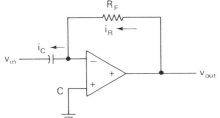

Figure 12-10 Differentiation OP-AMP

From a frequency response standpoint, we can write

$$\frac{V_{out}}{V_{in}} = \frac{R}{X_C} \underline{/180°} = \frac{R}{1/j2\pi fC} \underline{/180°} = 2\pi fCR \underline{/-90°}$$

Hence the gain goes up as frequency increases. This causes many problems.

12-5 COMPARATOR OPERATIONAL AMPLIFIER

Consider an open-loop OP-AMP, with inputs to the inverting and noninverting inputs as shown in Fig. 12-11. Suppose that the open-loop gain of the amplifier is 20,000. If V_A is more negative than V_B by 0.001 V, V_{out} should be equal to $-(-0.001 \times 20,000) = 20$ V. If the amplifier has $V_{CC} = +15$ V, the output will be limiting and at approximately $+14$ V.

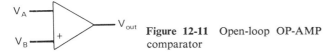

Figure 12-11 Open-loop OP-AMP comparator

Similar arguments hold if V_A is more positive than V_B by 0.001 V; V_{out} would be equal to -14 V. In this discussion no mention was made of the value of V_B. Hence we can use the OP-AMP to compare V_A to V_B. If V_A is slightly negative (within approximately 1 mV) with respect to V_B, the amplifier output will be positive and limiting; if V_A is slightly positive with respect to V_B, the OP-AMP output will be negative and limiting.

EXAMPLE 12-3 Voltages V_A and V_B are applied to the amplifier of Fig. 12-11. Plot V_{out}. The OP-AMP has $V_{CC} = \pm 10$ V and limits at 80 percent of V_{CC}.

Solution Whenever V_A is more positive than V_B, $V_{out} = -8$ V. Whenever V_A is more negative than V_B, $V_{out} = +8$ V (see Fig. 12-12).

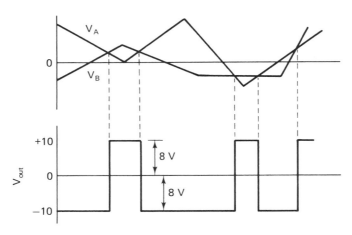

Figure 12-12 Solution to Example 12-3

12-6 DATA SHEET OF A COMMERCIAL OPERATIONAL AMPLIFIER

Many integrated-circuit OP-AMPS are commercially available. Typical is the μA741 whose data are shown in Fig. 12-13. In Fig. 12-13a, the ratings or the limiting values of voltages and temperatures that can be applied to it are given. Also given are the mechanical features and pin connections. Note that this OP-AMP is available in three packages, the common 14-pin dual-in-line A package, a metal-can T package, and an 8-pin dual-in-line V package. The internal circuit diagram is given. Note the differential inputs to Q_1 and Q_2 and the single-ended output through emitter resistors of transistors *NPN* Q_{14} and *PNP* Q_{20}. Note also the V^+ and V^- requirements.

The parameters are given in Fig. 12-13b of the data sheet. Typical voltage gain is 200,000, typical input resistance is 2.0 MΩ, and typical output resistance 75 Ω.

Figures 12-13c and 12-13d show how the parameters vary under different supply voltage and temperature conditions. In the center curve in Fig. 12-13d, note that the open-loop gain at very low frequency is 2×10^5. The open-loop gain begins to fall off at approximately 5 Hz, and the gain decreases by a factor of 10 to 1 (20 dB) for each decade increase in frequency up to 1 MHz, where the gain has fallen to unity. In the lower left corner of Fig. 12-13d, note that the output voltage capability is 28 Vp–p for a supply voltage of ± 15 V.

siynetics

µA741

LINEAR INTEGRATED CIRCUITS

The µA741 is a high performance operational amplifier with high open loop gain, internal compensation, high common mode range and exceptional temperature stability. The µA741 is short-circuit protected and allows for nulling of offset voltage.

- **INTERNAL FREQUENCY COMPENSATION**
- **SHORT CIRCUIT PROTECTION**
- **OFFSET VOLTAGE NULL CAPABILITY**
- **EXCELLENT TEMPERATURE STABILITY**
- **HIGH INPUT VOLTAGE RANGE**
- **NO LATCH-UP**

	µA741C	µA741
Supply Voltage	±18V	±22V
Internal Power Dissipation (Note 1)	500mW	500mW
Differential Input Voltage	±30V	±30V
Input Voltage (Note 2)	±15V	±15V
Voltage between Offset Null and V⁻	±0.5V	±0.5V
Operating Temperature Range	0°C to +70°C	-55°C to +125°C
Storage Temperature Range	–65°C to +150°C	–65°C to +150°C
Lead Temperature (Solder, 60 sec)	300°C	300°C
Output Short Circuit Duration (Note 3)	Indefinite	Indefinite

Notes
1. Rating applies for case temperatures to 125°C; derate linearly at 6.5mW/°C for ambient temperatures above +75°C.
2. For supply voltages less than ±15V, the absolute maximum input voltage is equal to the supply voltage.
3. Short circuit may be to ground or either supply. Rating applies to +125°C case temperature or +75°C ambient temperature.

A PACKAGE
(Top View)

1. NC
2. NC
3. Offset Null
4. Inv. Input
5. Non-Inv. Input
6. V⁻
7. NC
8. NC
9. **Offset Null**
10. Output
11. V⁺
12. NC
13. NC
14. NC

ORDER PART NO. µA741CA

T PACKAGE

1. Offset Null
2. Inverting Input
3. Non-Inverting Input
4. V⁻
5. Offset Null
6. Output
7. V⁺
8. NC

ORDER PART NOS. µA741T/µA741CT

V PACKAGE

1. Offset Null
2. Inv. Input
3. Non-Inv. Input
4. V⁻
5. Offset Null
6. Output
7. V⁺
8. NC

ORDER PART NO. µA741CV

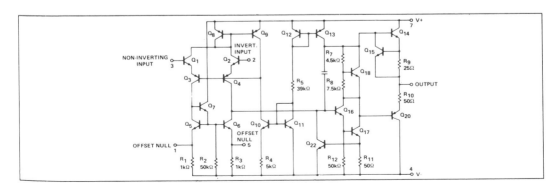

Figure 12-13 *µA741 OP-AMP. (Courtesy Signetics Corporation)*

(V_S = ±15V, T_A = 25°C unless otherwise specified)

PARAMETER	MIN.	TYP.	MAX.	UNITS	TEST CONDITIONS
μA741C					
Input Offset Voltage		2.0	6.0	mV	$R_S \leqslant 10k\Omega$
Input Offset Current		20	200	nA	
Input Bias Current		80	500	nA	
Input Resistance	0.3	2.0		MΩ	
Input Capacitance		1.4		pF	
Offset Voltage Adjustment Range		±15		mV	
Input Voltage Range	±12	±13		V	
Common Mode Rejection Ratio	70	90		dB	$R_S \leqslant 10k\Omega$
Supply Voltage Rejection Ratio		10	150	μV/V	$R_S \leqslant 10k\Omega$
Large-Signal Voltage Gain	20,000	200,000			$R_L \geqslant 2k\Omega$, V_{out} = ±10V
Output Voltage Swing	±12	±14		V	$R_L \geqslant 10k\Omega$
	±10	±13		V	$R_L \geqslant 2k\Omega$
Output Resistance		75		Ω	
Output Short-Circuit Current		25		mA	
Supply Current		1.4	2.8	mA	
Power Consumption		50	85	mW	
Transient Response (unity gain)					V_{in} = 20mV, R_L = 2kΩ, $C_L \leqslant$ 100pF
Risetime		0.3		μs	
Overshoot		5.0		%	
Slew Rate		0.5		V/μs	$R_L \geqslant 2k\Omega$
The following specifications apply for 0°C ≤ T_A ≤ +70°C					
Input Offset Voltage			7.5	mV	
Input Offset Current			300	nA	
Input Bias Current			800	nA	
Large-Signal Voltage Gain	15,000				$R_L \geqslant 2k\Omega$, V_{out} = ±10V
Output Voltage Swing	±10	±13		V	$R_L \geqslant 2k\Omega$
μA741					
Input Offset Voltage		1.0	5.0	mV	$R_S \leqslant 10k\Omega$
Input Offset Current		10	200	nA	
Input Bias Current		80	500	nA	
Input Resistance	0.3	2.0		MΩ	
Input Capacitance		1.4		pF	
Offset Voltage Adjustment Range		±15		mV	
Large-Signal Voltage Gain	50,000	200,000			$R_L \geqslant 2k\Omega$, V_{out} = ±10V
Output Resistance		75		Ω	
Output Short Circuit Current		25		mA	
Supply Current		1.4	2.8	mA	
Power Consumption		50	85	mW	
Transient Response (unity gain)					V_{in} = 20mV, R_L = 2kΩ, $C_L \leqslant$ 100pF
Risetime		0.3		μs	
Overshoot		5.0		%	
Slew Rate		0.5		V/μs	$R_L \geqslant 2k\Omega$
The following specifications apply for −55°C ≤ T_A ≤ +125°C					
Input Offset Voltage		1.0	6.0	mV	$R_S \leqslant 10k\Omega$
Input Offset Current		7.0	200	nA	T_A = +125°C
		20	500	nA	T_A = −55°C
Input Bias Current		0.03	0.5	μA	T_A = +125°C
		0.3	1.5	μA	T_A = −55°C
Input Voltage Range	±12	±13		V	
Common Mode Rejection Ratio	70	90		dB	$R_S \leqslant 10k\Omega$
Supply Voltage Refection Ratio		10	150	μV/V	$R_S \leqslant 10k\Omega$
Large-Signal Voltage Gain	25,000				$R_L \geqslant 2k\Omega$, V_{out} = ±10V
Output Voltage Swing	±12	±14		V	$R_L \geqslant 10k\Omega$
	±10	±13		V	$R_L \geqslant 2k\Omega$
Supply Current		1.5	2.5	mA	T_A = +125°C
		2.0	3.3	mA	T_A = −55°C
Power Consumption		45	75	mW	T_A = +125°C
		45	100	mW	T_A = −55°C

Figure 12-13 (continued)   μA741 OP-AMP. (Courtesy Signetics Corporation)

TYPICAL CHARACTERISTIC CURVES

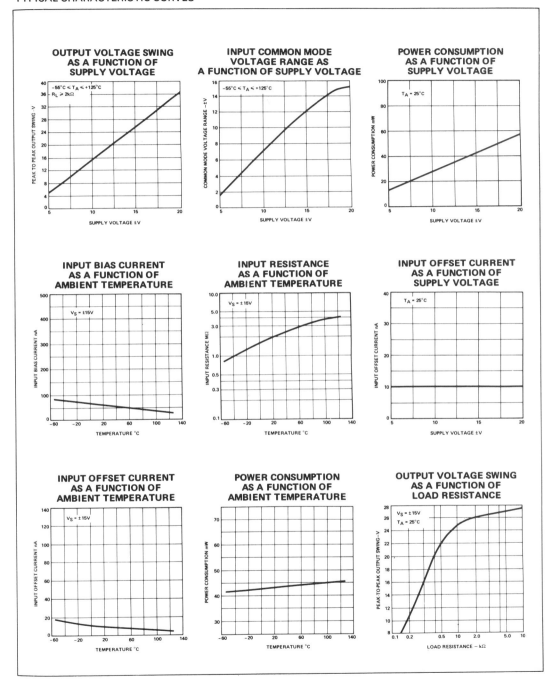

Figure 12-13 (continued)   μA741 OP-AMP. (Courtesy Signetics Corporation)

TYPICAL CHARACTERISTIC CURVES (Cont'd.)

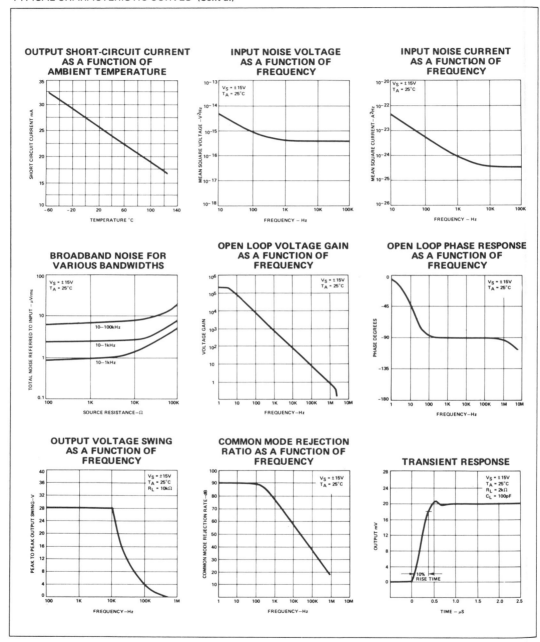

Figure 12-13 (continued)   µA741 OP-AMP. (Courtesy Signetics Corporation)

Problems

12-1. Find V_x.

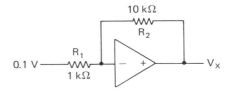

12-2. If R_2 in Prob. 12-1 equals 22 kΩ, what is V_x?

12-3. If R_1 in Prob. 12-1 equals 500 Ω, what is V_x?

12-4. Find V_x for the following values of V_A and V_B:

V_A	V_B	V_X
+0.1 V	0	
+0.1 V	+0.1 V	
+0.1 V	−0.1 V	

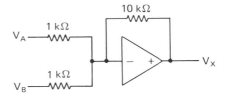

12-5. Complete the following table for V_x:

V_A	V_B	V_C	V_X
0	0	0	
0	0	+5	
0	+5	0	
0	+5	+5	
+5	0	0	
+5	0	+5	
+5	+5	0	
+5	+5	+5	

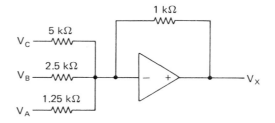

12-6. V_A changes from 0 to 0.2 V at time $t = 0$. What is the value of V_x after 10 ms?

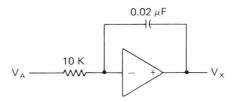

12-7. Plot a curve of V_x versus time for the values of V_A and V_B given in the graph.

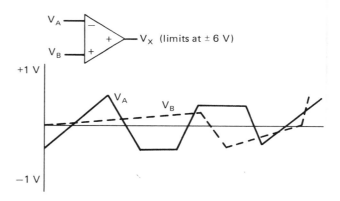

12-8. In the OP-AMP of Prob. 12-6, $V_A = 0.4 \cos 400t$. What is V_x?

12-9. In the OP- AMP of Prob. 12-6, $V_A = 0.4 \cos 400t + 0.4 \cos 800t$. What is V_x?

12-10. Show that the noninverting gain of the amplifier is given by

$$\frac{V_{out}}{V_{in}} = \frac{R_F + R_K}{R_K}$$

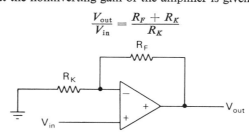

THIRTEEN

Digital-to-Analog (D/A) and Analog-to-Digital (A/D) Conversion[†]

13-1 INTRODUCTION

In all the previous chapters we have spent considerable time extolling the advantages of the digital world. Digital techniques are compatible with open and closed relays, with transistors cut off or in saturation, with holes or no holes in punched cards, etc. Digital electronics is simple; there are only two levels, a 1 and a 0.

However, the real or physical world is an analog one. Typical analog phenomena are voltage, current, pressure, temperature, sound level, light intensity, and distance. Analog quantities are continuous and have an infinite number of values, even negative. However, if we wish to transmit analog data, such as the light intensity from a picture in outer space, the noise level is frequently too high; this makes analog transmission of such data virtually impossible. However, it is possible to transmit digital data correctly, even in the presence of noise. It is much easier to store digital data than analog data. Hence the need exists to be able to take such analog data, encode it to a digital equivalent for transmission or storage, and then decode it back to its analog equivalent.

[†]M. E. Levine, *Digital Theory and Experimentation Using Integrated Circuits*. Englewood Cliffs, N.J.: Prentice-Hall, Inc., 1974, Expt. 16, Digital-to-Analog (D/A) and Analog-to-Digital (A/D) Conversion.

Some analog data, such as a voltage, may have a steady value; some analog data, such as the light level from a television picture, may be continuously changing. In that case, samples taken at regular intervals, using a sample and hold circuit, provide an excellent representation of the data. In digitizing, it does not take too many bits to satisfactorily represent analog data. For example in television, 6 bits are used for the gray scale. This gives 64 shades of gray between black and white. Digital voltmeters are very popular. Thirteen bits are used in a $3\frac{1}{2}$-digit voltmeter.

This chapter deals with the techniques of digital-to-analog (D/A) and analog-to-digital (A/D) conversion. In these techniques, the operational amplifier, whose basic operation was discussed in Chapter 12, plays a leading role.

13-2 DIGITAL-TO-ANALOG CONVERSIONS: BINARY WEIGHTED LADDERS

Consider a two-stage binary counter whose logic levels are $1 \equiv +5$ V and $0 \equiv 0$ V driving an OP-AMP with resistive feedback, as shown in Fig. 13-1. In considering Fig. 13-1a, the multiplying factor from Q_B to the output is $1/2.5$, and from Q_A to the output it is $1/5$. Figure 13-1b shows the resultant waveforms at the counter outputs and at V_{out} at the output of the OP-AMP. We can also tabulate the results, as in Table 13-1.

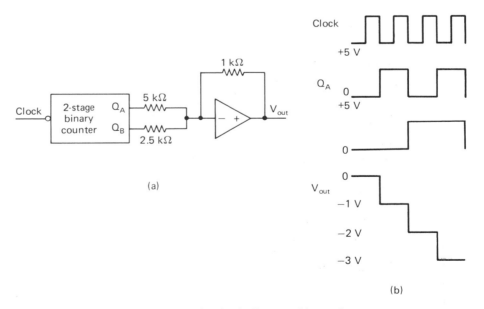

Figure 13-1   Two-bit D/A converter: (a) circuit diagram; (b) waveforms

Table 13-1   Two-bit D/A Converter

Clock Pulse	Q_B	Q_A	Count State	Q_B (V)	Q_A (V)	V_{out} (V)
0	0	0	00	0	0	0
1	0	1	01	0	+5	−1
2	0	0	10	+5	0	−2
3	1	1	11	+5	+5	−3

13-2.1   Three-Bit D/A Converter

Figure 13-2 gives the diagram and waveforms of a 3-bit D/A converter. In this case the multipliers are respectively 1/2.5, 1/5, and 1/10. Table 13-2 tabulates the results of this circuit.

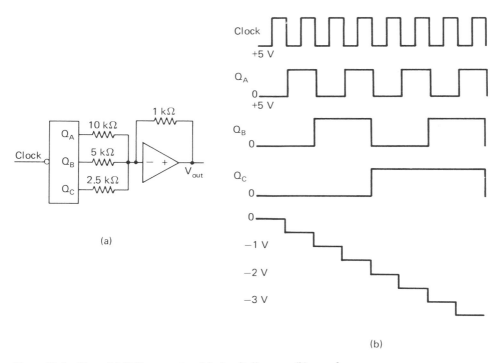

(a)

(b)

Figure 13-2   Three-bit D/A converter: (a) circuit diagram; (b) waveforms

Table 13-2 Three-bit D/A converter

Clock Pulse	Q_C	Q_B	Q_A	Count State	Q_C (V)	Q_B (V)	Q_A (V)	V_{out} (V)
0	0	0	0	000	0	0	0	0
1	0	0	1	001	0	0	+5	−0.5
2	0	1	0	010	0	+5	0	−1.0
3	0	1	1	011	0	+5	+5	−1.5
4	1	0	0	100	+5	0	0	−2.0
5	1	0	1	101	+5	0	+5	−2.5
6	1	1	0	110	+5	+5	0	−3.0
7	1	1	1	111	+5	+5	+5	−3.5

Now let us compare the results of these two D/A converters and try to arrive at some conclusions. In each case we have converted binary data, as represented by the *count state*, into an equivalent *analog voltage*, as given in the V_{out} column. In the 2-bit converter, the output changes in 1-V steps; in the 3-bit converter, in $\frac{1}{2}$ V steps. It is apparent that by going to a four-stage counter and with a 20-kΩ resistor for the fourth stage, the output voltage would change in $\frac{1}{4}$-V steps. By increasing the number of stages, we could make the steps as fine as we want and obtain any analog voltage as accurately as we desire.

Consider the V_{out} waveform, which is called a *staircase* waveform. In the 2-bit staircase, the steps are larger, being 1V. In the 3-bit staircase, the steps are $\frac{1}{2}$ V. In a 4-bit converter, the staircase would progress in steps of $\frac{1}{4}$V. Again it is apparent that by increasing the number of bits and steps we can make the staircase fine enough to approximate a straight line.

Again consider the 2-bit D/A converter. In going from 0 to +5 V, Q_A changes the output in 1-V steps, the least amount. We call this smallest amount the least significant bit (LSB). The largest amount here is the change in output due to a change in Q_B from 0 to +5 V. It changes the output by 2 V, which we call the most significant bit (MSB). In the three-stage counter, the MSB is still 2 V, but the LSB is now $\frac{1}{2}$ V. This D/A converter can distinguish between or resolve LSB differences of $\frac{1}{2}$ V. In a D/A converter the *resolution* is equal to the LSB.

Finally, the resistors, as we can see, go from 2.5 to 5 to 10 kΩ (to 20 kΩ for a four-stage counter). They increase in binary amounts. We call this combination a *binary weighted ladder. The accuracy* of the D/A converter is determined by the *accuracy* of the resistors of the ladder.

The preceding is tabulated for this system in Table 13-3.

Table 13-3   Binary weighted ladder D/A converters

Number of Bits	Smallest Resistor (kΩ)	Largest Resistor (kΩ)	LSB (V)	MSB (V)	Voltage Ranges (V)
2	2.5	5	1	2	0–3
3	2.5	10	$\frac{1}{2}$	2	0–3.5
4	2.5	20	$\frac{1}{4}$	2	0–3.75
5	2.5	40	$\frac{1}{8}$	2	0–3.875

13-3   DIGITAL-TO-ANALOG CONVERSION: R–2R LADDER

Digital-to-analog converters are available up to 16 bits. It is apparent by considering the extension of Table 13-3 to 16 bits that the largest resistor would have a very large value. This would cause problems for the following reasons:

1   Many different values of resistance are needed for the binary weighted ladder.
2   Very high values of resistance do not maintain their accuracy of resistance as temperature changes owing to unequal temperature coefficients.
3   It becomes difficult to manufacture high resistances to the precision required.
4   With high resistances, stray capacitance becomes important and a limiting factor if high sampling rates are to be used.

For these reasons the *R–2R* ladder of Fig. 13-3 is also used. It uses resistors of values *R* and 2*R* only.

Figure 13-3   Two-bit *R–2R* ladder D/A converter

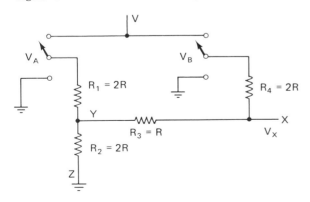

Referring to Fig. 13-3, let $V_B = 0$ (switch connected to ground). Let $V_A = V$. Break the network at point Y and apply Thevenin's theorem to the network to the left of point Y. $V_Y = V_{A/2}$ and $R_{\text{Th}} = R_1 \| R_2 = R$, as shown in Fig. 13-4.

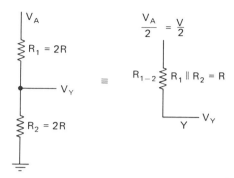

Figure 13-4   Network for V_A

We can now redraw the network to that of Fig. 13-5. It is apparent that this is again the same network as in Fig. 13-4, and $V_X = (V_{A/2}/2) = V_{A/4} = V/4$, with $R_{\text{Th}} = R$. V_A contributes either 0 or $V/4$ to V_X depending upon the switch position.

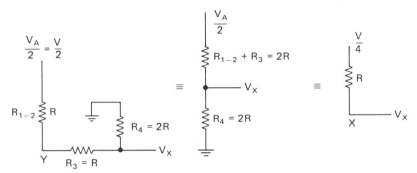

Figure 13-5   Equivalent network for V_A to V_X

Let $V_A = 0$ (switch connected to ground). Let $V_B = V$. If we look at Fig. 13-3, we see that $R_1 \| R_2 = R = R_{1-2}$. $R_{1-2} + R_3 = 2R$, and for V_B we can draw Fig. 13-6, from which we can see that a repetition of the voltage divider network gives us $V_X = V_{B/2} = V/2$ and $R_{\text{Th}} = R$.

By superposition of $V_A = V$ and $V_B = V$, $V_X = V/4 + V/2 = 3/4V$. We can write Table 13-4 for this 2-bit R–$2R$ ladder D/A converter.

We can now arrive at some general properties of the R–$2R$ D/A converter:

1   The resistance looking back toward the Z or ground terminal of the ladder junction point is always equal to R: R_Y toward $Z = R$ and R_X toward $Z = R$.

2   For each step that a voltage has to be processed up the ladder toward X, its contribution is decreased by a factor of 2.

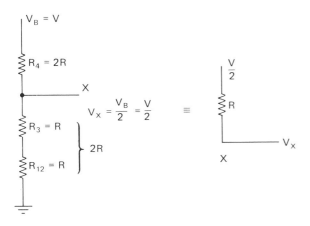

Figure 13-6 Network for V_B

Table 13-4 Two-bit R–$2R$ D/A converter

V_B	V_A	V_X
0	0	0
0	V	$\dfrac{V}{4}$
V	0	$\dfrac{V}{2}$
V	V	$\dfrac{3V}{4}$

Using these principles, a 4-bit D/A converter using the R–$2R$ ladder principle is shown in Fig. 13-7. The weights of the voltage contributions are as follows:

$$V_X = V_{D/2} = \frac{V}{2} = \text{MSB}$$

$$= V_{C/4} = \frac{V}{4}$$

$$= V_{B/8} = \frac{V}{8}$$

$$= V_{A/16} = \frac{V}{16} = \text{LSB}$$

We see that they follow a binary ratio as did those in the binary weighted ladder of Figs. 13-1 and 13-2.

The switches shown in Figs. 13-3 and 13-7 are known as analog switches and

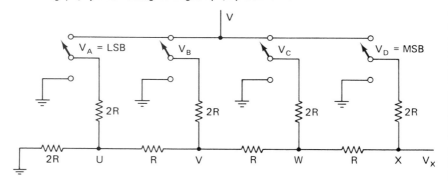

Figure 13-7   Four-bit *R–2R* ladder D/A converter

have stringent requirements. If any resistance is associated with them, it must be the same in going to *V* or to ground. They must have no offset voltage associated with them. This rules out bipolar transistors since they have an offset voltage (they do not begin to conduct) of approximately 10 mv. This offset can be improved with bipolar transistors by a factor of approximately 10 to 1, or to 1 mv, by using the transistor in the inverted mode, i.e., collector as emitter and emitter as collector. Field-effect transistors are used to a considerable extent since they have no offset voltage, but their resistance is high and must be considered in the design of the switch and ladder network.

13-4 ANALOG-TO-DIGITAL CONVERSION: GENERAL

The problem of A/D conversion is a complex one and still under active development. New techniques are continually appearing. We shall, therefore, only discuss some of the basic techniques.

Analog-to-digital converters can be classified into three groups:

1 Comparison type.
2 Capacitor charging type.
3 Voltage-controlled oscillator (VCO).

13-5 ANALOG-TO-DIGITAL CONVERSION: COMPARISON TECHNIQUE

In this method, an A/D converter is used to develop an analog voltage. This developed analog voltage is then compared to the voltage whose digital equivalent is desired. This type of A/D converter is illustrated in Fig. 13-8. It has a gated counter, a D/A converter–staircase generator and a voltage comparator.

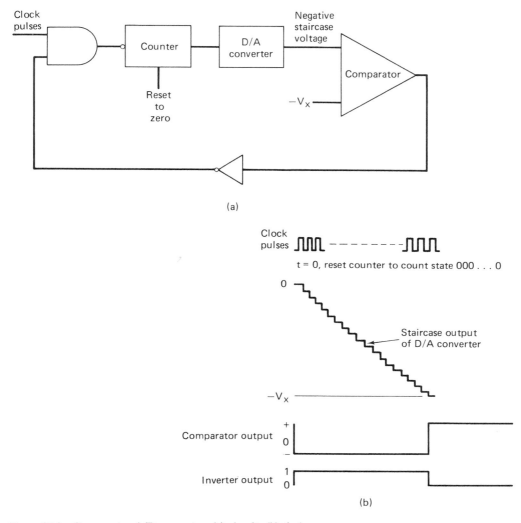

Figure 13-8 Comparator A/D converter: (a) circuit; (b) timing

1 At time $t = 0$, the counter is reset to zero.

2 This makes the staircase voltage equal to 0 and the comparator output negative.

3 The inverter output goes positive to logic 1 and enables the input gate so that clock pulses arrive at the counter.

4 The staircase voltage increases negatively until it becomes more negative than V_X.

5 The comparator output goes positive. This is inverted by the inverter to logic 0 and turns the counter off.

6 The counter count state is the digital equivalent of the analog voltage $-V_X$.

It should be noted that this method provides an approximation which is the equivalent of the next bit larger than V_x. Increased accuracy can be obtained by increasing the number of bits of the D/A converter to provide a more accurate comparison with V_x in the comparator. The accuracy of this method is limited by the ability of the comparator to resolve small differences in voltage. Open-loop OP-AMPs are used frequently for the comparator.

An interesting modification of Fig. 13-8 is to make the counter an up–down counter. The output of the inverter is connected to the up–down control of the counter. The counter will count up with a 1 applied and down with a 0 applied. The counter now cycles between the count states equivalent to $+1$ bit and -1 bit of the voltage V_x.

In the A/D converter of Fig. 13-8, if V_x is a large voltage, it takes a long time for the staircase to reach V_x. Much work has been done on developing techniques to reduce this time.

13-6 CAPACITOR CHARGING TYPES OF ANALOG-TO-DIGITAL CONVERTERS

13-6.1 Pulse-Width Converter: Single-Slope Method

A pulse is developed whose width is proportional to the voltage V_x. This time is measured and displayed with a series of count pulses. The number of pulses is chosen to display the unknown voltage directly. Consider Fig. 13-9, a pulse-width A/D converter.

Figure 13-9 Pulse width A/D converter: (a) circuit (b) timing

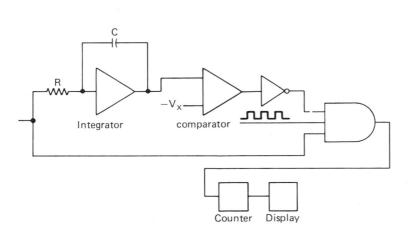

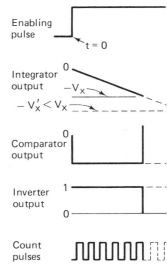

(a)

(b)

1   At time $t = 0$, a positive-going pulse is applied to the input of an integrator and to the input of a three-input AND gate.

2   At time $t = 0$, the integrator output equals 0, the comparator output is negative, and the inverter output is a 1. The AND gate is enabled.

3   Count pulses begin to be counted.

4   The integrator ramps down until $-V_X$. At $-V_X$, the comparator output changes sign, going positive, and through the inverter applies a 0 to the three-input AND gate, disabling it.

5   No more pulses get into the counter, which now displays the number of pulses passed through the AND gate.

6   For $-V'_X < -V_X$, it takes longer for the ramp to get to $-V'_X$ so more count pulses go through the gate and are counted.

7   From the timing diagram, it can be seen that the number of pulses that get into the counter is proportional to the time it takes the integrator to ramp down to $-V_X$.

This type of A/D converter is used quite frequently in digital voltmeters. When this is done, R, C, and the pulse frequency are selected to make the voltmeter direct reading. In these digital voltmeters the A/D conversion is repeated at regular intervals, frequently at a 60-Hz rate.

The accuracy and stability of this converter depends upon the stability of R and C of the integrator, since these determine the rate at which the integrator output voltage ramps down. This in turn determines the time it takes the ramp to get down to $-V_X$ and, therefore, the width of the gating pulse for the count pulses. Finally, the pulse generator repetition rate must remain fixed and not drift. Otherwise, the number of pulses counted will change, and the counter will display an incorrect count. The stability requirements necessitate a crystal-controlled pulse generator, which is complex and expensive.

13-6.2   Pulse Width A/D Converter: Dual-Slope Method

The dual-slope method eliminates the necessity for an expensive crystal-controlled oscillator and is very popular in digital voltmeters. As will be shown, many possible sources of error are eliminated.

Consider Fig. 13-10, the circuit and timing diagram of a dual-slope A/D converter.

1   At time $t = 0$, the switch (electronic) is connected to $-V_X$.

2   Also, at time $t = 0$, the counter is reset to a count of zero and begins to count pulses.

3   The integrator ramps in a positive direction for a fixed count, either 200 pulses for a $2\frac{1}{2}$-digit digital voltmeter (DVM) or 2000 pulses for a $3\frac{1}{2}$-digit DVM.†

†A $2\frac{1}{2}$-digit DVM can display from 0 to 199. A $3\frac{1}{2}$-digit DVM can display from 0 to 1999.

4 At the completion of the count capability of the counter (200 or 2000 pulses), the switch is operated to connect the integrator to $+V_{ref}$. The counter continues to count pulses, which again begin to fill up the counter.

5 The integrator ramps down until its output voltage returns to zero, which is detected by the zero (voltage) crossing detector.

6 The sensing of the zero voltage point is used to inhibit pulses from entering the counter. The pulse count is displayed and is proportional to the voltage $-V_X$.

Figure 13-10   Dual-slope A/D converter: (a) circuit; (b) timing

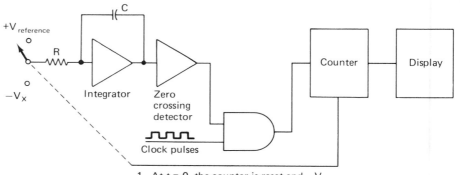

1. At t = 0, the counter is reset and $-V_X$ is connected to the integrator
2. At time when counter is filled, switch changes from $-V_X$ to $V_{reference}$

(a)

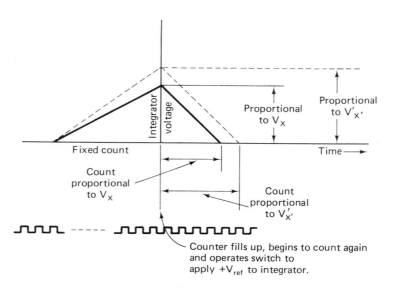

(b)

This A/D converter is a scheme that depends *only* upon the ratio of V_x to V_{ref}. The electronics is more complex, but it eliminates the need for stable R and C and for a very stable crystal-controlled oscillator. Let us see why this is so.

In Fig. 13-10a, suppose that integrator R were to double. This would halve the up-ramping rate, and the voltage attained by the integrator when the counter is filled would be half that shown in Fig. 13-10b. However, the down-ramp rate is also halved so that the time to reach 0 voltage is the same and does not depend upon R. Similar considerations hold for the stability of the integrator capacitor C.

Now consider the frequency stability of the pulse generator. Suppose that it were to drift down. The time for the up ramp would be increased, and the integrator voltage would be larger. It now would take longer for the down ramp to reach 0 V, but since the pulse generator frequency is lower, an identical number of count pulses would be counted during the down ramp. The pulse generator has to maintain its stability only over the time needed to complete any A/D cycle, and this is easily met by oscillators such as the UJT.

13-6.3   Voltage-Controlled Oscillator (VCO) Method

Figure 13-11a shows a frequency versus voltage plot of a VCO and block diagram of an A/D converter. Voltage-controlled oscillators operate by using a source of voltage to control the rate at which a capacitor charges to a triggering voltage. This triggering voltage initiates one cycle of the frequency. Many techniques have been developed to do this.

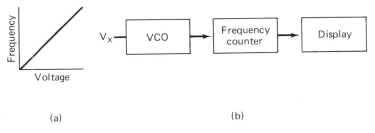

(a)   (b)

Figure 13-11   Voltage-controlled oscillator (VCO) A/D converter: (a) voltage-frequency plot; (b) block diagram

In the A/D converter shown in Fig. 13-11b, the frequency is measured by a frequency counter and displayed to give a visual indication.

Problems

13-1. What is meant by an analog voltage and how do analog voltages differ from digital voltages?

13-2. Define decoding and encoding and encoding with reference to D/A and A/D conversion.

13-3. What are the advantages of the R–$2R$ ladder compared to the binary weighted ladder?

13-4. If the 1 level equals $+10\,V$ and the 0 level equals $0\,V$, what is the MSB and LSB in the D/A converter?

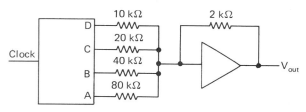

13-5. In Prob. 13-4, determine V_{out} for *DCBA* equal to (a) 1010; (b) 0101; (c) 1110.

13-6. It is desired to improve the resolution (LSB) of the D/A converter of Prob. 13-4 by adding two more stages E and F.
(a) What values of resistance are needed?
(b) What is the new value of the LSB?

13-7. It is desired to improve the resolution (LSB) of the D/A converter of Prob. 13-4 so that the converter has 12 bits. $R_A = 10{,}000\ \Omega$.
(a) What value of R is needed for the sixteenth bit?
(b) What is the magnitude of the LSB?

13-8. If the 1 level is 1 V and the 0 level is 0 V,
(a) What are the MSB and LSB?
(b) Find V_X for *CBA* = 011.
(c) Find V_X for *CBA* = 110.

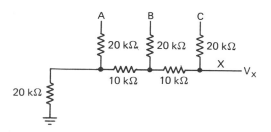

13-9. If the 1 level is 10 V and the 0 level is 0 V, what are the MSB and LSB?

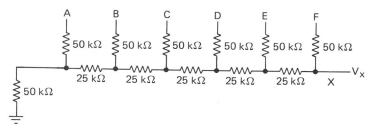

13-10. In Prob. 13-9, what is V_X for *FEDCBA* equal to (a) 101110; (b) 110010; (c) 111011?

13-11. How does a comparison type A/D converter operate?

13-12. What do we mean by a single-slope A/D converter? How does it function?

13-13. How does a dual-slope D/A converter operate?

13-14. Explain why the dual-slope A/D converter eliminates the need for accuracy in the R and C integrator values and why its pulse oscillator frequency does not have to be very stable.

13-15. Explain how the VCO type of A/D converter functions.

FOURTEEN

Illustrative Digital Techniques and Systems

14-1 INTRODUCTION

Digital electronics is penetrating into every phase of electronics and our everyday life. To attempt to cover all possible applications of digital electronics would require expanding this chapter alone into a multibook series. Therefore, we can only briefly illustrate with some simple systems a few of the areas in which digital electronics is being applied.

14-2 ALPHANUMERIC DISPLAY†

There are many different technologies available today for the display of numbers and letters. Each technology has its distinct advantages and disadvantages and these have to be considered in designing the display section of a digital system. The technologies available today are light-emitting diodes (LED), gas-discharge devices, vacuum fluorescent displays, vacuum incandescent displays, and liquid crystal display (LCD).

†David A. Laws, "Are you keeping pace with display technology?" *EDN*, Aug. 20, 1974, pp. 61–64.

14-2.1   Light-Emitting Diodes (LED)

Light-emitting diodes are made by fabricating a *PN* junction in gallium-based semiconductor materials. When these diodes are forward biased, they emit light. These diodes have a forward voltage drop of about 1.7 V and operate with currents of from 0.3 to 20 mA. For small calculators, sizes up to 0.1 in. are used; for larger displays, sizes up to 0.6 in. are available. GaAsP (Gallium–arsenic–phosphorus) diodes give red, orange, and yellow colors. Green display diodes (voltage drop = 2.2 V) are made from gallium and phosphorus. LED digital readouts are made as seven-segment displays (see Fig. 14-1).

Figure 14-1   Seven-segment display configuration

Single digit displays are packaged in either the 14-pin dual-in-line DIP style or in cases compatible with this 14-pin layout. LED displays come in two configurations, one with a common anode connection and the other with a common cathode configuration. To drive them, decoder–encoder ICs are available. To make up the difference between the diode voltage drop and V_{CC}, a voltage dropping resistor is needed. If needed, seven resistors are available in a single 14-pin DIP. Figure 14-2a shows the LED, resistors, and an encoding–decoding IC for the common-anode

Figure 14-2   Light-emitting diode with decoder-encoder driver IC: (a) common-anode display. Active low driver, TTL 7447; (b) common-cathode display. Active high driver, TTL 7448

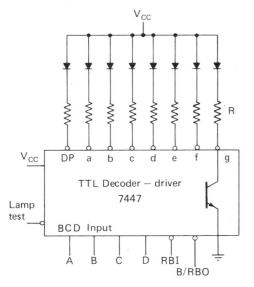

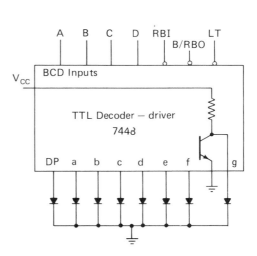

(a)

(b)

LED. Figure 14-2b shows the same components, but for the common-cathode LED. In the common-anode design (Fig. 14-2a), the output driver transistors are open collector and have to be driven into an *active low* state to allow current to flow through the LED segments. External resistors are needed, and these are chosen as a compromise between light output and currents used. In the common-cathode design, the IC output transistors are driven into an *active high* state to permit current to flow through the LED segments. (In the low state they shunt current away from the segments.) In this design, the dropping resistors are built into the IC, thus eliminating the need for external resistors.

It is annoying to have the extra zeros at the beginning and end of a numerical display. These decoder–driver ICs have RBI and RBO (ripple blanking input–output) inputs to blank these leading and trailing zeros. They also have a lamp test (LT) input to test if all the segments are operating properly. In addition, these decoder–drivers have a blanking (B) capability to permit intensity modulation or multiplexing from an external voltage source.

14-2.2   Gas-Discharge Devices

The gas-discharge display device is one of the oldest numeric display devices. In a gas tube, when a voltage greater than the ionization potential is applied between the anode and cathode, the gas will ionize and a glow appears at the cathode. The NIXIE† tube, a neon gas tube, glows orange. It requires an ignition voltage of between 170 and 200 V (high compared to the +5 V for TTL and LED systems) and sustains at about 140 V. The NIXIE tube has a common anode and 10 cathodes in the shape of the digits. The current required is approximately 0.5 mA. The high voltage needed can be obtained from the power line or from dc-to-dc converters from the 5 V needed for the ICs. Figure 14-3 shows the principle of the NIXIE tube. The NIXIE is driven frequently by an IC, typically the TTL type 74141. This is a BCD to decimal decoder–driver. It has 10 output driver transistors, one for

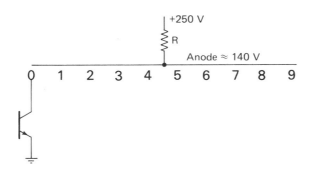

Figure 14-3   NIXIE display and transistor driver

†Burroughs Corporation.

each cathode. When the output transistor of the IC goes into its *active low* state, the voltage between the anode and cathode is high enough to ionize the display and the cathode glows at the digit displayed. When the transistor is cut off, the output transistor's breakdown voltage is in excess of 55 V. This keeps the anode–cathode voltage below the ionization voltage and there is no cathode glow. This 55-V breakdown for the output transistors is a rather severe requirement.

The Beckman display is a seven-segment gas-discharge device that operates on the same principle as the NIXIE tube.

When more digits are required, Panaplex II displays with 4 to 12 digits are available. They consist of a sandwich of two pieces of glass. They use seven-segment displays per digit and a gas mixture of neon and mercury.

14-2.3 Vacuum Fluorescent Displays

In these seven-segment displays,† electrons emitted by a heated filament are accelerated by a positive voltage and strike a phosphor-coated anode (segment). The energy of the electrons is converted into a green light. The spectrum of light is sufficiently broad so that different colors can be obtained with colored optical filters. This display requires an active high driver. The driver can be either an open collector TTL IC with an external load resistor or an MOS IC. Segment currents are approximately 1 mA, and the segment voltage is between +15 and +30 V. In addition, a filament within the display is always operating. The filament requirements are approximately 1.4 V at 50 mA. As can be seen, the voltages required are not compatible with the +5 V of TTL.

14-2.4 Vacuum Incandescent Displays

This is a seven-segment display in which each segment is a tungsten filament. This display is compatible with and operates off the +5 V of TTL. The filaments are connected across the 5-V supply through ICs that decode BCD information and have active low outputs. Tungsten is a metal with low cold resistance. Hence the ICs and the power supply must be able to handle the large inrush currents of the filaments. In the Numitron‡ the segment currents are 24 mA. The light output is white, quite high, and with a broad spectral response. Any color can be obtained with colored optical filters. Slow response makes it difficult to multiplex.

14-2.5 Liquid Crystal Display (LCD)

These seven-segment displays are causing much excitement because the power needed to operate them is extremely small. Thus they are potentially the solution to long-life, battery-operated displays, such as in the digital wristwatch. They

†DIGIVAC, Tung-Sol Division of Wagner Electric.
‡RCA Manufacturing Co.

can be made in comparatively large sizes. However, there are still major problems with life and temperature, and much work remains to be done in the development of LCDs.

Unlike the other displays, LCDs do not generate their own light, but modify the transmission of available light. No display can be seen when it is dark. To overcome this, a small light bulb is used in the dark to provide some light, but this detracts from the low power advantage of LCDs. They are made as transmissive or reflective devices.

LCDs consist of a thin layer of an organic liquid sandwiched between two layers of glass. Seven-segment patterns are coated on the glass. For displays that control the transmission of light, the patterns become transparent. In the reflective type one of the patterns is made highly reflective. There are two types of LCDs, the field-effect type and the dynamic-scattering type. With the field-effect type, application of voltages between the pattern rotates the plane of polarization of the light. Polarizing filters determine whether light is transmitted or reflected. In the dynamic-scattering type, application of voltage scatters the light, makes the liquid opaque, and prevents light from being transmitted.

LCDs do not tolerate dc voltages as it affects their life. Hence it is necessary to generate on ac voltage. This is done in an IC. LCDs can be driven directly by MOS and CMOS ICs and require voltages of the order of 20 V ac.

14-3 FREQUENCY COUNTER

Frequency counters measure the frequency of a periodic wave by counting the number of cycles in the wave over a period of one second or decimal fraction of a second. This counting is repeated at regular intervals. Different waveforms are processed into a standard pulse waveform, usually by a Schmitt trigger. These pulses are counted for the desired time and displayed on a digital display. To eliminate the annoyance of a visually changing count display during the count interval, most counters have a memory that is used to update the display. The accuracy of the measurement is determined by the accuracy of the count time interval. For inexpensive counters with a moderately accurate time base, the 60-Hz line is used as the timing source; for more accurate counting and at greater expense, crystal oscillators generate the time base.

Figure 14-4 is the block diagram of a three-digit frequency counter. The instrument counts for either 1s or decimal submultiples of 1 s. With a 1-s count time, it can measure frequencies from 0 to 999 Hz; with a 0.1-s count time, it can measure from 0 to 9990 Hz. To use a 1-s count time, a 10-s period is set up. Within this time, the third second is used as a count period, the seventh second as a strobe pulse to gate the count data into storage latches and update the display, and the ninth second as a pulse to reset the counter. These time slots are selected to minimize decoding pin count. Two possible time bases are available. One is a 60-Hz line-derived time base; the second is more precise and is derived from a crystal

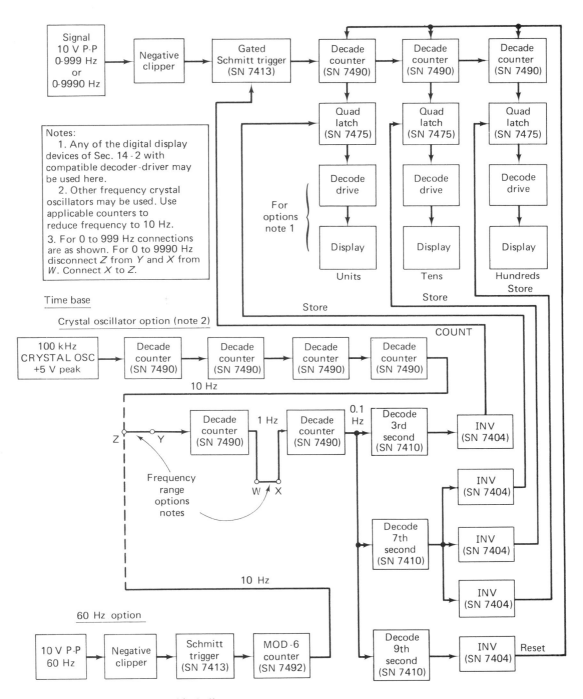

Figure 14-4 Frequency counter block diagram

oscillator. Schmitt triggers are used to convert and standardize the incoming signal and the 60-Hz signals to the fast rise and fall time signals needed by TTL ICs. The count Schmitt trigger is gated ON for the count interval.

The following digital principles are used in this frequency measuring instrument:

1 Schmitt trigger.
2 Crystal oscillator.
3 Decade counters for signal and time base.
4 Mod-6 counter to convert 60 Hz to 10 Hz.
5 Memory-strobed latch between decade counters and digital display.
6 Digital display.
7 Decoder–encoder for displays.
8 Decoder for count, strobe, and reset intervals.

14-4 DIGITAL TUNING INDICATOR

Magnavox radio receivers use a digital timing indicator in their AM/FM receivers. It makes use of many of the digital principles that have been developed in this text, specifically, the following:

1 Frequency counter.
2 Tens complement subtraction.
3 Seven-segment incandescent display.
4 Crystal oscillator time base.

Radio receivers operate on the superheterodyne principle. In this type of receiver a local oscillator is combined with an incoming signal to give an intermediate frequency, which is a fixed frequency and the difference between the signal and local oscillator frequencies. For example, in the broadcast AM (amplitude modulation) band (Fig. 14-5a),

Local oscillator	1455.0 kHz
Signal	1000.0 kHz
Intermediate frequency	455.0 kHz

The local oscillator is always 455.0 kHz higher in frequency than the signal frequency over the AM band of 530 to 1650 kHz. In the same way, in the broadcast FM (frequency modulation) band (Fig. 14-5b), the local oscillator is always 10.7 MHz higher in frequency over the FM band of 88.0 to 108 MHz.

At first thought, one would expect that an obvious method would be to measure the incoming signal in a frequency counter, but this is impractical for three reasons.

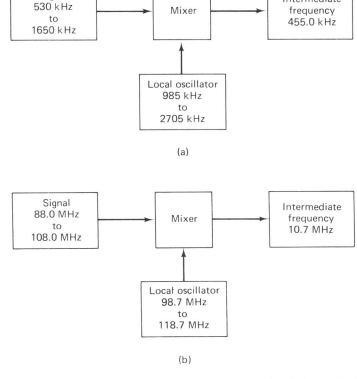

(a)

(b)

Figure 14-5 Broadcast radio receiver frequencies: (a) AM, amplitude modulation section of receiver; (b) FM, frequency modulation section of receiver

1 If the signal is very weak, it would have to be amplified separately to measurable levels.

2 With no signal, the display would either be nonexistent or noise would give spurious readings. The listener would not know at what frequency his receiver was operating.

3 Noise would interfere with the measurement.

Because of these problems, the display measures the local oscillator frequency and uses the tens complement method to subtract the intermediate frequency from it to arrive at the signal frequency. Since the local oscillator operates continuously over the band, it is always present and hence a display is always present. The voltage level of the local oscillator is sufficiently high so that only moderate amplification is required to raise its voltage to a level that is countable in a frequency counter.

We can write

$$\text{Intermediate frequency (IF)} = \text{local oscillator (LO)} - \text{signal (S)}$$

$$\text{IF} = \text{LO} - \text{S}$$

and
$$\text{S} = \text{LO} - \text{IF}$$

Suppose that we want to receive a station at 1000.0 kHz. We can then write

$$455.0 = \text{LO} - 1000.0$$

$$\text{LO} = 1455.0$$

$$1000.0 = 1455.0 - 455.0$$

To display 1000.0 kHz, we need a frequency counter that will measure the difference between 1455.0 and 455.0 kHz. However, using tens complement subtraction, we can write

```
     9999.9                    9544.9
   −  455.0                  +    .1
   ──────────                ──────────
     9544.9 = nines complement  9545.0 = tens complement
```

```
     1455.0
     9545.0
   ──────────
overflow one ──→ 1  1000.0
```

Tens complement subtraction lends itself readily to counter techniques. What has to be done now is to take a frequency counter and preset it to a count of 9545.0 instead of a reset to a count of zero as is usually done. Then apply the LO signal to the frequency counter. The counter begins with the tens complement of the intermediate frequency and counts up to the signal frequency, which can be displayed on a digital readout.

What about the overflow 1? Because of tens complement subtraction, no end-around carry is needed. If we use a five-stage counter, the overflow 1 is automatically discarded since it is the sixth digit.

We have now discussed the basic principles of the display. There are two other practical and simplifying steps.

1   Two possible adjacent signal channels might be 1000.0 and 1001.0 kHz. If the counter is preset to 9545.0 kHz and a most significant four-digit display is used that omits the digit to the right of the decimal point, the display reads 1000 for any signal frequency between 1000.0 and 1000.9 kHz. If, however, we split the channel difference and preset to 9545.5, the display reads 1000 between signals of 999.5 to 1000.4 kHz.

2 To preset the most significant digit to a 9 requires that the most significant stage have a decade counter. Practically, the most significant digit displayed is either a 1 or a 0, which is a binary count stage and much simpler electronically than a decade counter. Therefore, we can use a *T* flip-flop for the most significant bit and preset the counter to 1545.5. Let us try this for two stations, one at 900.0 kHz and one at 1100.0 kHz. The arithmetic is shown in Fig. 14-6.

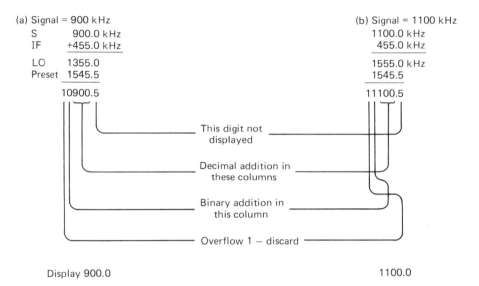

(a) Signal = 900 kHz

S	900.0 kHz
IF	+455.0 kHz
LO	1355.0
Preset	1545.5
	10900.5

(b) Signal = 1100 kHz

1100.0 kHz
455.0 kHz
1555.0 kHz
1545.5
11100.5

This digit not displayed

Decimal addition in these columns

Binary addition in this column

Overflow 1 — discard

Display 900.0   1100.0

Figure 14-6   Mathematical operations of display

The actual display whose operation is shown in the block diagram of Fig. 14-7 has the following additional features.

1 The FM LO is divided by a factor of 100. This gives frequencies for counting that are the same as those of AM signals. There is a divide by 4 using ECL and a divide by 25 with TTL.

2 The FM preset number is 9893.5, corresponding to the FM IF of 10.7 MHz.

3 A 204.8-kHz crystal oscillator is used for the time base. This is divided by binary stages to give two 10-ms pulses.

4 The LO is counted for 10 ms. During the next 10 ms, an update pulse is applied to storage latches that control display drivers and seven-segment incandescent displays; this is followed by the preset pulse. This gives a total cycle time of 20 ms, which means that the LO is counted 50 times per s.

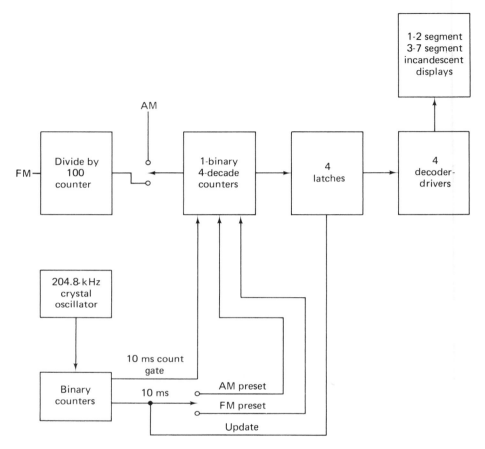

Figure 14-7 Radio receiver digital display

14-5 DIGITAL DATA TRANSMISSION: MODEMS

Transmission of digital data over telephone lines and other media is a very important factor in today's world. Typical examples are credit card verification, air line ticket reservations, and communication with remote computers.

Many of these data are transmitted over voice-grade telephone lines. These lines have a frequency range of 300 to 3200 Hz or a bandwidth of about 3000 Hz. Sometimes these lines can degrade to a 2100-Hz bandwidth. Digital data are defined by voltage levels, but to send such data it is necessary to be able to transmit down to direct current. Telephone lines cannot do this. Speed of data transmission is an important consideration. Practical limits of the rate of data transmission are of the order of the bandwidth. Thus for a 3000-Hz bandwidth the fastest rate is approximately 3000 bits per s (bps).

Digital data must be converted to a form suitable for transmission over telephone lines. This is done in a *MODEM*, short for modulator–demodulator. The Electronic Industries Association (EIA) has developed specification RS-232C, which specifies the voltage and impedance levels between *computers* and *modems* so that data from any computer can be converted for transmission, and at the receiving end data can be redigitized by the modem to computer logic levels. The two basic methods for digital data transmission are *frequency shift keying* (FSK) and *phase shift keying* (PSK).

For lower transmission rates on standard telephone lines, up to 1800 bps, FSK methods are used. At the sending modem, two audio frequencies are used for the 1 and the 0. These frequencies have not been standardized. Typical values might be $1 \equiv 1700$ Hz and $0 \equiv 1300$ Hz. The audio tone generators are transistor (OP-AMP) oscillators using resistance–capacitance networks or inductance–capacitance tuned circuits to set the frequency. At the receiving end, the modem has frequency selective filters to determine which of the low frequencies is being sent. It is quite common today to use active filters to obtain this frequency separation. Active filters are made with OP-AMPs that use resistance–capacitance feedback networks to achieve frequency selective gain amplification.

For higher bit rates, up to 4800 bps, PSK methods are used. Rates as high as this cannot be transmitted on the standard telephone line. The frequency response of the line can be improved by frequency compensation methods either at the modem or by the telephone company (dedicated lines). In PSK transmission, data are sent by means of pulses. Each pulse carries data for 2 bits of information and is called a *dibit*. The phase of a pulse compared to the phase of the previous pulse determines what data are being sent. At the sending end, digital circuits advance the phase of every dibit pulse. How much it is advanced is determined by the data of the 2 bits being transmitted. Table 14-1 gives a set of phase advance values that are being used, and the circuits for doing this are given in Fig. 14-8. At the receiving end, a pulse is delayed and its phase is compared with that of the next pulse. This is used to determine the data carried by the next pulse.

Table 14-1   DIBITs and phase

Data	00	01	11	10
Phase	45°	135°	225°	315°

The basic principles of dibit FSK generation are shown in Fig. 14-8.

1   A three-stage binary counter is shown in Fig. 14-8a. The phase of the output pulses of this counter can be set by means of a phase set pulse applied to the preset (*PR*) inputs of the stages. Phase set controls X_1 and

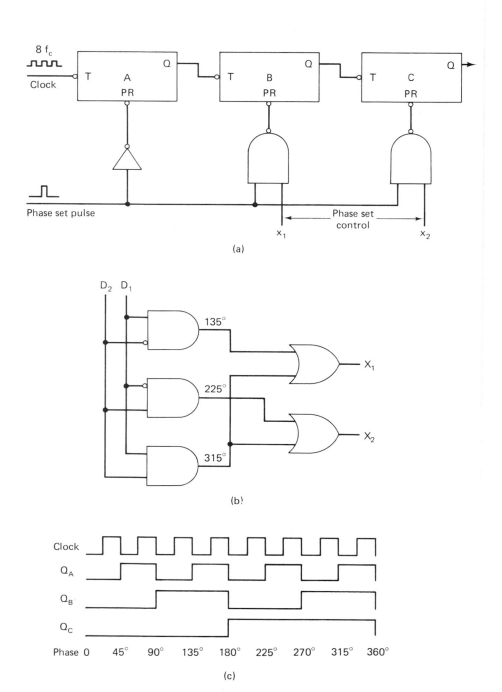

Figure 14-8 Dibit FSK generator: (a) counter circuit; (b) decoder-encoder; (c) waveforms

X_2 are used to determine which stages are preset. A clock is applied at a frequency eight times the output frequency.

2   Let phase set inputs X_1 and X_2 both be equal to 0. This disables gates G_1 and G_2.

3   At a time corresponding to a phase of $0°$ (Fig. 14-8c), apply the phase set pulse which is short. This is applied only to FF A (G_1 and G_2 disabled). This changes Q_A from a 0 to a 1. In effect, this shifts Q_A, Q_B and Q_C ahead by $45°$.

4   Suppose that $X_1 = 1$ and $X_2 = 0$. Now apply the phase set pulse again at a time corresponding to $0°$. This will set both Q_A and Q_B to 1. If we examine the waves of Fig. 14-8c, we see that the time when Q_A goes to 1 and Q_B is a 1 corresponds to an angle of $135°$. In effect, this shifts the waves of Q_A, Q_B, and Q_C ahead by $135°$. Comparing this to step 2, we see that $X_1 = 1$ advances the phase by $90°$.

5   Suppose that $X_1 = 0$ and $X_2 = 1$. A phase shift pulse sets Q_A and Q_C equal to 1. Q_A going to 1 and Q_C at a 1 corresponds to an angle advance of $225°$. $X_2 = 1$ advances the phase by $180°$.

6   We can now write a truth table (Table 14-2) between the dibit data of Table 14-1, phase angle shifts, and X_2 and X_1. To get from dibit data to the X_2 and X_1 inputs is a straightforward problem in decoding–encoding. The circuit diagram is given in Fig. 14-8b.

Table 14-2   Dibits to phase inputs

Dibit Data		Phase Angle	PR Inputs	
D_2	D_1		X_2	X_1
0	0	$45°$	0	0
0	1	$135°$	0	1
1	1	$225°$	1	0
1	0	$315°$	1	1

At the receiving modem, differential detectors are used to determine what data were sent. They operate in the following manner. When a signal is received, it is delayed by a time equal to the dibit or signaling interval. It is then used to generate two signals $90°$ out of phase. These are then compared in product detectors (like AND gates) with the next dibit pulse. Figure 14-9 shows how this is done. The two generated signals that are $90°$ out of phase are F (for the first bit detection) and S (for the second bit detector).

1   A signal at an angle of $45°$ is out of phase (different direction) with both F and S. Dibit $= 00$.

2   A signal at an angle of $135°$ is in phase with F, but out of phase with S. Dibit $= 01$.

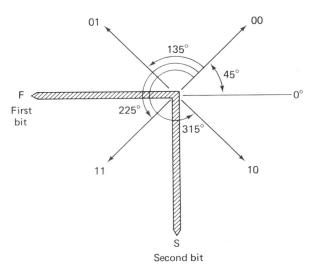

Figure 14-9   Differential detector for dibit recovery

3   A signal at an angle of 225° is in phase with both *F* and *S*. Dibit = 11.
4   A signal at an angle of 315° is in phase with *S*, but out of phase with *F*. Dibit = 10.

In the PSK method of Figs. 14-8 and 14-9, each signal pulse carried 2 bits of data. The signaling rate is expressed in *bauds*. This PSK system has a *bit* rate *two times* the baud rate.

More recent PSK systems send *tribits* as one of eight possible signals whose phases differ by 45°, as compared to the 90° of Fig. 14-9.

14-6   PHASE-LOCKED LOOP (PLL)

14-6.1   Frequency Synthesizer

Consider the problem of a multichannel communication system. To prevent adjacent channel interference, each channel must operate at a very precise and stable crystal-controlled frequency. Examples of such multifrequency systems occur in aircraft navigation, military communication systems, UHF television, amateur band radio, citizen's band radio, signal generators, etc. Suppose that a 50-channel system were needed. One would think that this would require 50 separate crystal oscillators. However, the PLL can generate the 50 channels of crystal stable frequencies with a *single crystal oscillator* reference frequency. The PLL is so important in modern technology that special ICs are available for PLL systems.

Figure 14-10 is a PLL system used to generate 1000 separate frequencies with crystal oscillator stability in 1-kHz steps over the frequency range of 1.000 to 1.999 MHz. The PLL operates in the following manner:

1 It is a frequency feedback system.

2 In an IC phase detector (MC4044), it compares the phase and frequency of two signals. One signal is generated by a crystal oscillator and in Fig. 14-10 is one hundredth of the crystal frequency. The second signal has its origin in a VCO-VCM (voltage-controlled oscillator/voltage-controlled multivibrator). Its frequency is also divided by a factor N, which is selectable in an IC programmable counter (MC74416).

3 The output of the phase detector goes into a low-pass filter and amplifier. Out of this comes a dc correction voltage.

4 Suppose that there were no crystal oscillator. The VCO-VCM oscillates at some free-running frequency. Now apply the crystal oscillator frequency divider output to the phase detector. If the VCO-VCM/N frequency is too high, the phase detector and filter will generate a dc correction voltage to decrease the VCO-VCM/N frequency until its frequency matches the crystal oscillator in frequency and phase with very little error. If VCO-VCM/N is too low, a correction voltage of opposite polarity will be applied to the VCO-VCM to increase its frequency.

5 In Fig. 14-10, N can be varied between 1000 and 1999, thereby generating 1000 different frequencies, all referenced to the original 100-kHz crystal oscillator and having almost the identical precision and stability as the 100-kHz oscillator.

Figure 14-10 Phase-locked loop frequency synthesizer. (Courtesy Motorola Semiconductor Products, Inc.)

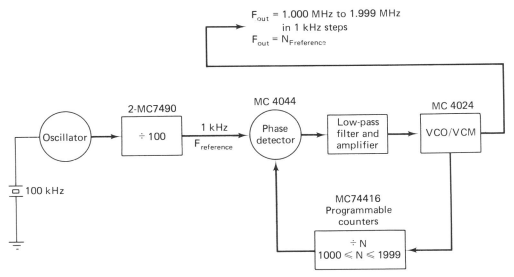

14-6.2 Frequency Shift Keying (FSK) Demodulator Detector

The PLL is used in FSK data transmission systems to directly demodulate the two frequencies and generate the 1 or 0 levels. Figure 14-11 is a diagram of an FSK demodulator. With no signal applied the VCO-VCM runs at its free-running frequency. Now apply an FSK upper frequency that is higher than the VCO-VCM free-running frequencies. The system will correct with a dc voltage level. Now apply an FSK lower frequency that is lower than the VCO-VCM free-running frequency. The system again corrects with an opposite (or different) dc voltage level. The two dc levels are used as the desired 1 and 0 logic levels.

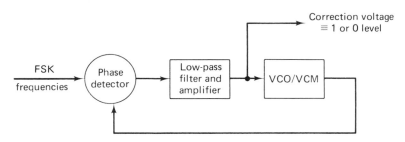

Figure 14-11   Frequency shift keying demodulator

14-7 ELECTRONIC CALCULATOR†

A digital electronic system that has attained enormous popular success is the hand-held electronic calculator. They are available with many different capabilities, from simple four-function units that can add, subtract, multiply, and divide to quite complex units directed toward specific applications, such as trigonometric and exponential function capability for engineers or handling problems in accounting or performing the many calculations involving interest in business.

Figure 14-12 shows the Hewlett-Packard HP-35 scientific calculator. It is a multifunction calculator with the ability to handle complex calculations both with conventional numbers and with trigonometric and exponential functions. The calculator contains five MOS/LSI circuits, three read-only memories (ROMs), and a control and timing circuit (C&T). The display consists of 15 seven-segment plus decimal-point LED numerals. Small numbers are displayed in conventional fashion, but very large and very small numbers are displayed in scientific notation with the power of 10 displayed by the two digits on the right of the display. Within the calculator the digits are expressed in BCD notation. Each word, consisting of 14 digits is therefore $4 \times 14 = 56$ bits long. Of these 14 digits, 10 are used for

†T. M. Whitney, F. Rode, and C. C. Tung, "The Powerful Pocketful: an Electronic Calculator Challenges the Slide Rule," *Hewlett-Packard Journal*, June 1972, Vol. 23, No. 10, pp. 1–9.

the value being displayed, 1 for the sign, 2 for the exponent, and 1 for the sign of the exponent.

The keyboard is arranged in a five-column, eight-row matrix. The matrix is continuously being scanned by signals from the C&T chip. When a key is pressed down, a contact is made between a row and a column. A code corresponding to this key is generated and transmitted to an ROM, and this starts a program stored in the ROM to service this key.

Since it would be impossible to store within a calculator of such small size all the values of the mathematical functions required, and since speed is not of great importance, the value of each mathematical function is calculated as needed. Three of the ROMs are used to store the necessary formulas (algorithms) required to do this.

The adder–subtractor computes the sum or difference between two numbers. It does this with a serial adder. This requires that numbers, stored in parallel form, be converted to serial form. The adder–subtractor functions in binary, but the digits are stored in BCD. Therefore, it is necessary to make corrections. In adding, the first three bits of each group are added as in binary. After the fourth bit, the sum is checked. If it is less than 9, it is correct, but if it is more than (1001) 9, the sum is corrected to decimal by adding (0110) 6.

Four of the registers form a four-register stack memory, the equivalent of 56 four-bit vertical shift registers. When new data are entered into the calculator,

Figure 14-12 Hewlett-Packard HP-35 scientific calculator. (Courtesy Hewlett-Packard Corp.)

previously stored data are moved up one bit in the stack. This feature permits complex calculations to be performed without the necessity of writing down intermediate steps of the calculation. The four sets of stored data can be recalled and viewed for checking; in addition, data in two of the stacks can be interchanged.

Power is provided by means of nickel–cadmium rechargeable batteries. These provide enough power for 4 hours of operation.

A block diagram of the calculator is given in Fig. 14-13.

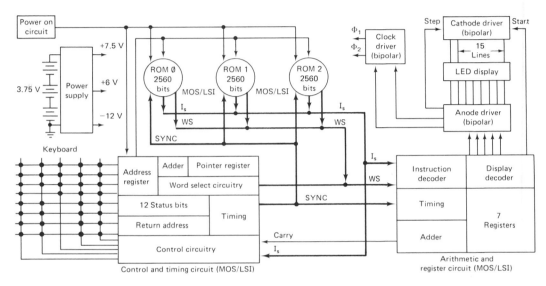

Figure 14-13 Block diagram of the HP-35 electronic calculator. (Courtesy Hewlett-Packard Corp.)

14-8 LOGIC PROBES AND KITS

Troubleshooting and fault location in digital systems in the past have required sophisticated equipment such as the cathode-ray oscilloscope (CRO) and demanded considerable skill on the part of the technician and engineer. Much of this has been simplified with the development of simple logic test instruments. These are the logic probe, logic clip, logic pulser, pulse memory, and logic comparator.

14-8.1 Logic Probe

A logic probe is shown in Fig. 14-14. It is a test probe that displays the logic level at a point in a logic system. This is done by means of a lamp indication on the probe. When the lamp is lit, the logic level is 1 (positive logic). A dark lamp repre-

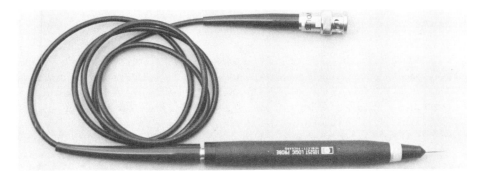

Figure 14-14   Logic probe. (Courtesy Hewlett-Packard Corp.)

sents a 0, and a dim lamp corresponds to an open circuit or a level in the region of the threshold voltage. Should a single pulse as short as 50 ns occur, it is stretched to 50 ms and displayed by the lamp. For continuous pulse trains as fast as 50 MHz, the lamp blinks at 10 times per second. This particular probe is compatible only with TTL, but the logic probe is very popular and is available for all the logic families.

14-8.2   Logic Clip

The logic clip (Fig. 14-15) is a 16-pin probe assembly that clips on the 14 pins or 16 pins of the dual-in-line (DIP) IC. On top of the clip are 16 LEDs that indicate the logic levels at each pin of the IC. Within the clip, there are logic circuits that automatically locate V_{CC} and ground, and use this V_{CC} and ground to activate circuits and LEDs within the clip. The logic clip is also TTL compatible.

Figure 14-15   Logic clip. (Courtesy Hewlett-Packard Corp.)

14-8.3 Logic Pulser

The logic pulser is a "single-pulse," TTL-compatible generating probe. The probe is connected to a point in a logic system and a switch on the probe is activated. This generates a single 0.3-μs pulse. Circuits within the probe determine if the logic level at the test point is high or low. If the level is high, the single pulse goes low for 0.3 μs; if the level is low, the single pulse goes high. Enough current is available in the pulse to overcome low impedance levels, and the pulse duration is short enough so that no damage is done to the IC. The impedance of the probe is normally high except for the brief pulse interval.

14-8.4 Pulse Memory

The pulse memory (Fig. 14-16) is used with the logic probe in determining the occurrence of a single pulse. It is connected between the logic probe and the power supply, and the logic probe is connected to the point under test. Should a single pulse occur, it is sensed by the logic probe and the pulse memory. In the pulse memory a cross-connected RS NAND FF is set when the pulse occurs, and the set position is indicated by an LED.

Figure 14-16 Pulse memory. (Courtesy Hewlett-Packard Corp.)

14-8.5 Logic Comparator

The logic comparator is a 16-pin probe clip that is clipped on 14- or 16-pin DIP ICs. It contains 16 LEDs, one for each IC pin. Within the probe is an identical reference-good IC. Signals for the IC under test are also applied to the reference IC, and the outputs from the IC under test and the reference IC are applied to a logic comparator (EXCLUSIVE-OR). Dissimilarities between the reference and test IC appear at the comparator outputs and are sent to pulse stretchers and the LED displays.

Digital electronics plays an important role in music generation, particularly in electronic organs. In music, identical notes differing by an octave have a 2-to-1 ratio in frequency. Hence a note one octave lower can be derived from the higher frequency note by means of a toggle flip-flop. The electronics providing this music capability is available as P-channel enhancement mode MOS/LSI ICs from American Microsystems, Inc. (AMI).

These ICs consist of four basic blocks: the frequency synthesizer, frequency divider, rhythm generator, and rhythm counter.

14-9.1 Frequency Synthesizer

The frequency synthesizer consists of two ICs that are basically multi-mod-N frequency dividers. This block generates the 13 highest frequencies used in electronic organs. It does this without any jitter and never requires tuning. This is possible since the frequencies are all derived from a single oscillator operating at a frequency of 2.00024 MHz. Figure 14-17 shows how the 13 highest notes are derived in two ICs, AMI parts S2555 and S2556.

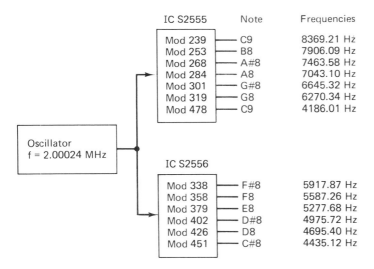

IC S2555	Note	Frequencies
Mod 239	C9	8369.21 Hz
Mod 253	B8	7906.09 Hz
Mod 268	A#8	7463.58 Hz
Mod 284	A8	7043.10 Hz
Mod 301	G#8	6645.32 Hz
Mod 319	G8	6270.34 Hz
Mod 478	C9	4186.01 Hz

Oscillator
f = 2.00024 MHz

IC S2556	Note	Frequencies
Mod 338	F#8	5917.87 Hz
Mod 358	F8	5587.26 Hz
Mod 379	E8	5277.68 Hz
Mod 402	D#8	4975.72 Hz
Mod 426	D8	4695.40 Hz
Mod 451	C#8	4435.12 Hz

Figure 14-17 Electronic organ frequency synthesizer

14-9.2 Frequency Divider

The frequency divider, AMI S2470, is a six-stage binary divider IC. By driving it with one of the outputs of the S2555 or S2556, all the lower octave notes can be obtained. For example, if the S2470 is driven by tone B8, the outputs of the IC will give B7, B6, B5, B4, B3, and B2. An S2470 is required for each note, a total of 12. This provides an 84-note tone generator.

14-9.3   Rhythm Generator

A rhythm system consists basically of a device capable of providing several measures of rhythm patterns for a given rhythm instrument. To do this properly in an electronic organ, one should have eight to ten rhythm-pattern outputs each capable of a 32- to 64-bit sequence. In addition to the basic output sequence, it should be possible to alter the sequence at the various outputs. With this capability it is then possible to have available rhythm patterns such as the waltz, fox trot, etc. All this can be done with a preprogrammed read-only memory (ROM), AMI IC S2566.

14-9.4   Rhythm Counter

The AMI S2567 resettable rhythm counter drives and provides the timing for the S2566. It is a six-stage, asynchronous binary counter with the ability to reset individual stages or combinations of stages. It is possible to use it to simultaneously generate several rhythm patterns.

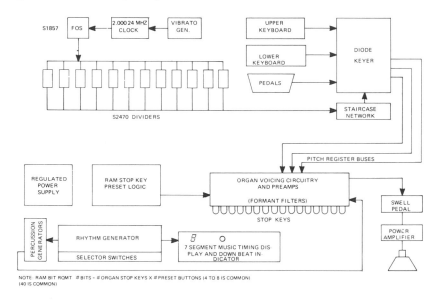

NOTE: RAM BIT RQMT # BITS = # ORGAN STOP KEYS X # PRESET BUTTONS (4 TO 8 IS COMMON)
(40 IS COMMON)

Figure 14-18   Typical electronic organ block diagram. (Courtesy American Microsystems, Inc.)

Figure 14-18 is a typical block diagram of an electronic organ. Several other sections are of digital interest. The outputs of T FFs are square waves and have only odd harmonies of the fundamental tone. To generate waves with even harmonies, one can add to the fundamental, at the right time in the cycle, sections of higher-frequency tones from its frequency divider and generate a staircase. This might be done in an operational amplifier. In another section the output of the rhythm generator is used to drive a seven-segment display device to indicate which rhythm pattern is in use.

14-10 AUTOMOTIVE APPLICATIONS FOR DIGITAL INTEGRATED CIRCUITS

There are many applications for digital ICs in the automobile. Some have already been put into service and some await further development. The CMOS family, with its low current requirement and high noise immunity, has a distinct advantage over the other logic families and much development work is underway with CMOS. The automotive market is so large that it will use custom ICs specially developed for these applications. Some control problems, particularly those relating to efficiency and pollution, are so complex that it will require a small computer or microprocessor in the car to resolve them.

The following are some applications for digital ICs in the auto:

1 Operation and pollution control
   Fuel injection
   Electronic ignition
   Catalytic converters
   Timing
   Spark advance
   Performance display
2 Safety
   Seatbelt controls and logic
   Antiskid
   Anticollision
   Flashers
   Fluid level sensors
3 Accessories
   Speed controls
   Indicators for speedometer–odometer–tachometer
   Antitheft
   Windshield wipers
   Temperature controls
   Headlight dimmers
   Solid-state clocks
   Electronic locks
4 Diagnosis
   Diagnostic checkouts at service centers

FIFTEEN

The Microprocessor

15-1 INTRODUCTION

The development in the early 1960s of the silicon planer process, which permitted making many transistors simultaneously, in addition to improving their high-speed characteristics and switching speed, brought with it the capability of interconnecting these transistors and forming integrated circuits. At first it was difficult to do this with more than a few transistors (small-scale integration, SSI) because of the chemical and metallurgical problems involved, but improvements in the technology and manufacturing processes have permitted more and more transistors to be combined. It is now possible to build large-scale-integrated circuits (LSI) with as many as 10,000 transistors in a single IC.

Simultaneously, the availability of LSI ICs permitted the development of compact complex systems, which continued to increase in complexity as more and more LSI circuits and functions became available. Those systems for the most part made use of available standard ICs. It would have been simpler in many cases to use ICs developed specially for the specific system, but the startup costs involved in developing such ICs were extremely high; only in cases where large quantities of such special ICs would be used was it economically justifiable to manufacture such ICs.

The problem confronting the IC manufacturer and system designer was to develop a universal digital IC, even if quite complex, that could be used to solve a multitude of system problems and which could be manufactured in such large quantities that the cost would be quite low. The solution to this problem resulted in a very complex digital IC system called a *microprocessor*. It is composed of a few multifunction ICs. These ICs are capable of being controlled to perform many different logic functions by means of a set of instructions in digital format called an *instruction set*. Among these functions are AND, OR, exclusive-OR, add, subtract, shift, address memory, fetch data from memory, store data in memory, and make decisions.

In applying these ICs to system problems the solution is accomplished by means of an organized series of commands called a *program*. The microprocessor, or minicomputer, thereby changes the solving of system problems from a specific dedicated hard-wired system to one using generalized ICs and a program. Prior to the advent of the microprocessor problems were solved with hard-wired components.

With the microprocessor, one uses a program and a common logic system for the solution of many problems. As a result, the microprocessor based system has the following advantages over the hard-wired system.

1. The same logic components are used to solve many system problems. Only the program has to be changed. A hard-wired system can solve only a specific problem.
2. Since the same components are used for many systems, it becomes less expensive to develop new systems and add additional features to old systems.
3. New products can be developed more rapidly.
4. Reliability is improved. The same components are used and there are fewer different elements needed.
5. The individual ICs cost less since they are used in many systems.
6. The individual ICs become standardized.
7. Greater familiarity with components is attained by engineering, production, and service personnel because of the greater use of the same components.
8. The IC manufacturer is able to improve product reliability and reduce cost because of increased production.
9. There are potential reductions in size, weight, and power consumption.

15-2 FUNCTIONS IN A MICROPROCESSOR

As we have discussed in the previous section, the microprocessor functions by changing the technique from the hardware of a logic system to the software of a program. To do this, the microprocessor system must provide some basic logic functions. These are shown in block form in Fig. 15-1, in which the system and blocks illustrate the operations of a microprocessor or minicomputer. Micro-

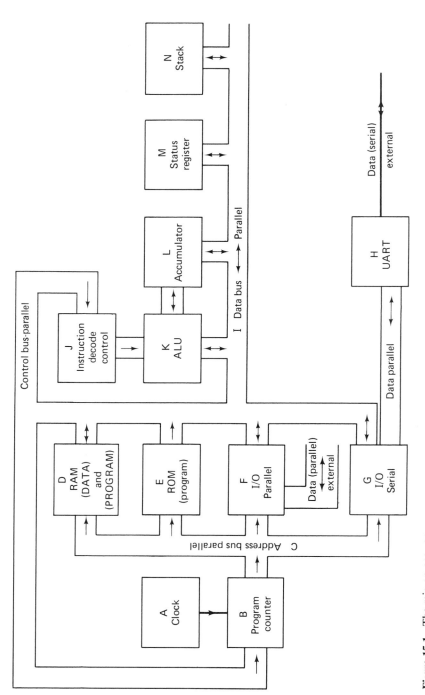

Figure 15-1 The microprocessor

processors are available from many semiconductor manufacturers, and each manufacturer combines functions in different arrangements or architecture, depending upon his understanding of the optimum way to solve system problems.

In Fig. 15-1, the blocks function and are interrelated in the following manner:

A. Clock

The clock is an oscillator. Its frequency can be determined either by a resistor–capacitor (*RC*) network or by a crystal. Its function is to generate the system basic timing. It determines when commands are executed and when data are moved. It is coupled to and provides the timing for the program counter (B). The clock frequency is usually of the order of 1 MHz.

B. Program Counter

The Program Counter is a multi-stage binary counter. Its primary function is to address memory. The number of stages in this counter determines the memory-addressing capability of the miroprocessor. Many microprocessors have 16 stages; with this capability it becomes possible to address 65,536 (65K) memory locations. This counter has presetting capability and therefore can be started from any count upon command.

C. Address Bus

The addressing outputs of the program counter are connected by means of a parallel set of connections to the memory elements D, E, F, and H of the micro-processor.

D. Random Access Memory (RAM)

The random access memory is addressed by the program counter. If this is the only memory element in the system, it will be used for both the program and the data. When there is an ROM in the system, the RAM is used for data and the ROM for the program. If the RAM is used for programming, modifications in the program or the introduction of new programs is done quite easily. RAM memories are used during the design stages of a program. This makes it possible to debug errors in programming. Programs can be tested and easily modified until they are correct.

E. Read-Only Memory ROM

The read-only memory is addressed by the program counter. Its primary function is to store the control elements of a specific program. Once a program is debugged and correct, it can be stored in ROM. There are three types of ROM. They are:

1   Mask programmable. A metal mask is designed inside the ROM and applies shorts and opens inside the ROM to get 1s and 0s at the desired locations. This is expensive and is used when large production runs justify the cost.

2   PROM-Programmable. Fusible links are blown out in the ROM to create 0s at the desired locations. This is less expensive than mask programmable.

3   EROM-Erasable. This memory can be programmed electrically and, if necessary, can be erased by exposure to ultra-violet light for about 10 minutes. This is quite useful for program evaluation and debugging and for small production runs.

To simplify memory referencing and location it has become common practice to divide the memory into small blocks of 256 bytes called a *page*. With 65,576 (= 256 × 256) memory locations, this becomes 256 pages (starting from page zero) of memory. For example, on page zero, the memory locations are as follows:

Binary	Decimal	Hexadecimal
0000 0000 0000 0000	000 000	00 00
to	to	to
0000 0000 1111 1111	000 255	00 FF

On page 255, the memory locations are as follows:

Binary	Decimal	Hexadecimal
1111 1111 0000 0000	255 000	FF 00
to	to	to
1111 1111 1111 1111	255 255	FF FF

F.   Input/Output (I/O) : Parallel

The function of this I/O block is to provide means for transferring data in and out of the microprocessor in parallel form. For example, data from a parallel memory element, such as a magnetic or punched tape, can be transferred into this element. When incoming data are stored (on a temporary basis) in this element, an interrupt command is generated. This will interrupt the main program, and the microprocessor will act on the data in this I/O port. The reverse is also true. Parallel data from the microprocessor will be routed through this I/O device to the outside world.

G.   Input/Output (I/O) : Serial

A microprocessor also must be capable of handling serial data from the outside world. This is done in blocks G and H. Block G functions in an identical manner to block F, but processes parallel data serialized in block H.

H. Universal Asynchronous Receiver Transmitter (UART)

The UART is a logic element that takes input serial data from the outside world (e.g., MODEM data) and converts them into parallel data, and, in reverse, converts from parallel to serial form for outputting. This is a typical shift register function.

I. Data Bus

Data are moved around the microprocessor in parallel form. At first, microprocessors handled 4 bits of data. Now most microprocessors operate with 8 bits of data, and some are beginning to use 16 bits. An 8-bit data group is known as a byte, and a 4-bit group is called a nibble. The data bus is bidirectional; data can be transferred in either direction.

J. Instruction Decode Control

The program contains the commands in binary. In this block, they are decoded and translated into control functions for operation on the data. The binary control data are obtained from the memory elements by way of the data bus.

K. Arithmetic Logic Unit (ALU)

The ALU is the fundamental operating element in the microprocessor. Under instructions from the control unit, it performs arithmetic and logic functions on data.

L. Accumulator

The accumulator is a temporary storage register (clocked *RS* flip-flops) whose main function is to act as the communication path between the data bus and the ALU. For example, when two numbers are to be added, the procedure is to fetch the first number from memory and store it in the accumulator. When the control command to add is received, the second number is fetched from memory and applied to the ALU. The "add" command is now executed and the sum appears in the accumulator, replacing the first number. The timing for this is generated by the clock. The sum can now be held in the accumulator for additional operations or can be returned to memory. Shift operations (multiply or divide by 2) can be performed on data held in the accumulator. This is done in conjunction with the status register, block M.

M. Status Register

The status register is a group of storage flip-flops called status bits or status flags that give information about the results of the last instruction. An important factor in programming is the ability of the program to make decisions. This is done by examining the levels of the status flags. As a result of such a decision, the program may be modified by changing the sequence of the program counter. This program modification is called a *jump* or a *branch*. The following are typical status flags:

1 Overflow: in performing arithmetic operations, the result may be too large to be handled by the available number of bits in the accumulator. This is indicated by the level of the overflow flag.
2 Negative: this flag is used in signed arithmetic to indicate whether the result of an addition is positive or negative.
3 Zero: this flag will indicate whenever there is a zero in the accumulator as a result of data movement or calculation.
4 Carry: there are two operations that require the use of a carry bit. Although the number of bits that can be added at any one instruction in the ALU is limited by the design of the microprocessor, it is possible to add larger numbers by a suitable program. To do this, the lower-order bits are added together and may generate a carry for addition to the next higher order of bits. The second case when the carry bit is used is in the shift (multiply by or divide by 2) instruction. In this case, the carry bit acts like an extension of the accumulator, increasing its size by 1 bit.
5 Decimal mode: this flag is set when it is desired to perform arithmetic operations using numbers expressed in BCD coding.

N. Stack Register

The stack register is an area of memory used for temporary storage. Although shown as a separate block in Fig. 15-1, it can be implemented by assigning a section of RAM memory to it. Two uses for it are as follows:

1 Handling of interrupts
2 Handling of subroutines. A subroutine is a set of instructions that is continually repeated. For example, the program might call for the value of the sine to be calculated many times.

In the handling of an interrupt request, the main program of the microprocessor is interrupted, but the status of the program must be preserved since the program will be continued after the interrupt request is completed. The status of the program is preserved in a stack and is handled on a last-in, first-out (LIFO) basis. Upon completion of the current instruction, the following information is loaded into the stack in order: status register, accumulator, and program counter (address of next instruction). After the interrupt request has been serviced, the main program status is retrieved from the stack in reverse order, with the program counter address first, followed by the accumulator and then the status register. The main program can now continue.

In the handling of a subroutine, the main program is effectively interrupted by a "branch to subroutine" instruction. To do this, the address of the next main instruction must be preserved; this is done in the stack. When the subroutine program is completed, an instruction in it causes a return to the main program. This retrieves the address of the main program, which now can continue.

15-3 CENTRAL PROCESSING UNIT (CPU)

Although not shown in this manner in Fig. 15-1, it is common·practice in designing a microprocessor system to lay out the architecture in such a way that there is a CPU, which consists of the program counter, the instruction decode control unit, the ALU, the accumulator, the status register, and the stack, all in a single IC. This puts all the control and operating functions into a single unit.

15-4 ARITHMETIC IN THE MICROPROCESSOR

Addition in the microprocessor is performed with an ADD instruction to the ALU. The first number is transferred from memory to the accumulator. The second number from memory is then added to the first number and the sum appears in the accumulator. Symbolically, we can write

$$A + M \longrightarrow A \tag{15-1}$$

or if we are considering carries and there is a 1 in the carry flag from a previous addition, we can write

$$A + M + C \longrightarrow A + C \tag{15-2}$$

Arithmetic is performed using 2s complements. The reason for this can be seen if we add 1 and -1, using both 1s and 2s complement methods, as follows:

		1s complement	2s complement
	1	0000 0001	0000 0001
$+$	-1	1111 1110	1111 1111
	Sum	1111 1111	1 0000 0000

In the 1s complement method, the sum (without recomplementing) is equal to 1111 1111, whereas in the 2s complement method the sum is equal to 0000 0000, if we ignore the overflow.

Arithmetic operations can be performed either as unsigned or as signed arithmetic. In dealing with unsigned numbers, if we add or subtract and the result is too large for the number of bits in the accumulator, the overflow can be interpreted as either an overflow or a carry. For example,

$$\begin{array}{r} 1100\ 1000 \\ +1110\ 1011 \\ \hline \text{Overflow or carry} \rightarrow 1\ \ 1101\ 0011 \end{array}$$

In signed arithmetic, the interpretation of the results is not so evident because we have to consider both sign and overflow. In signed arithmetic (assuming that we are using 8-bit words), the most significant bit, bit 7, is the sign bit. A 0 in bit 7 indicates a positive number and a 1 in bit 7 is a negative number. Negative numbers are represented in 2s complement form. For example,

$$\begin{array}{r} +27 = 0001\ 1011 \\ -27 = 1110\ 0101 \\ \nearrow \\ \text{Sign bit} \end{array}$$

Consider the following examples of signed arithmetic addition with 8 bits. Bits 0 through 6 are bits of the number and bit 7 is the sign bit. In performing this addition, we have to consider the "overflow" from bit 6 to bit 7 and the "carry" from bit 7.

EXAMPLE 15-1   Add two positive numbers with no overflow.

$$\begin{array}{lll} 0011\ 0010 & (A) & +50 \\ +0011\ 0111 & (M) & +55 \\ \hline 0110\ 1001 & (A) & +105 \end{array}$$

The overflow is 0 and the carry is 0. The correct positive answer is shown by the 0 in bit 7.

EXAMPLE 15-2   Add a positive and negative number with positive result.

$$\begin{array}{lll} 0011\ 0010 & (A) & +50 \\ +1111\ 1010 & (M) & -6 \\ \hline 1\,0010\ 1100 & (A) & +44 \end{array}$$

The overflow is 1 and the carry is 1. The correct positive number is shown by the 0 in bit 7. The 1 in the carry position has no significance in signed arithmetic.

EXAMPLE 15-3   Add a positive and negative number with negative result.

$$\begin{array}{lll} 1111\ 0010 & (A) & -14 \\ +0000\ 0101 & (M) & +5 \\ \hline 1111\ 0111 & (A) & -9 \end{array}$$

The overflow is 0 and the carry is 0. The correct negative answer (-9 in 2s complement) is shown by the 1 in the bit 7 position.

EXAMPLE 15-4   Add two negative numbers without overflow.

$$
\begin{array}{llll}
& 1111\ 1001 & (A) & -7 \\
+ & 1111\ 0100 & (M) & -12 \\
\hline
1\ & 1110\ 1101 & (A) & -19 \text{ (2s complement} \\
& & & \text{neglecting carry bit)}
\end{array}
$$

The carry is 1 and the overflow flag from bit 6 is 1. Negative correct answer is shown by the 1 in bit 7.

EXAMPLE 15-5   Add two positive numbers with overflow.

$$
\begin{array}{llll}
& 0101\ 0010 & (A) & 82 \\
+ & 0110\ 0101 & (M) & 101 \\
\hline
& 1011\ 0111 & (A) & -73 \quad ?
\end{array}
$$

The overflow is 1 and the carry is 0. The apparent negative sum shown by the 1 in bit 7 is incorrect. How can the computer know this? If we compare the overflow and carry bits in this example, we see that they are *different*, whereas in Examples 15-1 to 15-4 they were *alike*. The computer will recognize this and make a correction. When overflow and carry are *alike*, the answer is correct. When *different*, the sum is incorrect.

EXAMPLE 15-6   Add two negative numbers with overflow.

$$
\begin{array}{llll}
& 1011\ 0110 & (A) & -74 \\
+ & 1010\ 1011 & (M) & -85 \\
\hline
1\ & 0110\ 1001 & (A) & +105
\end{array}
$$

The carry is 1 and the overflow flag from bit 6 is 0. They are not alike and therefore an incorrect result is indicated.

15-5 LOGIC OPERATIONS

15-5.1 AND

This operation is performed on any bit by ANDing it with a 0 when it is desired to convert that bit to a 0.

EXAMPLE 15-7

$$
\begin{array}{lll}
& 1011\ 1101 & A \\
\text{AND} & 0000\ 1111 & B \\
\hline
& 0000\ 1101 & A \cdot B
\end{array}
$$

The first 4 bits have been converted to 0s and the last 4 bits are unchanged.

15-5.2 OR

This operation is performed on any bit by ORing it with a 1 when it is desired to convert that bit to a 1.

EXAMPLE 15-8

$$
\begin{array}{r}
1010\ 1010 \quad A \\
\text{OR} \quad \underline{1111\ 0000} \quad B \\
1111\ 1010 \quad A + B
\end{array}
$$

The first 4 bits have been converted to 1s and the last 4 bits are unchanged.

15-5.3 EXCLUSIVE OR

This operation is performed on any bit by EXCLUSIVE-ORing it with a 1 when it is desired to invert that bit.

EXAMPLE 15-9

$$
\begin{array}{r}
1010\ 1010 \quad A \\
\text{EXCLUSIVE-OR} \quad \underline{1111\ 0000} \quad B \\
0101\ 1010 \quad A \oplus B
\end{array}
$$

The first 4 bits have been inverted and the last 4 bits are unchanged.

15-6 PROGRAMMING PRINCIPLES†

A microprocessor/minicomputer solves problems by means of a program. A program is a sequence of instructions, written by a programmer, which direct and control the movement of data within (and external to) the microprocessor in order to arrive at the solution to the problem. In writing a program, the programmer has additional factors to consider. He will probably find that there is more than one way to write his program. He should write it so that it uses as little memory as possible. Memory is expensive and since in many problems time is a factor, he should explore program options which minimize the time.

The basic operation of any computer can be expressed by the phrase *Fetch and Execute*. The operation of the microprocessor is repetitive. It fetches an instruction and executes it. It then continues on to fetch and execute the next instruction and continues in the same manner until the program is completed. In the fetch phase of the instruction, the program counter points to a location in memory which contains an instruction code. This instruction code is sent to an

†The material in this section is based upon the MOS Technology, CPU part numbers MCS 6502 through 6505. They are 8 bit (1 byte) data microprocessors. The 6502 can address 65 k memory location (65 k bytes of memory) while the 6503 through 5 have lower addressing capability.

instruction register (block J Fig. 15-1) where it is decoded to determine which operation will be now executed.

The program is stored in memory, to be acted upon by the microprocessor. Memory words can be interpreted in different ways

1 OP CODE (Operation code, instruction-code) The computer identifies the OP CODE as an instruction to perform an operation on data words. Typical operations are add, subtract, shift, compare, AND, OR.

2 Operand. Data on which an operation is performed is an Operand. For example, in addition it might be the addend.

3 Memory address. The memory location where data are stored is a memory address.

4 Data.

There are many ways of writing programs. We shall discuss a few.

15-6.1 Machine Language—Object Code

Written using the 1s and 0s that the computer needs, these programs are extremely difficult to remember and the instruction codes are difficult to recognize; errors are likely to occur. A simplification step frequently employed is to write the program in hexadecimal notation. For example, an instruction (OP CODE) to transfer data from memory to the accumulator might be 1010 0101 or in hexadecimal A5. A program written in machine language is said to use an object code. Each step in the program represents a separate location in memory. The memory address must be specified by the programmer.

15-6.2 Mnemonic—Source Code

To simplify program writing and to minimize errors, programs are written using symbolic terms, which the programmer remembers much easier than binary or hexadecimal. This uses three- or four-letter mnemonic symbols, which have the appearance of a shorthand way of writing the instruction. For example, load accumulator from memory becomes LDA. Translators or compilers using assembly programs convert them into binary code for the microprocessor. A mnemonic instruction is called a source code. Each step in the program represents a separate location in memory.

15-6.3 High-Level Languages

High-level language programs are written following rules that make programming even simpler. The programs are written using ordinary language and algebra-like statements. A statement in the program can result in the usage of many positions in memory, and the programmer does not have to assign specific locations in memory for the program since this is done automatically. Programs written in high-level languages are converted by a compiler into machine language object codes. Typical languages are FORTRAN (arithmetic, scientific oriented), BASIC (arithmetic oriented), and COBOL (business oriented).

Table 15-1 lists in alphabetical sequence the instruction set of the MOS Technology MCS 6502–6505 CPUs in mnemonics. The actual OP CODE used depends on how data are to be stored in memory.

Let us consider the instruction

LDA: Load Accumulator with Memory

When this instruction is completed, data from memory will have been transferred into the accumulator while leaving the data still in memory. There are three basic ways of storing data in memory for this instruction. These are called addressing modes.

Absolute Addressing Mode

This is a 3-byte instruction and takes four clock cycles to complete. This is considered the most normal mode for data addressing. The total instruction is the OP CODE followed by the memory location where the data are stored. For example, suppose that we want to put the number 23_{10} ($= 17_{16}$) in the accumulator, and it is to be located in memory location 1011 0110 0001 1110. When the CPU sees the OP CODE for this instruction, which is AD and is the first byte, it knows that this must be followed by 2 bytes of memory. The first byte following the OP CODE is the low-order byte of the data address, and the second byte is the high-order byte of the data address. A program with the instruction will look like this. Memory locations are arbitrary and can be anywhere within successive available memory. The succession is in binary in order to be compatable with the program counter which counts in binary.

Memory Location of Program (HEX)	Program (Binary)	Program (Hex)	Program (Mnemonic)	Comments
⋮	⋮	⋮	⋮	⋮
OB 24	1010 1101	AD	LDA	Load acc.–Absolute
OB 25	0001 1110	1E		Low-order–Address bus
OB 26	1011 0110	B6		High-order–Address bus
⋮	⋮	⋮	⋮	⋮
Memory B6 1E	Data (Binary) 0001 0111	Data (Hex) 17		

Page Zero Addressing Mode

This is a 2-byte instruction and takes three clock cycles to complete. The first byte is the OP CODE and the second byte is the location in memory, on page

Table 15-1 Instruction set, MOS technology MCS 6502–6505 CPUs

ADC	Add Memory to Accumulator with Carry
AND	"AND" Memory with Accumulator
ASL	Shift Left One Bit (Memory or Accumulator)
BCC	Branch on Carry Clear
BCS	Branch on Carry Set
BEQ	Branch on Result Zero
BIT	Test Bits in Memory with Accumulator
BMI	Branch on Result Minus
BNE	Branch on Result Not Zero
BPL	Branch on Result Plus
BRK	Force Break
BVC	Branch on Overflow Clear
BVS	Branch on Overflow Set
CLC	Clear Carry Flag
CLD	Clear Decimal Mode
CLI	Clear Interrupt Disable Bit
CLV	Clear Overflow Flag
CMP	Compare Memory and Accumulator
CPX	Compare Memory and Index X
CPY	Compare Memory and Index Y
DEC	Decrement Memory by One
DEX	Decrement Index X by One
DEY	Decrement Index Y by One
EOR	"Exclusive-or" Memory with Accumulator
INC	Increment Memory by One
INX	Increment Index X by One
INY	Increment Index Y by One
JMP	Jump to New Location
JSR	Jump to New Location Saving Return Address
LDA	Load Accumulator with Memory
LDX	Load Index X with Memory
LDY	Load Index Y with Memory
LSR	Shift One Bit Right (Memory or Accumulator)
NOP	No Operation
ORA	OR Memory with Accumulator
PHA	Push Accumulator on Stack
PHP	Push Processor Status on Stack
PLA	Pull Accumulator from Stack
PLP	Pull Processor Status from Stack
ROL	Rotate One Bit Left (Memory or Accumulator)
RTI	Return from Interrupt
RTS	Return from Subroutine
SBC	Subtract Memory from Accumulator with Borrow
SEC	Set Carry Flag
SED	Set Decimal Mode
SEI	Set Interrupt Disable Status
STA	Store Accumulator in Memory
STX	Store Index X in Memory
STY	Store Index Y in Memory
TAX	Transfer Accumulator to Index X
TAY	Transfer Accumulator to Index Y
TSX	Transfer Stack Pointer to Index X
TXA	Transfer Index X to Accumulator
TXS	Transfer Index X to Stack Pointer
TYA	Transfer Index Y to Accumulator

zero, where the data are located. The OP CODE for this, which is A5, is different from that for absolute addressing. It is not necessary to use memory and time to specify page zero. Let us again load the accumulator with data. 23_{10} is to be in location 1E on page zero. A program with this instruction will look like this.

Memory Location of Program (Hex)	Program (Binary)	Program (Hex)	Program (Mnemonic)	Comments
OB 24	1010 0101	A5	LDA	Load acc.–zero page
OB 25	0001 1110	1E		Zero page address

Memory (page zero)	Data (Binary)	Data (Hex)		
00 1E	0001 0111	17		

Immediate Addressing Mode

This is a 2-byte instruction and takes two clock cycles to complete. The first byte contains the OP CODE for the instruction, which also specifies the addressing mode. The second byte contains the data. The OP CODE A9 is different from these in A and B. In effect, there is no need to go first to another memory location for the data. Suppose that we again want to load 23_{10} into the accumulator. The program with these instructions will look like this.

Memory Location of Program (Hex)	Program (Binary)	Program (Hex)	Program (Mnemonic)	Comments
OB 24	1010 1001	A9	LDA	Load acc.-Immediate
OB 25	0001 0111	17		Data

As we can see, these are choices in programming. Absolute addressing can use all the memory but takes 3 bytes and four clock cyles. Page zero addressing is faster, takes 2 bytes and three cycles but is limited to 256 memory locations. Immediate addressing takes 2 bytes and two cycles, and is used when a fixed number is incorporated in a program.

Implied Addressing Mode

This is a 1-byte instruction and is used for control commands. It does not reference any other memory location.

Relative Addressing

This is used with decision controls. It is used with the branch (decision) instruction and determines whether the program is to continue in normal sequence or is to be skipped forward or backward. It is a 2-byte instruction. The first byte is the condition test, and the second instruction is a number that gives the number of program steps which the program is to be changed. The number is added or subtracted from the memory location of the next control instruction of the program.

15-7 PROGRAMMING EXAMPLES

Table 15-2 lists OP CODES that will be applicable to the writing of programs in this section and to the solution of problems at the end of this chapter. In the examples, all memory locations and programs are in hexadecimal.

EXAMPLE 15-10   Write a program that will move data from location 02BA to location 04C7.

Solution

Memory Location	Program	Mnemonic	Comments
0000	AD	LDA	Load-absolute
0001	BA		⎰Original data
0002	02		⎱ address
0003	8D	STA	Store-absolute
0004	C7		⎰Final data
0005	04		⎱ address

EXAMPLE 15-11   Write a program that will convert the last 4 bits of any byte to zeros and store the result in location F6 on page zero. The original number is in memory location F307.

Solution

Memory Location	Program	Mnemonic	Comment
0001	AD	LDA	Load-absolute
0002	07		⎰Original data
0003	F3		⎱ address
0004	29	AND	AND-immediate
0005	F0		ANDing data
0006	85	STA	STORE-page zero
0007	F6		Data address-page zero

Table 15-2 Operation codes

		Addressing Modes				
Bytes	3	2	2	1	2	
Mne-monic	Abso-lute	Zero Page	Imme-diate	Implied or Accu-mulator	Rela-tive	Comments
ADC	6D	65	69			Add memory to acc. with carry
AND	2D	25	29			AND memory to acc.
ASLA				0A		Shift left 1 bit acc.
ASL	0E	06				Shift left 1 bit memory
BCC					90	Branch-carry clear (C = 0)
BCS					B0	Branch-carry set (C = 1)
BEQ					F0	Branch on zero (Z = 1)
BM1					30	Branch on result minus (N = 1)
BNE					D0	Branch on result not 0 (Z = 0)
BPL					10	Branch on result plus (N = 0)
BVC					50	Branch on overflow clear (V = 0)
BVS					70	Branch on overflow set (V = 1)
CLC				18		Clear carry flag
CLD				D8		Clear decimal mode
CLV				B8		Clear overflow flag
CMP	CD	C5	C9			Compare memory and acc (A–M)
DEC	CE	C6				Decrement memory by one
EOR	4D	45	49			Exclusive-or memory with acc.
INC	EE	E6				Increment memory by one
JMP	4C					Jump
LDA	AD	A5	A9			Load accumulator
LSRA				4A		Shift right 1 bit acc.
LSR	4E	46				Shift right 1 bit memory
ORA	0D	05	09			OR memory with accumulator
ROLA				2A		Rotate left accumulator
ROL	2E	26				Rotate left memory
RORA				6A		Rotate right accumulator
ROR	6E	66				Rotate right memory
SBC	ED	E5	E9			Subtract mem. from acc. with borrow
SEC				38		Set carry flag
SED				F8		Set decimal mode
STA	8D	85				Store acc. in memory

EXAMPLE 15-12 Write a program to add two binary numbers together and store the result in location A6 of page zero. The numbers are in locations 00 and 01 of page zero. In location A7, page zero, store the carry (1 or 0) in bit 0 with all other bits equal to 0.

To solve this problem we have to introduce some new concepts and instructions.

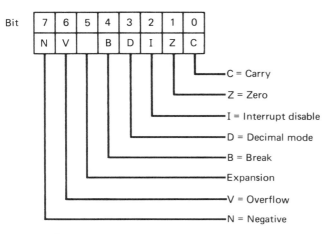

Figure 15-2 Status register

1. The Status Register is organized as in Fig. 15-2. It is 8 bits long. In the register the flags (status indicators) are equal to 1 when there is a carry, when the accumulator is equal to zero, when performing arithmetic in the decimal mode, and when there is an overflow. The N flag is the level of bit 7 of the accumulator. Branch tests test the status of these flags.

2. There are shift and rotate instructions that involve the carry flag.

a. ASL: shift left one bit (memory or accumulator).

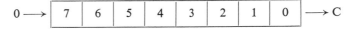

This instruction shifts data one bit position to the left. A 0 appears in bit 0, and the level of bit 7 is transferred into the carry.

b. LSR: logical shift right (memory or accumulator).

This instruction shifts data one bit position to the right. A 0 appears in bit 7, and the level of bit 0 is transferred into the carry.

c. ROL: rotate left (memory or accumulator).

The level of bit 7 is shifted into the carry, the carry is shifted into bit 0, and all other bits are shifted one bit position to the left.

   d. ROR: rotate right (memory or accumulator).

The level of the carry is shifted into bit 7, the content of bit 0 is shifted into the carry, and all other bits are shifted one bit position to the right.

Solution

Memory Location	Program	Mnemonic	Comment
0000	Data		First number
0001	Data		Second number
0002	01		Data for ANDing
0003	18	CLC	Clear carry
0004	D8	CLD	Clear decimal mode (set binary mode)
0005	A5	LDA	Load acc. zero page
0006	00		Data location
0007	65	ADC	Add with carry zero page
0008	01		Data
0009	85	STA	Store acc. zero page
000A	A6		Address of sum
000B	2A	ROL	Rotate left acc.
000C	25	AND	AND-zero page
000D	02		ANDing data (in location 02)
000E	85	STA	Store Accumulator (carry)
000F	A7		Address

Subtraction

Subtraction is accomplished by setting the carry equal to 1 (SEC instruction). A carry equal to 1 is a no-borrow condition. An SBC (subtract memory from accumulator with borrow) command is used. In the microprocessor, the number to be subtracted is converted into its 2s complement form and added. If, as the result of this addition, $C = 1$, the result is positive. If $C = 0$, the result is negative in 2s complement form.

EXAMPLE 15-13 Write a program for multiplying two binary numbers. There are three methods of performing such a computation.

Solution: Method A In this method, the multiplicand is added to itself as many times as it takes to complete the multiplication. A specific program must be written for each problem. The program is repetitive.

Memory Location	Program	Mnemonic	Comments
0000	XX		Multiplicand
0001	04		Multiplier
0002	18	CLC	Clear carry flag
0003	D8	CLD	Clear decimal mode
0004	Data		⎰Address of
0005	Data		⎱ product
0006	A5	LDA	
0007	00		Multiplicand
0008	65	ADC	
0009	00		
000A	65	ADC	
000B	00		
000C	65	ADC	
000D	00		

From this point the program follows steps 0009 to 000F of Example 15-12. The programmer must leave adequate space in memory for the product. Note that steps 0008 through 000D were repetitive and were based upon a multiplier equal to 4.

Solution: Method B   This is a general method for the solution of this problem. It sets up a loop, which is tested for the completion of the problem. The instruction DEC

Memory Location	Program	Mnemonic	Comments
0000	0X	Data	Multiplicand
0001	0Y	Data	Multiplier
0002	00	Data	Product
0003	18	CLC	Clear carry
0004	08	CLD	Binary mode
0005	A5	LDA	⎰Load acc. zero page
0006	00		⎱Multiplicand
0007	C6	DEC	⎰Decrement. M–1 ⟶ M
0008	01		⎱Multiplier
0009	F0	BEQ	⎰Branch if zero (multiplier)
000A	05		⎱No. of branch steps from 000B (= 000B + 05)
000B	65	ADC	⎰Add with carry
000C	00		⎱Multiplicand
000D	4C		⎰Jump-loop back to
000E	07		⎰ decrement multiplier-
000F	00		⎱ absolute to location 0007
0010	85	STA	⎰Store accumulator
0011	02		⎱Product location
End			

decrements the multiplier (in memory) by 1 and checks to see if it is equal to 0 with a branch instruction BEQ. If the decremented multiplier is equal to 0, the program ends. If it is not equal to 0, the multiplicand is added to the accumulator. This is shown in a flowchart (Fig. 15-3). The BEQ instruction can be used after any operation. It is followed by a number that locates the step in the program where the branch is activated. This number is added to the program step of the next control instruction following BEQ.

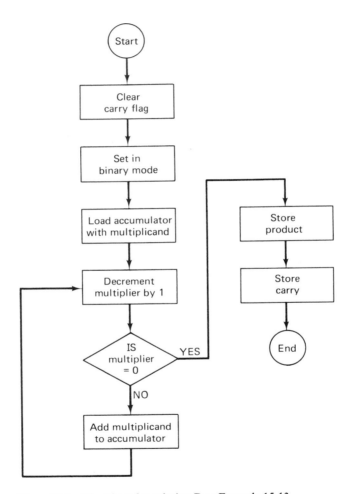

Figure 15-3   Flowchart for solution B to Example 15-13

Solution: Method C   There is still a third method of performing multiplication, which follows the method we normally use. It is the shift and add procedure shown in Fig. 15-4. In this method, we add if the bit of the multiplier is equal to 1. A bit status can only be tested in the status register. If we shift to the right (LSR), the least significant bit can be tested in the carry position. At the same time, the multiplicand can be shifted to the left for addition. The procedure comes to an end when the multiplier is equal to 0. Figure 15-4 is a flowchart for this procedure. The actual program is left for a problem.

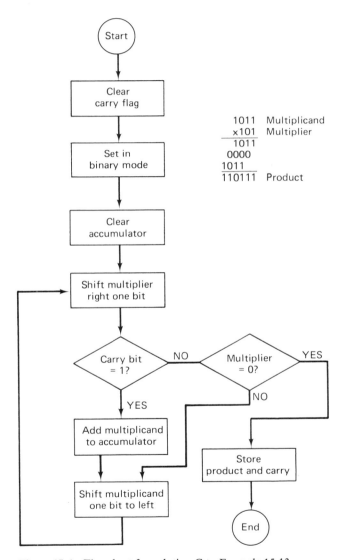

```
1011    Multiplicand
×101    Multiplier
1011
0000
1011
110111  Product
```

Figure 15-4 Flowchart for solution C to Example 15-13

15-8 A MICROPROCESSOR PROGRAMMED ADJUSTABLE MONOSTABLE

As we have previously discussed, the microprocessor changes the solution of problems from a hard-wired component system designed to solve a specific problem to one using a program. Let us consider a specific problem, namely the adjustable monostable. This problem was discussed in Sec. 9-10 using a two-gate monostable and using ICs specifically designed for this function, types 9601/8601 and the 555 timer. In these circuits, the monostable time was controlled and changed by adjusting R or C or both.

Now let us do this with a programmed microprocessor. To do this we have to introduce a new OP-CODE namely the mnemonic NOP (HEX-EA). This is a nooperation code but it does take 2 clock cycles to complete so that it can be used to consume time. In this program we will use the zero-status flag as our time indicator.

Monostable Program

Memory Location	Program	Mnemonic	Comments
0000	01	Data	
0001	C6	DEC	⌠Decrement data to zero.
0002	00		⌡Zero status flag goes to a 1
0003	EA	NOP	⌠Use as many NOP
0004	EA	NOP	│ operations as
0005	EA	NOP	│ needed to generate
0006	EA	NOP	│ the monostable
0007	EA	NOP	⌡ time.
·	·	·	
·	·	·	
·	·	·	
·	EA	NOP	
·	EE	INC	⌠Increment data. Zero
·	00		⌡ status flag will go to a 0

As can be seen from the program, initially the value 1 is stored in memory location 0000. Where it is decremented, it goes to a value of 0 and this makes the Zero status flag go to a 1. Then a series of NOP instructions are used. Each requires 2 clock cycles. If a 1 MHZ clock is used this is 2 microseconds. If a crystal is used, this 2 μs is quite accurate. As many NOP instructions are used as are required to generate the needed time. At the conclusion of the series of NOP instructions an increment instruction removes the 0 from memory location 0000 and the Zero status flag goes to a 0.

The monostable time in this illustrative example is limited by the number of available memory locations. As can be seen on Fig. 15-1, the status register is connected to the data bus so that the Zero status flag level can be transmitted to the outside world through an I/O part.

Problems

15-1. Perform the following additions in binary signed arithmetic and check to see if the sign of the sum and the sum are correct. Tabulate the carry bit and the bit 6 overflow and check if a correct answer does occur where the carry bit and overflow bits have identical levels. The numbers given are to the base 10: (a) $+27 + (+54)$; (b) $+84 + (-38)$; (c) $-58 + (+18)$; (d) $+92 + (+190)$.

15-2. Repeat Prob. 15-1 for (a) $+60 + (-28)$; (b) $+(-104) + (+18)$; (c) $+(-184) + (-147)$.

15-3. Write a program that will interchange the memory locations of two numbers. The numbers are at page zero locations 37 and A6. Write the program in hexadecimal binary, mnemonics, and comments.

15-4. Write a program that will interchange the memory locations of two numbers. The numbers are at locations 24A6 and 36DB. Write the program in hexadecimal, binary, mnemonics, and comments.

15-5. A byte is stored in memory location 0000. Write a program in hexadecimal, mnemonics, and comments to store its 1s complement in location 0001 and its 2s complement in location 0002.

15-6. Write a program to interchange the first 4 bits of 2 bytes.

15-7. Write a program to add three numbers. The numbers are located at A210, A211, and A212. The sum is to be at location A213 and the carry at location A214.

15-8. Write a program for the multiplication of two numbers according to solution method C and flowchart of Fig. 15-4.

15-9. Write a program to check the decrement function of a microprocessor by storing the numbers 0 to 5 in memory locations 0000 to 0005. Do this with 5 separate decrements.

15-10. Repeat Prob. 15-9, but use a loop with a final check to see when the number has been decremented to 0.

15-11. It is desired to store the squares of the numbers 1 to 5 in memory locations 0101 to 0105. Using the formula $(a + 1)^2 = a^2 + 2a + 1$, draw a flowchart and write a program.

SIXTEEN

The Integrated-Circuit Data Sheet and Its Use

16-1 INTRODUCTION

The purpose of this chapter is to provide guidelines for using IC specifications and for using ICs. Before you can properly use an IC, you have to look at its data sheet. Frequently, individual IC data sheets do not give the complete specification for an IC. They give specific information for the IC in question, but general information about the IC family may be given in a separate section of an IC manual, and this also must be consulted.

The function of a data sheet is to provide the user with the following:

1 Defining specifications for the IC.
2 Limiting conditions and characteristics.
3 Applications or user-oriented information.
4 Information for interchangeability.

On the data sheet or in the data manual you will find the following:

16-1.1 Part Number

Each IC has an identifying part number. Interchangeable ICs are also made by other manufacturers. Unfortunately, IC part numbers have not been standardized, but cross-reference lists between the IC manufacturers are available. In many cases, part numbers from different sources are similar, but similarity does not always ensure interchangeability. For example, the TTL quad two-input NAND gate can be obtained from different manufactuers as a MC7400P, MC7400L, 7400DC, SN7400N, SJ7400J, or DM7400N. These are all electrically the same, with the only difference being either a plastic or ceramic 14-pin DIP case. However, types MC830F and MC830P, while electrically the same, have quite different cases and are not mechanically interchangeable. The MC830F is in a flat-pack case and the MC830P is in a DIP case.

16-1.2 Logic Function and Logic Diagram

The logic function and diagram indicate the basic function of the IC. Typical examples are NAND, NOR, adder, and counter.

16-1.3 Mechanical Specifications

Mechanical specifications are needed for printed circuit layout or socket design. Frequently, these specifications will be found in the general section of the data manual.

16-1.4 Maximum Ratings

Maximum ratings specify maximum conditions that can be applied to an IC. Failure to observe these ratings can lead to catastrophic failure. Typical ratings are for voltage, current, temperature, shock, vibration, humidity.

16-1.5 Terminal Pin (Pinout) Information

Terminal pin information tells how to make electrical connections to the IC.

16-1.6 Operating Characterisitcs

1. Supply voltage operating range.
2. Operating temperature range.
3. Fan out: the maximum number of gates that can be driven by the gate.
4. Fan in: the limiting number of inputs that can be connected to a gate.
5. Impedance levels.
6. Logic voltage swing.
7. Threshold voltage.
8. Noise immunity.
9. Dissipation.
10. Current requirements.
11. Thermal resistance.
12. Switching speeds.
13. High-frequency performance.

16-1.7 Applications Information

Applications information shows how the IC is used. It may be simple, such as a truth table, or it may be quite detailed and complex with complete information on how the IC is used in a digital system.

A typical IC data sheet for the *JK* flip-flop, type MC5472–MC7472 is discussed in the next section.

16-2 MC5472-MC7472 DATA SHEET

The defining data for this IC are given in Figs. 16-1a to 16-1d. Since it is an IC of the TTL 54/7400 family, some of its data are given in the general section of a TTL IC data manual. Items in the general section of the manual and not on this data sheet are mechanical data, maximum ratings, and noise immunity.

16-2.1 Figure 16-1a

The following information is given on this data sheet:

1   Part number: type MC5472–MC7472
2   Supplementary part number information. Suffixes added to the part number are used to indicate the type of package
3   Function description: *JK* FF.
4   Logic diagram and pinout information. The pinout information is given next to the logic diagram. For the dual-in-line (DIP) package, the pin connections are indicated as shown and for the flat pack the pin connections are given in brackets.

   Directly below the logic diagram is a statement of the *JK* AND input property of the IC ($J = J1 \cdot J2 \cdot J3$, $K = K1 \cdot K2 \cdot K3$).

   Below this is a typical *JK* FF truth table.

   To the left of the truth table is a logic diagram for the internal construction of the IC.

   Completing this block is information on the loading factors, dissipation, and high-frequency performance.
5   On the bottom of the page is a circuit schematic for this IC corresponding to the logic diagram. It is quite complex. The typical TTL multiple-emitter construction is used for gating everywhere within the IC. TTL is very fast and short leads should be used for wiring to the ICs. If the leads are too long, the inductance of the leads will resonate with the stray capacitance and cause ringing. To minimize the ringing effect, the IC has built-in reverse-biased diodes at all inputs. This can be seen at the *J* and *K*, the $\overline{\text{SET}}$–$\overline{\text{RESET}}$, and the clock inputs.

MC5472 • MC7472

APRIL 1971

Add Suffix F for TO-86 ceramic package (Case 607).
Suffix L for TO-116 ceramic package (Case 632).
Suffix P for TO-116 plastic package (Case 605) MC7472 only.

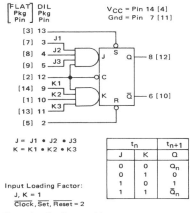

V_{CC} = Pin 14 [4]
Gnd = Pin 7 [11]

$$J = J1 \bullet J2 \bullet J3$$
$$K = K1 \bullet K2 \bullet K3$$

	t_n	t_{n+1}
J	K	Q
0	0	Q_n
0	1	0
1	0	1
1	1	$\bar{Q}_n$

Input Loading Factor:
J, K = 1
$\overline{\text{Clock}}, \overline{\text{Set}}, \overline{\text{Reset}}$ = 2

Output Loading Factor = 10
Total Power Dissipation = 40 mW typ/pkg
Propagation Delay Time = 30 ns typ
Operating Frequency = 20 MHz typ

This negative-edge-clocked J-K flip-flop operates on the master-slave principle. Three K inputs are ANDed together, and three J inputs are ANDed together. $\overline{\text{SET}}$ and $\overline{\text{RESET}}$ inputs are also available. The device helps minimize package count in J-K flip-flop applications requiring AND gating into the J or K inputs.

LOGIC DIAGRAM

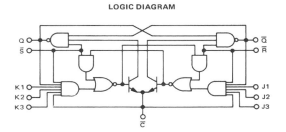

CIRCUIT SCHEMATIC

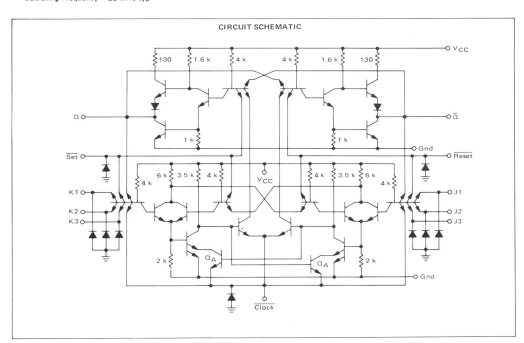

Figure 16-1a MTTL MC5400/7400 series JK flip-flop data sheet. (Courtesy Motorola Semiconductor Products, Inc.)

The purpose of this sheet is to specify the dc characteristics of the IC. It specifies the test conditions of current, voltage, and pin connections that must be applied to the IC. With these conditions applied, this data sheet specifies the limiting currents and voltages that will be measured at any IC type MC5472–MC7472. The definition of the symbols (e.g., I_{OL}) are given in the general section of the data manual. The characteristics specified are as follows:

1 Input forward current: the maximum input current from an IC when an input is connected to $V_{IL} = 0.4$ V (logic 0). Note that this current is negative. It has to be sinked.

2 Input leakage current: the maximum input current into an IC when an input is connected to $V_{IH} = 2.4$ V (logic 1).

3 Output voltages: the maximum (V_{OL}) low-level and minimum (V_{OH}) high-level voltages.

4 Short-circuit current: the maximum and minimum values of current obtained when an output terminal is shorted to ground under normal high output level conditions. This gives output impedance data.

5 Power requirements.

16-2.3   Figures 16-1c and 16-1d

These two pages in combination give high-frequency performance data. The table on the top of Fig. 16-1c gives the test conditions and limits for toggling frequency and switching time. On the top of Fig. 16-1d is a discussion on how and when data must be applied to these FFs to be transferred to the output.

The bottom of Figs. 16-1c and 16-1d relate to the measurement of switching times. The applied waveforms and their definitions are found on Fig. 16-1c, and the circuit used and connection to a 50 Ω oscilloscope are given on the lower half of Fig. 16-1d.

16-3 GUIDE TO USE OF INTEGRATED CIRCUITS

Although each logic family has its own requirements, the following general comments apply to all the families. All the IC manufacturers provide applications data for the various logic families.

16-3.1 Grounding and Ground Plane

The ground leads should be made as heavy as practical. This reduces noise caused by the $I \times R$ drop. Ground planes reduce the resistance of the ground line, add bypassing capacitance, and act as shields, and should be used whenever possible.

ELECTRICAL CHARACTERISTICS

Test procedures are shown for only one J and one K input, plus the Set, Reset, and Clock inputs. To complete testing, sequence through remaining J and K inputs in the same manner.

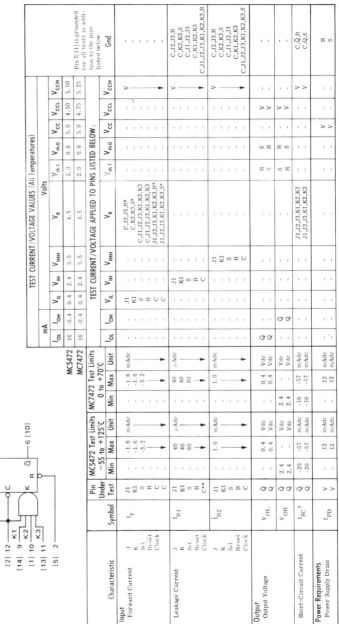

Figure 16-1b dc characteristics. (Courtesy Motorola Semiconductor Products, Inc.)

TEST PROCEDURES

(Numbers shown in test columns refer to waveforms.)

TEST	SYMBOL	INPUT				Q	$\bar{Q}$	LIMITS		
		$\bar{C}$	J, K	$\bar{R}$	$\bar{S}$			Min	Max	Unit
Toggle Frequency	f_{Tog}	1	1	2.4 V	2.4 V	†	†	15	—	MHz
Turn-On Delay	t_{pd-}	2	2	2.4 V	2.4 V	3	3	10	40	ns
Turn-Off Delay	t_{pd+}	2	2	2.4 V	2.4 V	4	4	10	25	ns
Turn-On Delay	t_{sd-}	2.4 V	2.4 V	5	6	7	8	—	40	ns
Turn-Off Delay	t_{sd+}	2.4 V	2.4 V	5	6	7	8	—	25	ns
Enable Voltage	V_{EN}	2	2.0 V	2.4 V	2.4 V	†	†	†	—	—
Inhibit Voltage	V_{INH}	2	0.8 V	2.4 V	2.4 V	‡	‡	‡	—	—

†Output shall toggle with each input pulse.
‡Output shall NOT toggle.

VOLTAGE WAVEFORMS AND DEFINITIONS

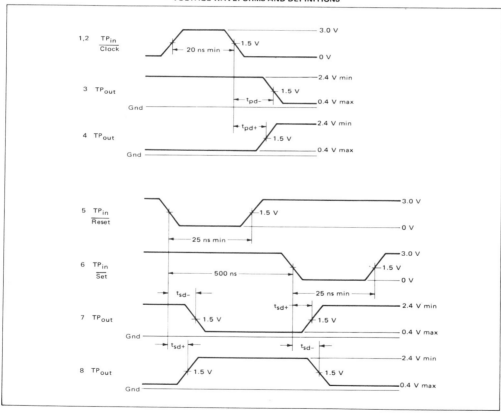

Figure 16-1c High frequency performance data. (Courtesy Motorola Semiconductor Products, Inc.)

OPERATING CHARACTERISTICS

Data must be applied to the J-K inputs while the clock is low. When the clock input goes to the positive logic "1" state, the data at the J and K inputs is transferred to the master section, where it is stored until the clock changes to the positive logic "0" state. Data at the J and K inputs must not be changed while the clock is high. When the clock returns to the positive logic "0" state, information in the master section is transferred to the slave section.

Application of a logic "0" to the $\overline{\text{Reset}}$ input will force the Q output to the logic "1" state. The $\overline{\text{Reset}}$ input overrides the clock.

Since no charge storage is involved in this flip-flop, rise and fall times are not important to its operation. Clock fall times as long as 1.0 μs will not adversely affect the operation of the flip-flop. The clock pulse need only be wide enough to allow the data to settle in the master section. This time, which is the setup time for a logic "1", is 20 ns minimum.

Transistors Q_A have been added to the standard flip-flop circuit to protect the device against negative clock transients. This addition prevents both outputs from changing to the logic "1" state when transients in excess of –2.0 V appear at the clock.

SWITCHING TIME TEST CIRCUIT

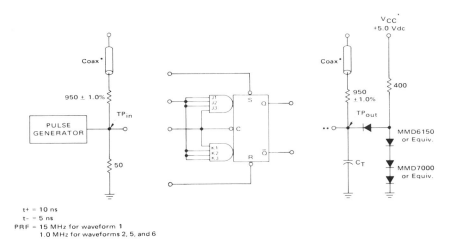

t+ = 10 ns
t– = 5 ns
PRF = 15 MHz for waveform 1
1.0 MHz for waveforms 2, 5, and 6

Two pulse generators are required and must be slaved together for testing $\overline{\text{Set}}$ and $\overline{\text{Reset}}$. Only one pulse generator is required for J, K, and Clock tests.

*The coax delays from input to scope and output to scope must be matched. The scope must be terminated in 50-ohm impedance. The 950-ohm resistor and the scope termination impedance constitute a 20:1 attenuator probe. Coax shall be CT-070-50 or equivalent.

**A load is connected to each output during the test.

C_T · 15 pF = total parasitic capacitance, which includes probe, wiring, and load capacitances.

MOTOROLA *Semiconductor Products Inc.*

BOX 20912 • PHOENIX ARIZONA 85036 • A SUBSIDIARY OF MOTOROLA INC

Figure 16-1d High Frequency performance data. (continued.) (Courtesy Motorola Semiconductor Products, Inc.)

16-3.2 Bypassing

Use large values of capacitance where power enters a PC board, and use smaller values of capacitance distributed throughout a system. At high speeds the series inductance of large values of capacitance makes them ineffective.

16-3.3 Power Supplies

In general, power supplies should be well regulated and bypassed to reduce noise. One must be careful that under no conditions is the output voltage permitted to exceed the maximum voltage ratings of the ICs. In addition, they must be carefully checked for transient response to make sure that a transient voltage spike does not exceed IC voltage ratings and cause catastrophic failure. In general, the dc current requirements of ICs increase as the operating frequency increases. Both the power supplies and the ICs must be capable of handling this increase in current, dissipation, and temperature.

16-3.4 Unused Inputs

Unused inputs should not be left unconnected lest they pick up noise. For connection to logic 1, perhaps the best place is to connect them to the output of an unused gate or inverter. If this is not convenient, IC data books or the manufacturer should be consulted. For connection to logic 0, grounding an unused input is satisfactory.

APPENDIX

Standard Logic Symbols
ANSI Y32.14–1973

There has been considerable activity by industry and the military service for many years to develop a standard set of logic symbols. The most recent set of standards is the American National Standard ANSI Y32.14–1973, Graphic Symbols for Logic Diagrams (two-state devices). The industry groups involved were the Institute of Electrical and Electronics Engineers and the American National Standards Institute. This standard has been adopted by the Department of Defense. The following pages show the ANSI Y32.14–1973 symbols and compare them with previous standards. This is Appendix C† of the standard.

†Reprinted from IEEE Std 91-1973, Graphic Symbols for Logic Diagrams (Two-State Devices) (ANSI Y32.14-1973), by permisson of the Institute of Electrical and Electronics Engineers, Inc.

APPENDIX C

Composite Chart Showing Relationship of Graphic Symbols for Logic Diagrams from ANSI Y32.14-1973 (IEEE Std 91-1973) and the Superseded Standards

Column 1: Description of Logic Functions

Column 2: ANSI Y32.14-1973
    2A: Rectangular-shape symbols
    2B: Distinctive-shape symbols

Column 3: Other industry standards

Column 4: ANSI Y32.14-1962
    IEEE Std 91-1962
    MIL-STD-806C (Navy)
    4A: Uniform-shape symbols
    4B: Distinctive-shape symbols

Column 5: MIL-STD 806B

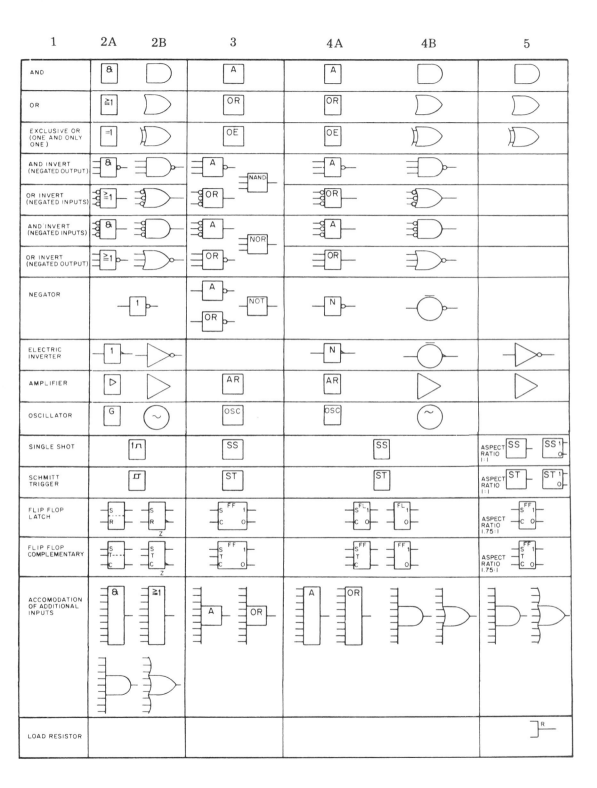

	1	2A	2B	3	4A	4B	5
AND							
OR							
EXCLUSIVE OR (ONE AND ONLY ONE)							
AND INVERT (NEGATED OUTPUT)							
OR INVERT (NEGATED INPUTS)							
AND INVERT (NEGATED INPUTS)							
OR INVERT (NEGATED OUTPUT)							
NEGATOR							
ELECTRIC INVERTER							
AMPLIFIER							
OSCILLATOR							
SINGLE SHOT							
SCHMITT TRIGGER							
FLIP FLOP LATCH							
FLIP FLOP COMPLEMENTARY							
ACCOMODATION OF ADDITIONAL INPUTS							
LOAD RESISTOR							

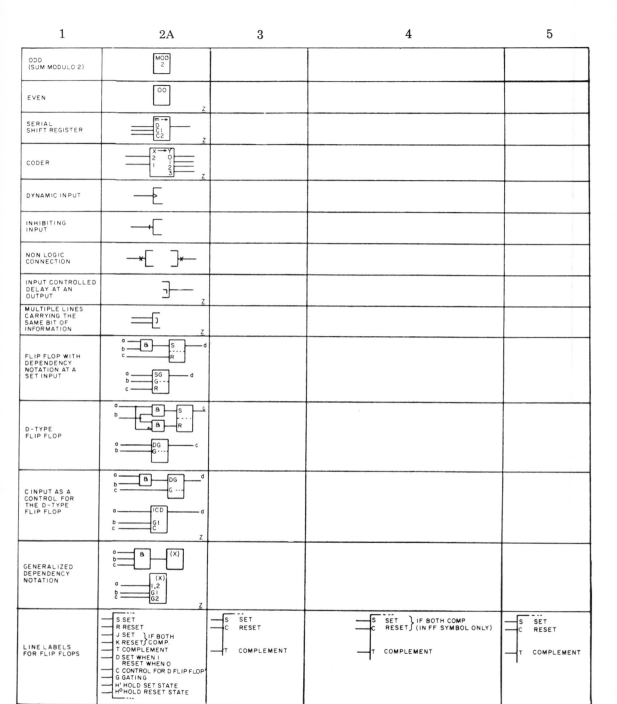

	1	2A	3	4	5

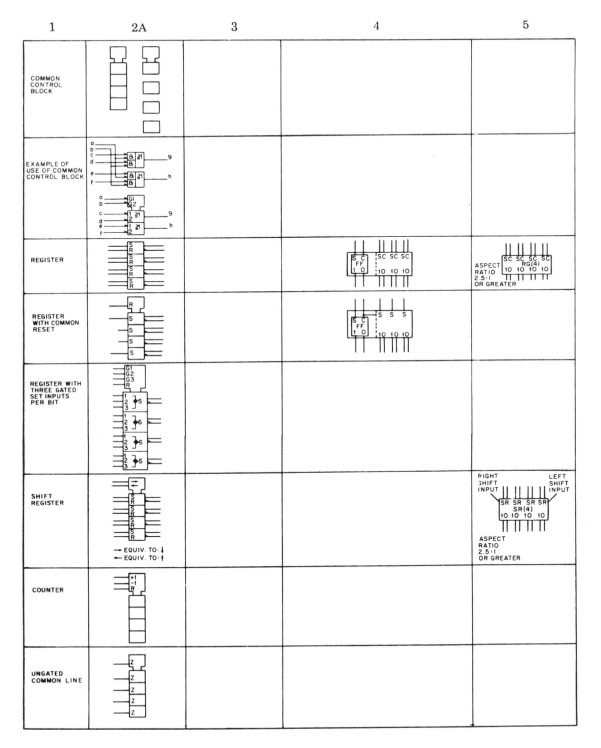

	1	2A	2B	3	4A	4B	5
DOT AND			&◇		(x) A / (x) A	(x) / (x)	(x) / (x)
DOT OR			≧1 / ≥1 / ◇		(x) OR / (x) OR	(x) / (x)	(x) / (x)
TIME DELAY		(t)	(t)	TD (t)	TD (t)	(t)	(t)
TIME DELAY (ZERO TRAILING EDGE DELAY)		(t) O	(t) O	TR (t)			
TIME DELAY (UNEQUAL LEADING AND TRAILING EDGE DELAYS)		(t1)(t2)	(t1) (t2)	TR (t1) TR (t2)			
NEGATIVE POLARITY							
POSITIVE POLARITY							
NOT							
EXTENDER CONNECTION		& E / & E			A E / A	E	E
GENERAL LOGIC SYMBOL FOR FUNCTIONS NOT ELSEWHERE SPECIFIED							ASPECT RATIO 2:1 OR GREATER
THRESHOLD		≧ m					
m AND ONLY m		= m					
INPUT IDENTITY		=					
MAJORITY		>n/2					

Answers to Selected Problems

1-1. $Z = 0$ when X and $Y = 0$

1-2. $T = 1$ when R and $S = 1$

1-3. $0\ \text{V} \equiv 0$, $120\ \text{V} \equiv 1$

1-4. $M = 0$ when $K = 0$ and $L = 1$

1-5. $M = 1$ when $K = 0$ and $L = 1$

1-6. $P = 1$ when $X = 1$, $Y = 1$, $Z = 0$

1-7. $R = 1$ when $X = 1$, $Y = 0$, $Z = 0$

1-8. (a) $C = 0$ when $A = 1$ and $B = 0$; (b) $C = 1$ when $A = 1$ and $B = 0$

1-9. $N = 0$ when $K = 1$, $L = 1$, $M = 0$

1-10. $D = 1$ when $A = 0$, $B = 0$, $C = 1$

1-15. (a) $\overline{\overline{AB\bar{C}D\bar{E}}}$; (b) $(\overline{X\bar{Y}})(\overline{Y\bar{Z}\bar{X}})(\overline{RS\bar{T}\bar{U}})$

1-16. (a) $\overline{X\bar{Y}Z}$; (b) $\bar{A}(\bar{\bar{B}}E)(K\bar{L}M)$

1-17. (a) $\overline{\bar{R} + \bar{S} + T}$; (b) $\bar{K} + L + \overline{\bar{A} + B + \bar{D}}$;

(c) $\overline{\bar{A} + B + \bar{D}} + (\bar{U} + \bar{V}) + \overline{A + B}$

1-18. (a) $\overline{K + \bar{L} + M + \bar{P}}$; (b) $(\overline{\bar{X} + Y + \bar{Z}}) + \bar{B}$;

(c) $(\overline{X + \bar{Y}}) + L + N + (\overline{A + B})$

1-19.

1-20.

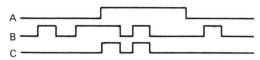

1-21.

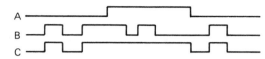

1-22.

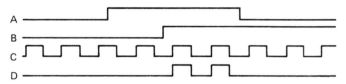

1-23. $Z = \overline{(A + \bar{B} + C)(A + \bar{B} + \bar{C})} + \bar{A}B$

1-24. $Z = A + \bar{B} + (A + \bar{B} + C + D)(\overline{AB\bar{C}})(\bar{A}\bar{B})$

1-27. (a) A; (b) AB

1-28. (a) $\bar{X} + \bar{Z}$; (b) AB

1-29. (a) $B + \bar{C}$; (b) $AB + \bar{C}(A + B)$

1-31. B

1-32. No

1-33. $\bar{B}C$

1-34. $B\bar{C} + AB$

1-35. (a) $\bar{T} + RST$ or $\bar{T} + RS$; (b) $\overline{\bar{R}T + \bar{S}T}$

1-36. S

1-38. $B\bar{C}D + AC\bar{D}$

1-40. $\bar{B}CD$

1-42. $\bar{L}$

1-44. $AB + AC + BC$

1-45. Use $\frac{1}{4}$ SN7400

1-46. Use $\frac{3}{4}$ SN7400

1-47. Use SN7430 $+ \frac{4}{6}$ SN7404

1-48. Use $\frac{3}{4}$ SN7400 $+ \frac{1}{3}$ SN7410 $+ \frac{1}{6}$ SN7404

1-49. Use $\frac{1}{3}$ SN7410 $+ \frac{3}{4}$ SN7404 $+ \frac{4}{6}$ SN7404

1-50. Use $\frac{1}{2}$ SN7420 $+ \frac{2}{6}$ SN7404

2-1. (a) 8.5 V; (b) 5.0 V; (c) 3.5 V; (d) 2.7 V; (e) 2.0 V

2-4. (a) 10.75; (b) 10; (c) 10; (d) 5

2-6. 4

2-7. (a) 0.52; (b) 0.42.

2-9. 0.775 V

2-11. At $V_{BB} = 0.715$ V, $V_{CE} = 2.88$ V

2-12. (a) 1.06 V; (b) 0.62 V
2-14. 1.56 V
2-16. 1.2 V
2-17. (a) 1.05 V; (b) 0.97 V
2-19. 6.5 mA
2-20. (a) 0.26 mA; (b) 0.4 mA
2-21. (a) -3 V; (b) $+7.4$ V
2-22. (a) 0.147 mA; (b) 1.47 mA; (c) 1.62 mA; (d) 1.01 V
2-24. 14.8 mA
2-25. 12

3-7. $X = \overline{AB + CD}$
3-8. $X = \overline{A + B + C}$
3-9. (a) High-level noise immunity $= 5.2$ V; (b) low-level noise immunity $= 3.2$ V

4-3. (a) 13; (b) 103; (c) 21
4-5. (a) 0.875; (b) 0.1875
4-7. (a) 57.25; (b) 85.3125
4-9. (a) 11101; (b) 111101; (c) 101110; (d) 10011100
4-11. (a) 0.101; (b) 0.10110$^+$; (c) 0.00111
4-13. (a) 111001.01; (b) 11101.011
4-15. (a) 1101010; (b) 10111100; (c) 101000
4-17. (a) 36; (b) 432; (c) 695; (d) -36; (e) -432
4-19. (a) 110; (b) 1001; (c) 1110; (d) -1100
4-21. (a) 11111; (b) -1110
4-23. (a) 10001111; (b) 10101111
4-25. 1111101
4-27. (a) 110.01; (b) 101.101
4-29. (a) 11; (b) 10110; (c) 1001.00$^+$
4-31. (a) 185; (b) 124
4-33. (a) 43; (b) 1354; (c) 423
4-35. (a) 6736.33; (b) 100010.011111
4-37. (a) 12; (b) 26; (c) 8F; (d) 202
4-39. (a) CF; (b) 1010110001000111; (c) 5.E

5-1. $EO = X\bar{Y} + \bar{X}Y$
5-3. $B = (X + Y)(\bar{X} + \bar{Y}) = X\bar{Y} + \bar{X}Y$
5-4. $B = \bar{X}\bar{Y} + XY$
5-6. 1 HA, 15 FA
5-8. (b) Four three-input AND, one four-input OR
5-9. (a) $D = \bar{X}\bar{Y}\bar{B_i} + \bar{X}Y B_i + XY\bar{B_i} + X\bar{Y}B_i$
     (c) Four three-input AND, one four-input NOR
5-11. (a) $B_o = \overline{\bar{X}\bar{Y}\bar{B_i} + XY\bar{B_i} + X\bar{Y}\bar{B_i} + X\bar{Y}B_i}$
     (b) $B_o = \bar{Y}\bar{B_i} + X\bar{B_i} + X\bar{Y}$
     (c) Three two-input AND, one three-input NOR

6-13. (a) 2000 Hz
(b) 1000 Hz

6-17. For clock, data appears at output when clock pulse returns to 0. For C_D and S_D, activate by making them go to a 0.

6-18. X, don't care; U, undefined. The output can arrive at a level determined by transistor parameters and not by data.

7-1. $f_D = 2.5$ kHz

7-4. 0 to 1111 or 0 to 15

7-5. 0000 to 1111

7-6. 1111 to 0000

7-7. For five stages divisions by $\frac{1}{32}$; count capability 0 to 31

7-8. 90 ns

7-12. 4, 5, 7

7-14. Feedback to FF A and FF C

7-17. Mod-6

7-18. Mod-11

7-19. 000, 001, 010, 011, 100, 101

7-20. 000, 001, 010, 100, 101, 110

7-21. 0100

7-26. At X, 333 Hz

7-31. 10

7-32. Two mod-5, one mod-7, one mod-3

8-7. (a) 4; (b) left; (c) 80 μs

8-8. (a) 5; (b) right; (c) 100 μs

8-9. 40 μs

8-10. 128 μs

8-11. 51.2 ms

8-14. (b) $Q_A Q_B Q_C = 000, 100, 110, 111, 011, 001$; (c) 120 Hz; (e) $\bar{Q}_C, Q_B, \bar{Q}_A$ is one possible solution

8-15. (a) 1200 Hz; (b) 1000 Hz

8-17. $Q_A Q_B = 00, 10, 01$, mod-3

8-18. $Q_A Q_B Q_C = 000, 100, 010, 101, 110, 011, 001$, mod-7

8-19. $Q_A Q_B Q_C = 100, 110, 111, 011, 101, 010, 001$, mod-7

9-1. (a) 14 μs; (b) 42 μs; (c) 56 μs; (d) 17,900 Hz

9-4. (d) 35,700 Hz

9-5. 20

9-7. Good for Q_1, not good for Q_2

9-8. (c) 2.2τ

9-9. (a) 1 μs; (b) 2.2 μs

9-10. (a) 0.7τ or 14 μs; (b) 0.84τ or 16.9 μs

9-12. $\approx 29,600$ Hz

9-13. $\approx 26,100$ Hz

9-14. $\approx 30,600$ Hz

9-15. (a) 19.3 Hz; (b) 21.9 Hz

9-17. 1.33×10^5 ns

9-18. (a) 2.827 V; (b) 2.474 V
9-19. (a) 28.6 and 167.3°; (b) 0.96 ms

10-1. (a) 0010 0111
10-2. (a) 785
10-3. (a) 1100 0110 1110
10-4. (a) 368
10-5. (a) 1011 0101 0100
10-6. (a) 0110 1010
10-7. (a) 10 00010 01 00100
10-8. (a) 010000100100
10-11. (a) 0110 0100 0
10-12. (a) 0101 0111 1
10-13. Error is in fourth row and third column.
10-14. 111000
10-16. (a) $\approx 1.4°$
10-17. (a) 101 0000 100 0001 100 0111 100 0101 011 0010 011 1000
10-19. Decoding gates are $\bar{B}\bar{A}$, $\bar{B}A$, $B\bar{A}$, and BA.
10-21. Decoding gates are $0 = \bar{D}\bar{A}$, $1 = D\bar{C}$, $2 = C\bar{B}$.
10-23. Decoding gates are $0 = \bar{C}\bar{B}\bar{A}$, $3 = \bar{C}BA$, $8 = DB\bar{A}$ or $CB\bar{A}$.
10-25. $C = A$, $D = \bar{A}\bar{B}$, $E = \bar{A} + \bar{B}$

11-3. (a) 64; (b) 34; (c) 20; (d) 16
11-4. (a) 6; (b) 6; (c) 6; (d) 6
11-5. (a) 8; (b) 32; (c) 64; (d) 128
11-6. (a) 4; (b) 8; (c) 10; (d) 12
11-12. (a) 42; (b) 50
11-13. Nine planes of 32×32 square matrix, 66 wires per plane
11-15. 12.5 ms
11-16. 25,100
11-17. 4,520,000
11-18. 180×10^6 bps
11-19. (a) 23.04×10^6; (b) 207×10^6; (c) 60 s; (d) 120s; (e) 192,000; (f) 1.73×10^6

12-1. -1 V
12-4. At $V_A = 0.1$ V and $V_B = 0.1$ V, $V_X = -2$ V
12-5. At $V_A = +5$ V, $V_B = 0$ V, $V_C = +5$ V, $V_X = -5$ V
12-6. -10 V
12-8. $-5 \sin 400t$

13-4. MSB $= 2$ V, LSB $= 0.25$ V
13-5. (a) 2.5 V
13-6. (a) 320 kΩ; (b) $\frac{1}{16}$ V
13-7. (a) 20.48 MΩ; (b) ≈ 1 mV
13-8. (a) MSB $= 0.5$ V, LSB $= 0.125$ V; (b) 0.375 V
13-9. MSB $= 5$ V, LSB $= 0.15625$ V
13-10. (a) 7.1875 V

Index